Lecture Notes in Physics

Springer
Berlin
Heidelberg
New York
Barcelona
Hong Kong
London
Milan
Paris
Tokyo

Physics and Astronomy ONLINE LIBRARY

http://www.springer.de/phys/

Editorial Policy

The series *Lecture Notes in Physics* (LNP), founded in 1969, reports new developments in physics research and teaching -- quickly, informally but with a high quality. Manuscripts to be considered for publication are topical volumes consisting of a limited number of contributions, carefully edited and closely related to each other. Each contribution should contain at least partly original and previously unpublished material, be written in a clear, pedagogical style and aimed at a broader readership, especially graduate students and nonspecialist researchers wishing to familiarize themselves with the topic concerned. For this reason, traditional proceedings cannot be considered for this series though volumes to appear in this series are often based on material presented at conferences, workshops and schools (in exceptional cases the original papers and/or those not included in the printed book may be added on an accompanying CD ROM, together with the abstracts of posters and other material suitable for publication, e.g. large tables, colour pictures, program codes, etc.).

Acceptance

A project can only be accepted tentatively for publication, by both the editorial board and the publisher, following thorough examination of the material submitted. The book proposal sent to the publisher should consist at least of a preliminary table of contents outlining the structure of the book together with abstracts of all contributions to be included.
Final acceptance is issued by the series editor in charge, in consultation with the publisher, only after receiving the complete manuscript. Final acceptance, possibly requiring minor corrections, usually follows the tentative acceptance unless the final manuscript differs significantly from expectations (project outline). In particular, the series editors are entitled to reject individual contributions if they do not meet the high quality standards of this series. The final manuscript must be camera-ready, and should include both an informative introduction and a sufficiently detailed subject index.

Contractual Aspects

Publication in LNP is free of charge. There is no formal contract, no royalties are paid, and no bulk orders are required, although special discounts are offered in this case. The volume editors receive jointly 30 free copies for their personal use and are entitled, as are the contributing authors, to purchase Springer books at a reduced rate. The publisher secures the copyright for each volume. As a rule, no reprints of individual contributions can be supplied.

Manuscript Submission

The manuscript in its final and approved version must be submitted in camera-ready form. The corresponding electronic source files are also required for the production process, in particular the online version. Technical assistance in compiling the final manuscript can be provided by the publisher's production editor(s), especially with regard to the publisher's own Latex macro package which has been specially designed for this series.

Online Version/ LNP Homepage

LNP homepage (list of available titles, aims and scope, editorial contacts etc.):
http://www.springer.de/phys/books/lnpp/

LNP online (abstracts, full-texts, subscriptions etc.):
http://link.springer.de/series/lnpp/

W. Plessas L. Mathelitsch (Eds.)

Lectures on Quark Matter

Springer

Editors

Willibald Plessas
Leopold Mathelitsch
Universität Graz
Institut für Theoretische Physik
Universitätsplatz 5
8010 Graz, Austria

Supported by the Österreichische Bundesministerium für Bildung, Wissenschaft und Kultur, Vienna, Austria

---

Cover Picture: (see figure 10, page 300, contribution by L. McLerran in this volume)

Library of Congress Cataloging-in-Publication Data applied for.
Die Deutsche Bibliothek - CIP-Einheitsaufnahme

Lectures on quark matter / W. Plessas ; L. Mathelitsch (ed.). - Berlin ; Heidelberg ; New York ; Barcelona ; Hong Kong ; London ; Milan ; Paris ; Tokyo : Springer, 2002 (Lecture notes in physics ; 583)
(Physics and astronomy online library)
ISBN 3-540-43234-5

ISSN 0075-8450
ISBN 3-540-43234-5 Springer-Verlag Berlin Heidelberg New York

Springer-Verlag Berlin Heidelberg New York
a member of BertelsmannSpringer Science+Business Media GmbH

http://www.springer.de

Printed in Germany

Typesetting: Camera-ready by the authors/editor
Camera-data conversion by Steingraeber Satztechnik GmbH Heidelberg
Cover design: *design & production*, Heidelberg

Printed on acid-free paper
SPIN: 10860999 54/3141/du - 5 4 3 2 1 0

Preface

This volume contains the written versions of the lectures delivered at the "40. Internationale Universitätswochen für Theoretische Physik" in Schladming, Austria. The 40th "Schladming Winter School" took place during the period March 3rd–10th, 2001. The topic of the School was "Dense Matter".

After the establishment of quantum chromodynamics as the fundamental gauge field theory of strong interactions it soon became an intriguing question whether a new form of matter consisting just of the ultimate constituents of hadrons, i.e. quarks and gluons, would be possible. Could nuclear matter undergo a phase transition and transform to quark matter? What would be the necessary conditions for the creation of the so-called quark-gluon plasma? Did such a state exist at the beginning of the universe and could it still be found somewhere in our cosmos? These were only a few of the questions that could be posed on the issue of a deconfined state of quark matter. Theoreticians rapidly came up with a variety of answers. Sometimes the corresponding predictions were rather speculative but gradually they gained a more quantitative nature. Experimental physicists started to think about ways of realizing the new form of matter in the laboratory. Soon the idea of letting heavy nuclear systems collide at high energies was born. Thereby, possibly, conditions could be reached such that the hadron constituents could get deconfined over a reasonably large local domain and one could observe quark matter. The discipline of heavy-ion physics developed rapidly at the interface between nuclear and particle physics. A lot of effort went into the theoretical and experimental investigations of heavy-ion reactions. In particular, experimentalists had a hard time reaching a stage where they could manage head-on collisions of heavy nuclei at energies large enough so that a quark-gluon plasma could be formed. After many years and a long series of experiments, in early 2000 sufficient and convincing enough experimental data were accumulated so that physicists at CERN could announce the observation of quark matter. Evidently, this brought new enthusiasm to the field of heavy-ion physics. Also, one could then expect exciting new evidence of the quark-gluon plasma from RHIC, which started data taking later on in 2000. Through these developments one certainly had enough reason to devote the 2001 Schladming Winter School to the topic of "Dense Matter".

We are happy that we got some of the most renowned experts in the field to lecture at Schladming. Thus the meeting not only became a respectable jubilee Winter School – the 40th in a continuous series since 1962 – but was also very successful scientifically. Practically all relevant topics relating to heavy-ion physics and the quark-gluon plasma were dealt with in the lectures presented. C. Lourenço summarized the modern experimental evidence on quark matter formation as they were achieved at CERN. M. Gyulassy complemented them with the most recent data from RHIC, along with exposing the theory of ultra-relativistic heavy-ion reactions, and M. Alford reviewed aspects of quark matter in compact stars. The general theory of the quark-gluon plasma was presented by J.-P. Blaizot, while A. Rebhan explained the treatment within thermal gauge field theories. The evidence on the properties of the quark-gluon plasma so far obtained from lattice QCD calculations were reviewed by F. Karsch. Finally E.V. Shuryak and L. McLerran opened exciting views on a variety of new phenomena that can be studied through quark matter, for example, color superconductivity or the formation of a color glass condensate. We should also mention that all of these lectures were accompanied by a number of seminars given on related topics by the participants of the School.

Here we would like to express our sincere gratitude to the lecturers for all their efforts in preparing, presenting, and finally writing up their lectures. Our thanks are also due to the main sponsors of the School, the Austrian Federal Ministry for Education, Science, and Culture and the Government of Styria, for providing generous support. We also appreciate the contributions from the University of Graz and the valuable organizational and technical assistance from the town of Schladming, Ricoh Austria, and Hornig Graz. Furthermore, we thank our secretaries, S. Fuchs and E. Monschein, a number of graduate students from our institute, and, last but not least, our colleagues from the organizing committee for their valuable assistance in preparing and running the school.

Graz,
October 2001

Leopold Mathelitsch
Willibald Plessas

Contents

List of Contributors

M. Alford
Dept. of Physics and Astronomy
Glasgow University
Glasgow G12 8QQ
United Kingdom
m.alford@physics.gla.ac.uk

J.-P. Blaizot
Service de Physique Théorique
CEA Saclay
F-91191 Gif-sur-Yvette Cedex
France
blaizot@spht.saclay.cea.fr

M. Gyulassy
Department of Physics
Pupin Lab
Columbia University
New York, NY 10027
USA
gyulassy@nt3.phys.columbia.edu

F. Karsch
Fakultät für Physik
Universität Bielefeld
D-33615 Bielefeld
Germany
karsch@physik.uni-bielefeld.de

C. Lourenço
EP Division
CERN
CH-1211 Genève 23
Switzerland
carlos.lourenco@cern.ch

L. McLerran
Nuclear Theory Group
Brookhaven National Laboratory
Upton, NY 11793
USA
mclerran@bnl.gov

A. Rebhan
Institut für Theoretische Physik
Technische Universität Wien
Wiedner Hauptstrasse 8 – 10
A-1040 Vienna
Austria
anton.rebhan@tuwien.ac.at

E.V. Shuryak
Department of Physics
SUNY
Stony Brook, NY 11794-3800
USA
shuryak@dau.physics.sunysb.edu

Quark Matter Production in Heavy-Ion Collisions

Carlos Lourenço

EP Division, CERN
CH-1211 Genève 23, Switzerland

1 Introduction

The study of high-energy heavy-ion collisions is presently a very active field in experimental particle physics, with the RHIC collider at BNL in operation since summer 2000 and with the ALICE experiment being prepared to study this kind of physics at LHC energies. The first goal of these experimental attempts, which started in 1986, with the AGS and SPS fixed-target programs, is the discovery of the phase transition from confined hadronic matter to deconfined partonic matter. The idea that such a phase transition should exist, between hadronic and quark matter, has been around since the first models of the quark structure of hadrons. It is presently studied in detail in the framework of lattice QCD calculations, which predict its occurrence when the temperature of the system exceeds a critical threshold at around 170 MeV, corresponding to a critical energy density of around 600 MeV/fm^3 [1]. Fig. 1 illustrates how the energy density (in units of T^4) depends on the temperature of the medium (in units of T_c), increasing by an order of magnitude within a very small temperature range. At the critical temperature, two phenomena should occur: the color degrees of freedom become deconfined and chiral symmetry (spontaneously broken in the hadronic world) gets restored. Both should lead to observable effects, to be looked for in properly designed experiments. The proof of existence of the quark matter phase and the study of its properties are key issues in QCD, for the understanding of confinement and chiral symmetry.

When this new state of matter was postulated, some signatures of its formation in high-energy nuclear collisions were proposed, on the basis of theoretical arguments, among which we can highlight the enhancement of strange particle production, the suppression of charmonia states (J/ψ, χ_c and ψ'), due to the screening of the $c\bar{c}$ binding potential in the QGP colour soup, and the production of thermal dileptons, electromagnetic radiation emitted by the 'free' quarks. The results obtained by the SPS experiments, after 15 years of collecting data with proton and ion beams, provide "compelling evidence for the existence of a new state of matter, in which quarks roam freely", produced in central Pb-Pb collisions at the highest SPS energies. Among the most exciting observations are the enhanced production of multistrange hyperons, the centrality dependence of the J/ψ suppression pattern and the

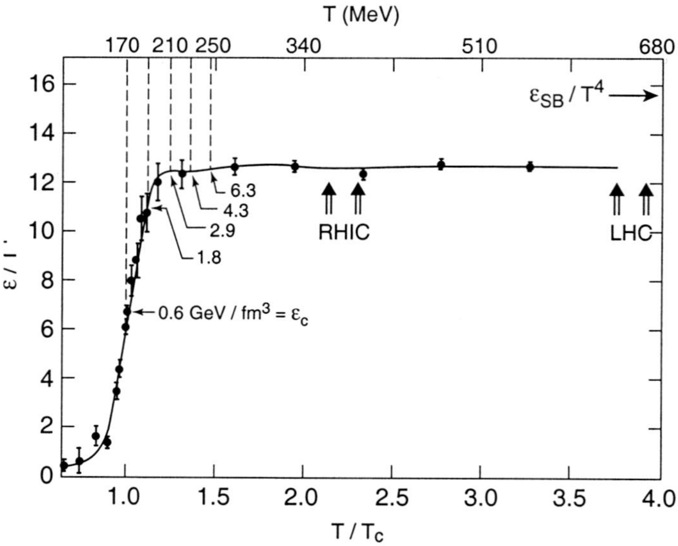

Fig. 1. Energy density versus temperature, showing a phase transition at a critical temperature $T_c \sim 170$ MeV

enhancement of intermediate mass dimuon production, a possible indication of thermal dimuons. Besides, the enhanced production of low-mass dileptons may be an indication of approach to chiral symmetry restoration.

In view of these exciting results and in order to clarify important questions remaining open, a new experiment, NA60, has been approved at CERN, extending the SPS runs with heavy-ion beams and bridging the gap between the original SPS program and the future high-energy wonderland of ALICE. At the other end of the energy scale, the NA49 experiment will have a ten day extension with Pb ions at 20 and 30 GeV per nucleon, to complete the energy scan of global strangeness production. So far it has collected data at 158, 40 and 80 GeV per nucleon. This extension of the SPS heavy-ion running time will bring to a proper conclusion the program started in 1986, whilst providing valuable information, complementary to the studies underway at RHIC.

Here I review some of the most interesting results obtained by the CERN SPS experiments, with particular emphasis on the progress made in the last two years (since the Quark Matter 1999 conference). I will also present some pertinent questions that still remain open and explain how the future SPS program will address those issues.

The very large amount of experimental results obtained by the CERN SPS experiments since 1986 is so vast and diversified that a proper review would require a much more extensive article, jointly prepared by several of the active players in the field. The Quark Matter 1999 conference, the last one before the RHIC experiments started collecting data, ended with two summary talks that reviewed in detail the status of the field in terms of hadronic [2] and dilepton [3] signals. A few attempts have also been made to see in a coherent way some of the most significant results, in particular those obtained with the lead beam at the SPS [4]. A tentative summary has been proposed [5], basically saying that "the combined results provide compelling evidence for the existence of a new state of matter, featuring many of the characteristics of the primordial soup in which quarks and gluons existed before they clumped together as the universe cooled down".

We can discuss at length the scientific meaning (and opportunity) of these words, but it is certainly appropriate to say that all the CERN SPS experiments have been successful in delivering significant information, many of them having seen "what they were looking for".

However, these 15 years have also confirmed that heavy-ion collisions lead to very complicated (and fastly evolving) systems, and that it is difficult to extract clear messages from the observations. Central collisions between two Pb nuclei, at the highest SPS energies, lead to the production of hundreds of final-state particles widely emitted without discernible structures. These complex, and apparently chaotic, final states can be studied applying statistical concepts, attempting descriptions based on (non-perturbative) QCD thermodynamics, and summarized by macroscopic variables like temperature, pressure, etc. But such studies present formidable challenges, requiring complex (and quite expensive) experimental techniques, demanding huge amounts of computing time, and leading, after a major effort of many people, to a few points on a figure, not always easy to interpret. In spite of the strong indications that very interesting phenomena occur in the early stages of a Pb-Pb collision, at the highest SPS energies, we still do not know the final answer to the critical question that motivates this field: can we convince ourselves and the community at large that we have formed quark matter in the laboratory?

The final clarification of the present SPS results requires a careful and systematic approach, to establish beyond reasonable doubt that the QCD phase transition from hadronic to quark matter happens in central Pb-Pb collisions at the highest SPS energies. Further work on the available data remains to be done and, in some cases, where information is obviously missing, new measurements should be urgently performed. Models that claim to explain the available results must provide specific predictions for future measurements, with appropriate and carefully explained uncertainty bands. When and if the new observations validate those predictions, we will have made substantial progress in our understanding.

2 Overview of Heavy-Ion Collisions at the SPS

To recognise specific features of heavy-ion collisions with respect to a simple superposition of nucleon-nucleon interactions, it is very important to make a 'scan' in the centrality of the events. While the most peripheral collisions should be similar to conventional physics, the most central and violent events, where the two nuclei collide head-on, are expected to reveal some kind of 'anomalous' behaviour.

The centrality variables, like the impact parameter, b, or the number of participant nucleons, $N_{\rm part}$, can be determined from directly measured quantities, such as the charged hadron multiplicity, $N_{\rm ch}$, the transverse energy, $E_{\rm T}$, and the zero-degree (forward) energy, $E_{\rm ZDC}$. Assuming that $N_{\rm ch}$ and $E_{\rm T}$ are directly proportional to $N_{\rm part}$ or, similarly, that $E_{\rm ZDC}$ scales linearly with the number of spectator nucleons from the projectile ion (wounded-nucleon model [6]) it is possible to describe with good accuracy the measured distributions in the framework of the Glauber model of nucleus-nucleus collisions. A typical example [7], using $N_{\rm ch}$, can be seen in Fig. 2.

Figure 3 illustrates an alternative way to measure the centrality of the collisions, using the forward energy measured in a hadronic calorimeter placed in the beam line (at zero degree). If the incident Pb beam ion traverses the traget without interacting, all the energy (208×158 GeV, or 33 TeV) is deposited in the calorimeter, resulting in the 'beam peak' seen in the figure. This figure also illustrates the fact that the experiments can select the collected events, at the trigger level, according to the centrality of the collisions. This is particularly useful to increase the relative fraction of central events in the collected data. Without this feature, many experiments would almost only collect the much more frequent peripheral collisions, lacking statistics in the most interesting region.

Figure 4 shows the baryon rapidity distributions, in the centre-of-mass system, for central Pb-Pb collisions, compared to the S-S and pp distributions, scaled up to match the number of nucleons participating in the Pb-Pb collisions. The broad peaks at rapidities around 1.5 correspond to the projectile and target nucleons, shifted to mid rapidities from the beam and target rapidities due to the loss of energy induced by their mutual traversing (an effect commonly known as 'stopping'). The S beam had an energy of 200 GeV per incident nucleon, while the Pb beam was less energetic, 158 GeV per nucleon, therefore having a somewhat smaller 'beam rapidity'. However, the reason why the Pb peaks are closer to mid rapidity than the S peaks is mostly due to the fact that the Pb nuclei are much bigger, thereby being much more effective in 'stopping' each other.

Most of the particles newly produced in a heavy-ion collision, like pions and kaons, for instance, are produced at mid-rapidity, after the two colliding nuclei crossed each other. Figure 5 shows the pseudo-rapidity distributions of charged particles produced in Pb-Pb collisions, at 158 GeV per nucleon (top)

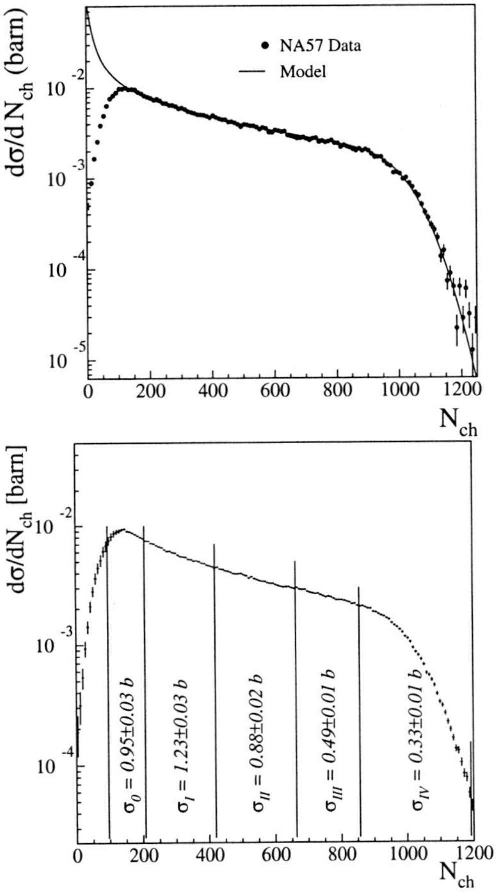

Fig. 2. The charged hadron multiplicity distribution measured by NA57 adjusted by the wounded-nucleon model (top) and illustrating the splitting of the event sample in different centrality classes (bottom)

and at 40 GeV per nucleon (bottom), for several different centrality classes, tagged by the E_{ZDC} energy of the events.

Besides the number (or rapidity density) of produced charged particles and the total forward energy measured in a zero-degree calorimeter, also the flux of energy released in the tranverse plane, E_{T}, provides a good estimator of the geometry of the heavy-ion collision. In fact, the measurement of E_{T} is essentially equivalent (but much easier to do experimentally) to the measurement of the total multiplicity of produced particles (mostly pions), from the point of view of sampling the events in different centrality classes. It is actually a remarkable observation that the ratio between E_{T} and the number of produced charged particles remains essentially constant from the

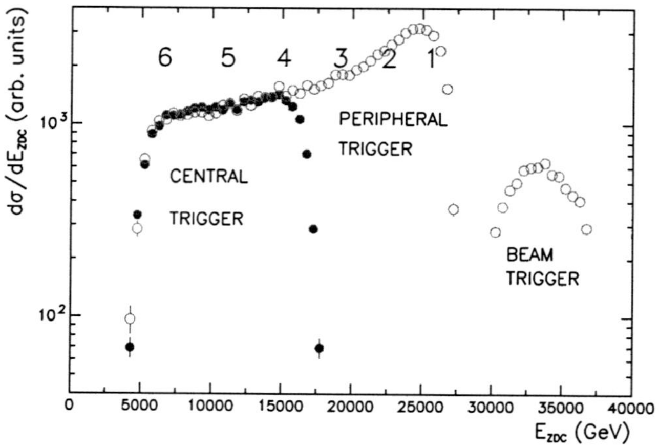

Fig. 3. Forward energy distributions measured by the NA50 zero-degree calorimeter, for three different event samples selected at the trigger level

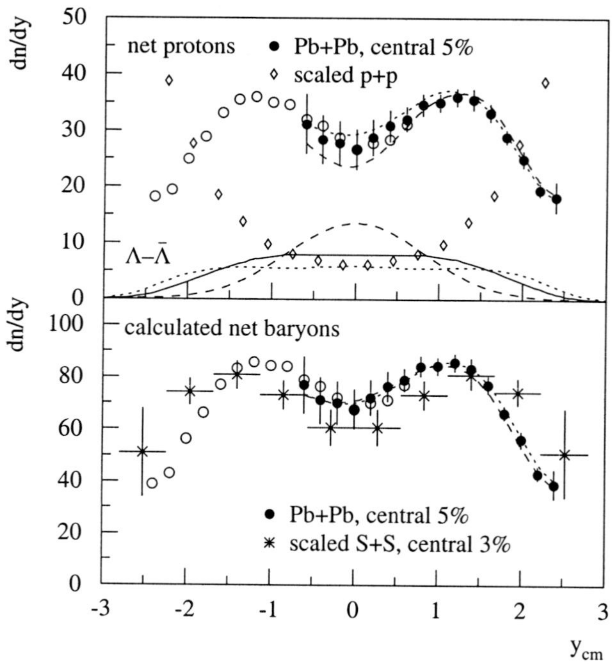

Fig. 4. Rapidity distributions of baryons for central Pb-Pb collisions, compared to scaled S-S and pp distributions

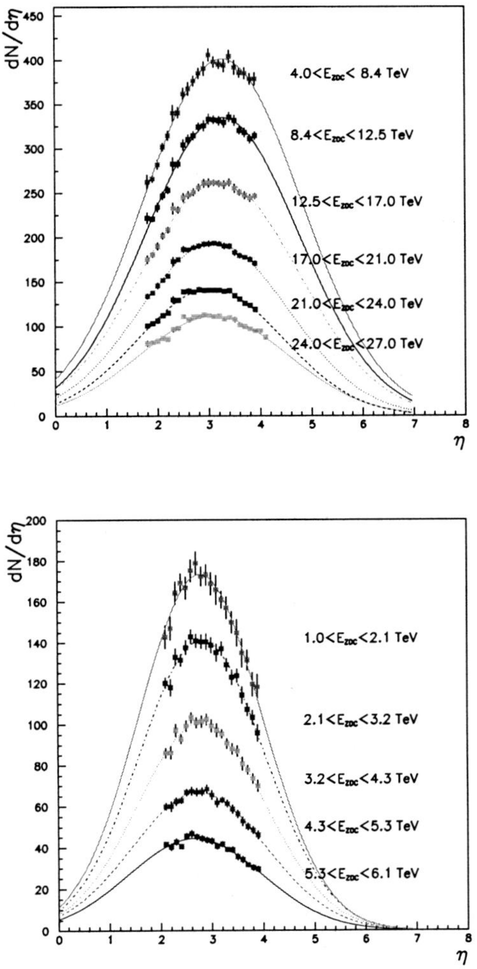

Fig. 5. Pseudo-rapidity distributions of charged particles produced in Pb-Pb collisions of different energies and centralities, as measured by the NA50 experiment. The data points are fitted with gaussians

most peripheral to the most central collisions (this remains valid at the higher energies of RHIC [8]).

Figure 6 shows the transverse-energy distributions of S-Au and Pb-Pb collisions, as measured by the NA35 and NA49 experiments, with calorimeters placed at mid-rapidity. Besides allowing to tag the events from peripheral to central collisions, the measurement of the transverse-energy rapidity density, at mid-rapidity, is very important for the estimation of the energy density, ϵ,

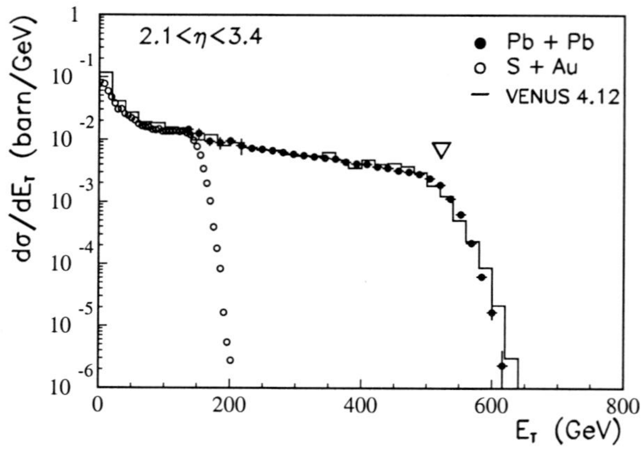

Fig. 6. Transverse-energy distributions of S-Au and Pb-Pb collisions, from NA35 and NA50. The Pb data are compared to a calculation done with the VENUS Monte-Carlo event generator

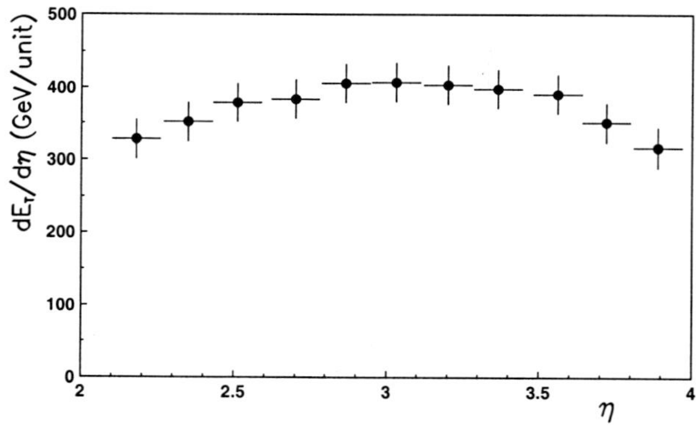

Fig. 7. Rapidity density of the transverse energy emitted in Pb-Pb collisions, as measured by NA49. At mid-rapidity, 2.92, the value 400 GeV per unit rapidity is observed

reached in these collisions. Figure 7 shows that the value 400 GeV is reached in central Pb-Pb collisions at the highest SPS energies [9].

From the measured values of mid-rapidity $\mathrm{d}E_{\mathrm{T}}/\mathrm{d}y$, the energy density can be estimated according to the formulation proposed by Bjorken almost 20 years ago [10], assuming a boost-invariant longitudinal expansion. The estimates give $\epsilon \sim 3.2$ GeV/fm^3 for central Pb-Pb interactions [9], significantly higher than the value 600-700 MeV/fm^3 calculated in lattice QCD for the occurrence of the phase transition [1]. Table 1 shows the corresponding estimates for smaller collision systems, S-S and S-Au.

Table 1. Values of the energy density, ϵ, reached in some collision systems studied at the SPS, estimated according to the Bjorken model

System	E_{lab}/A (GeV)	N_{part}	ϵ (GeV/fm^3)
S-S	200	58	1.3
S-Au	200	113	2.6
Pb-Pb	158	390	3.2

3 Strangeness Production

One of the earliest predictions in the field of high-energy heavy-ion physics is that particles containing strange quarks should be produced more often if the produced system goes through a quark-gluon plasma phase. An increase of around a factor 2 has indeed been observed [11], in global strangeness production, when comparing heavy-ion to elementary collisions, at around the same colliding energies, as can be seen in Fig. 8.

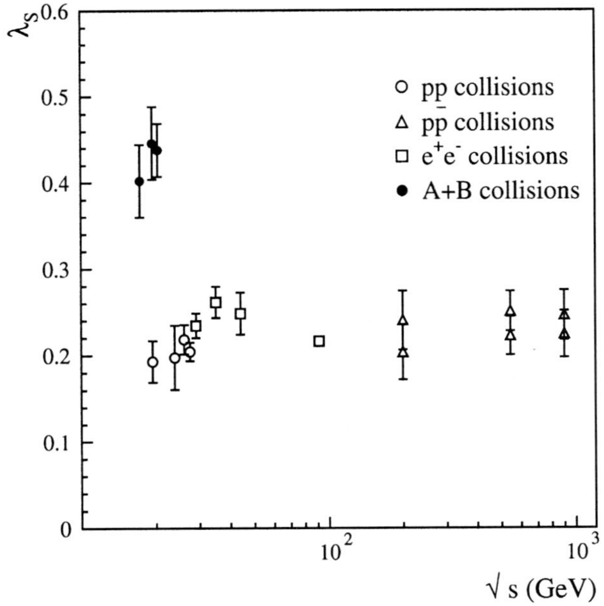

Fig. 8. The strangeness suppression factor, λ_s, for the production of strange quarks with respect to the production of u and d quarks, as a function of the collision energy, for elementary and nuclear collisions

Global strangeness yields are dominated by kaon production (around 75 % of all the produced strange particles). Figure 9 shows how the kaon multiplicity, per produced pion, evolves with the system size, from peripheral to central Pb-Pb collisions, including points from some other (smaller) colliding systems. These measurements were done, in particular, by the NA49 large-acceptance experiment [12].

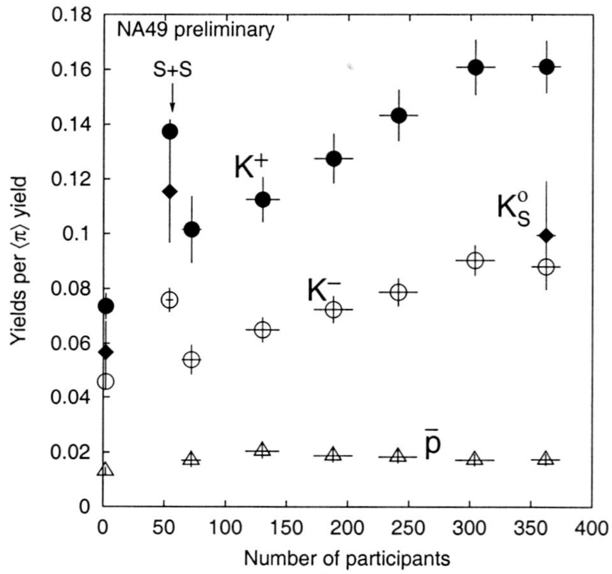

Fig. 9. Production yield of kaons (and anti-protons), normalized to pion production, versus the number of nucleons involved in the pp, S-S and Pb-Pb collisions

The most spectacular observations have been made, however, in the multi-strange hyperon sector. The very large enhancement factors in particle yields per participating nucleon (see Fig. 10), reaching a factor around 17 for the Ω, a triple-strange hyperon, and the fact that these factors are significantly higher for the states with more strange quarks, i.e. $E_\Omega > E_\Xi > E_\Lambda$, where E_i is the enhancement of the particle i with respect to p-A interactions, are naturally explained if the particle yields are determined from statistical hadronization of a strangeness-enhanced plasma phase. On the contrary, such enhancement levels cannot be reproduced in conventional (final-state hadronic rescattering) scenarios, given the short lifetime of the expanding hadronic system (see, however, the recent work mentioned in Ref. [13]). When these results were presented at the Quark Matter 1999 conference [14], a question was left in the air: is there a threshold behaviour in the enhancement pattern, between the p-Be and p-Pb points and the Pb-Pb values? The flat pattern observed in the Pb-Pb data indicated very little dependence on the centrality of these collisions, for $N_{\text{part}} > 100$. Where was the transition?

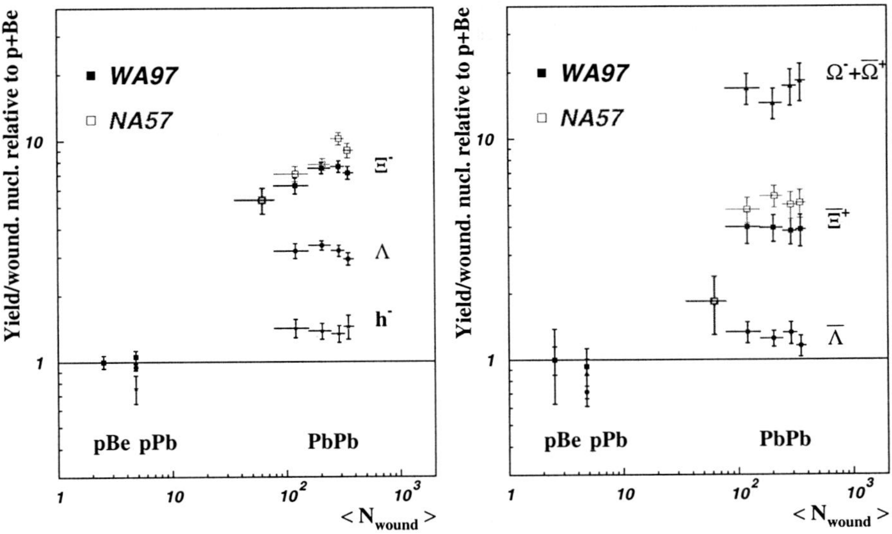

Fig. 10. Multi-strange hyperon production yields, at mid-rapidity, per wounded nucleon, normalized to the p-Be system, as a function of the number of wounded nucleons. The right panel collects the particles with no valence quark in common with the colliding nucleons

The NA57 experiment has continued these studies, making a special effort to collect peripheral Pb-Pb collisions. The first results were presented at the Quark Matter 2001 conference [15]. They show, as can be seen in Fig. 10, that the enhancement of $\bar{\Xi}^+$ production increases by a factor 2.6, from $N_{\text{part}} = 62$ to 121. The confirmation of a threshold behaviour in the strangeness enhancement pattern may come from the data analysis of the other strange hyperons. Unfortunately, having only five bins in centrality, and four of them showing a flat behaviour, the Pb-Pb pattern measured by the NA57 experiment will fall short of showing a clear transition, with a characteristic threshold, if it exists. A better understanding of the apparent onset of the enhancement would require studying 'intermediate mass' nuclear collisions, like In-In, for instance, in small centrality steps. Unfortunately, the effort required by the preparation of the ALICE experiment seems to prevent the realisation of such future studies.

Another major addition to the strangeness chapter is being provided by the NA49 experiment, when running with proton beams. The very large acceptance of this experiment makes it particularly appropriate in the study of asymmetric collision systems, as p-A interactions, where the reflection of the probed phase space window around midrapidity cannot be performed. First results of "NA49-hadrons" have also been shown at the Quark Matter 2001 conference [16], raising some questions on the "p-A reference baseline" used by WA97/NA57 in the extraction of the enhancement factors. Fortunately, we will see further data on this issue in the near future, since "NA49-hadrons" has been approved for further running in the next few years, and NA57 will collect more data on proton induced collisions in 2001.

Still in the strangeness sector, long standing questions concerning ϕ production remain unclear. NA50 sees, in the dimuon decay channel, a strong increase in the yield of ϕ mesons produced in heavy-ion collisions and a transverse-mass spectrum with a rather low 'inverse slope' [17], contrary to the observations of the NA49 experiment [18], in the K^+K^- decay channel. Future measurements of low-p_T ϕ production in the dilepton channel, by NA60, should help clarifying the source of discrepancy.

4 Evolution of the Final State

The understanding of the particle multiplicities (or relative production yields) and of their kinematical distributions gives significant information on the properties of the system that, by hadronization, resulted in the observed final states. The hadronic data collected at the SPS, in particular with high-energy Pb-Pb collisions, has been studied in the framework of statistical models of the hadronization process. The data collected in the large acceptance NA49 detector have been particularly useful for these studies, and to fix the free parameters of the models, such as the chemical freeze-out temperature, the baryon chemical potential, etc.

Figure 11 shows several measurements of ratios of particles yields (points), compared to values (lines) calculated in a particular statistical model [19], where the hadrons are produced as a gas in complete chemical equilibrium, with a chemical freeze-out temperature of 168 MeV and a baryon chemical potential of 266 MeV. Given the fact that this temperature is very close to the value expected for the phase transition to the QGP phase, it is tempting to imagine that the system crosses the phase boundary, from the partonic to the hadronic phases, shortly before the chemical freeze-out point. Furthermore, the almost complete strangeness saturation assumed in these models indicates that the observed strangeness enhancement comes essentially from the partonic phase.

Once the hadrons are produced, the chemistry step is over but the particles continue to collide with each other, influencing their kinetics. It is only later, once the system has expanded further, that the particles stop interacting and fly through to the detectors. This 'thermal freeze-out' point can be probed by looking at the transverse mass spectra of identified particles, for instance. The flux of particles in the transverse plane is purely due to the production mechanisms taking place during the collision, while in the longitudinal direction things are made more complicated by the very high

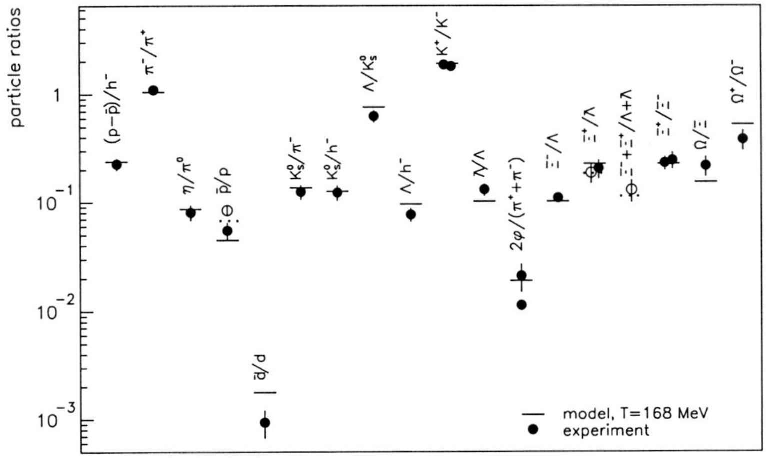

Fig. 11. Comparison of particle yields measured in Pb-Pb collisions, at the SPS, with the values expected in a thermodynamical model assuming statistical particle production from a thermal bath

initial-state energy of the colliding nuclei. The NA44, NA49 and WA97 experiments, among others, have shown that all the hadrons exhibit exponential transverse mass spectra, that can be simply characterized by the temperature of the medium at thermal freeze-out, T_f, and by the mean transverse flow velocity of the medium, $\langle v_\perp \rangle$. These two values determine the inverse slope of the transverse-mass distributions, for each particle species, roughly as $T_f + 0.5 \cdot m_0 \cdot \langle v_\perp \rangle^2$. This linear dependence with the mass of the produced particles, m_0, can be seen in Fig. 12 [20].

These observations show that the thermal motion of each produced hadron is superimposed on the ordered collective flow of the whole system, due to the radially expanding fireball, looking like a microscopic version of the Big Bang [21]. The departure of the Ω from the linear trend may be due to an early decoupling (freeze-out) of this particle, probably because, having zero isospin, it cannot form resonances with the copious pions, that substantially contribute to the equilibration of the other hadrons [22].

To separately determine the radial flow velocity and the freeze-out temperature of the system, it is very important to have other, independent, sources of information. For instance, the transverse flow of the system can also be seen through its influence on the Bose-Einstein pion correlations [23]. By following the dependence of the HBT transverse radius on the transverse momentum of the pion pair (see Fig. 13), it is possible to extract the correlation between the radial flow velocity and the freeze-out temperature of the system. Fortunately, the combination of this correlation with the information obtained from the study of the hadronic transverse mass spectra, allows to

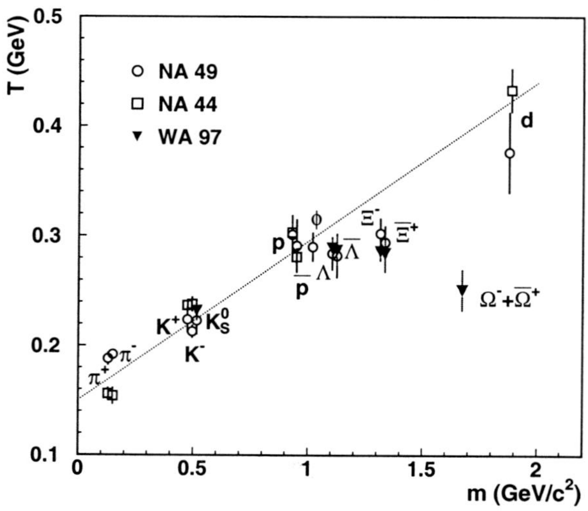

Fig. 12. Inverse slope, T, of the exponential m_{T} spectra, as a function of the mass of the produced particle, for Pb-Pb collisions at the highest SPS energies

separate the random thermal motion from the collective flow, as can be seen in Fig. 13, leading to the values $T_f \sim 100$ MeV and $\langle v_\perp \rangle \sim 0.55\,c$ [24].

5 Low-Mass Dilepton Production

The CERES experiment has observed [25] that the yield of low mass e^+e^- pairs measured in p-Be and p-Au collisions is properly described by the expected "cocktail" of hadronic decays, while in Pb-Au collisions, on the contrary, the measured yield, in the mass region 0.2–0.7 GeV, exceeds by a factor of 2.5 the expected signal [26], as shown in Fig. 14.

The dileptons from $\pi^+\pi^-$ annihilation would increase the expected yield around the mass of the ρ meson, not reproducing the measured shape. The excess dileptons are concentrated at low p_{T} (Fig. 15) and their yield seems to scale with the square of the charged particle multiplicity (Fig. 16).

These observations are consistent with the expectation that the properties of vector mesons should change when produced in dense matter. In particular, near the phase transition to the quark-gluon phase, chiral symmetry should be partially restored, making the vector mesons indistinguishable from their chiral partners, thereby inducing changes in their masses and decay widths [27]. The short lifetime of the ρ meson, shorter than the expected lifetime of the dense system produced in the SPS heavy-ion collisions, makes it a sensitive probe of medium effects and, in particular, of chiral symmetry restoration.

The present measurements are not accurate enough to clearly distinguish between a change in the ρ mass (signaling the restoration of chiral

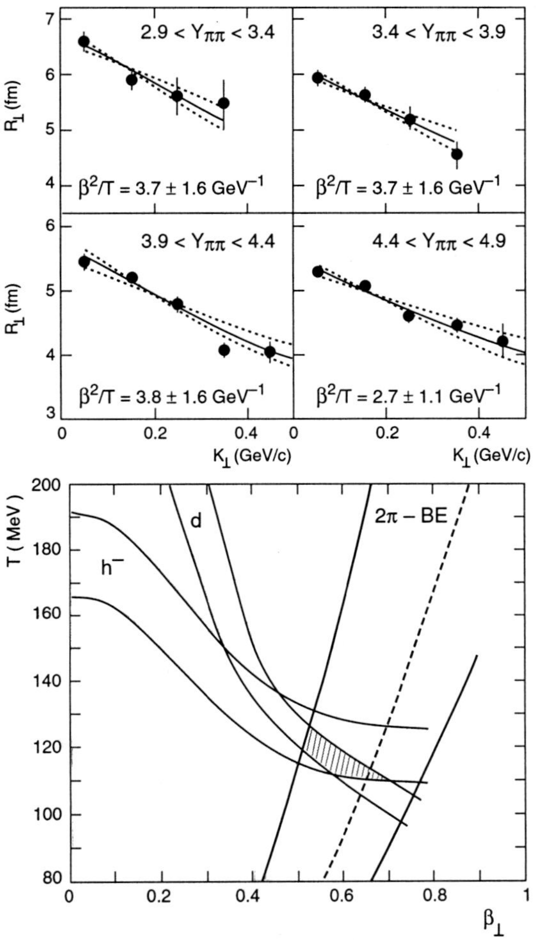

Fig. 13. Dependence of the transverse radius of the produced system on the transverse momentum of the pion pair, for several rapidity ranges (top) and correlation between thermal freeze-out temperature and radial flow velocity for central Pb-Pb collisions (bottom)

symmetry) and a broadening due to conventional hadronic interactions [28]. Already E. Shuryak [29] and B. Müller [30], in their Quark Matter 1999 papers, emphasized the importance of a considerable improvement in the CERES measurements of low-mass dilepton production, in terms of signal to background ratio, mass resolution, and statistics. A TPC was added to the CERES setup [31], to improve the momentum resolution of the dielectron measurement. Unfortunately, problems in the data taking during the year 1999 have prevented the CERES experiment from collecting a significant sample of dilepton events [32]. Those problems were solved in time for

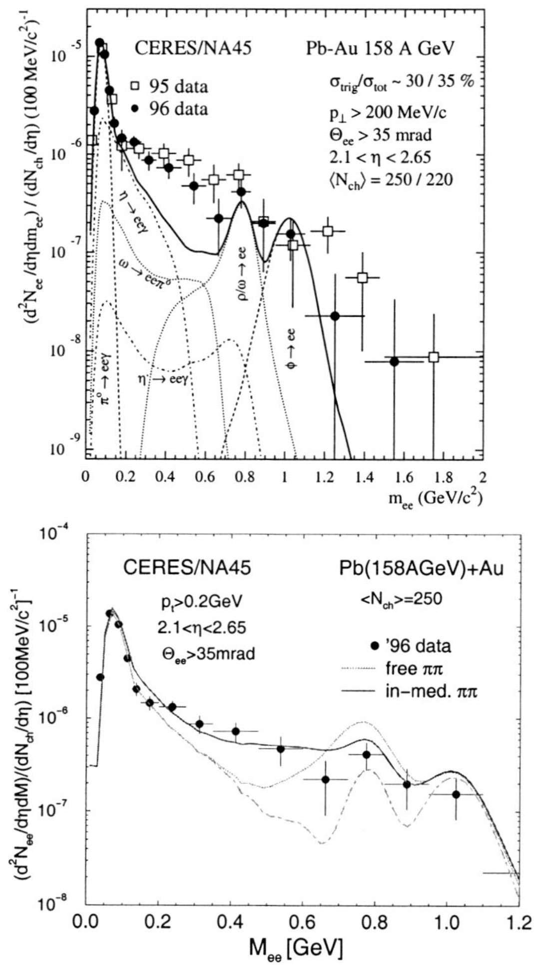

Fig. 14. Dielectron mass distribution measured by CERES in Pb-Au collisions, compared to the expected hadronic decays (top) and to the contribution from $\pi\pi$ annihilation with and without in medium effects (bottom)

the run of the year 2000 and we are eagerly waiting for the results from this new data set. A first look into this new data sample indicates that a mass resolution close to the expected value of around 2 % may be within reach. Unfortunately, the collected statistics will probably not be enough to have an accurate measurement of the ω resonance, which would be very helpful in the studies of the in-medium modifications apparently affecting the ρ.

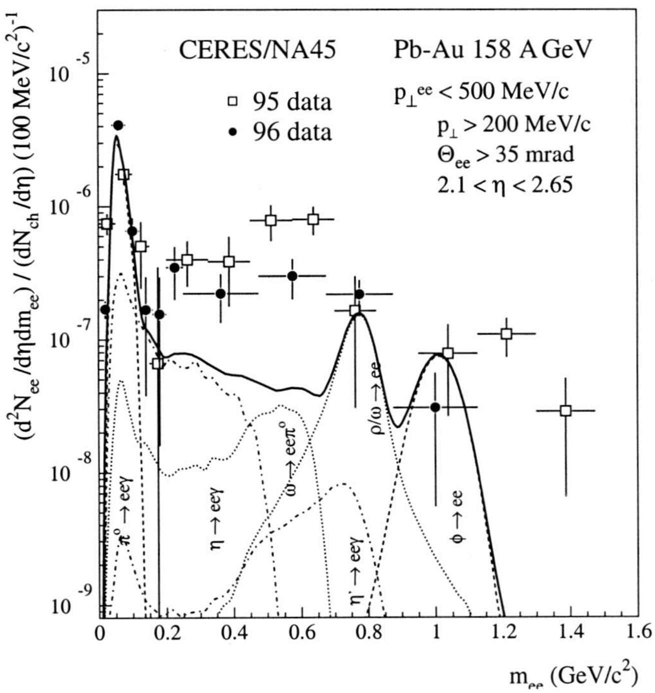

Fig. 15. Mass distribution of low-p_T dielectrons produced in Pb-Au collisions, as measured by CERES

6 Intermediate-Mass Dilepton Production

The NA38 and NA50 experiments have studied [33] the production of dileptons in the mass window between the ϕ and the J/ψ peaks, as a superposition of Drell-Yan dimuons and simultaneous semileptonic decays of D and $\bar{D}$ mesons, after subtraction of the combinatorial background from pion and kaon decays [34].

The Drell-Yan and open-charm contributions were calculated with the PYTHIA event generator [35] with the MRS A set of parton distribution functions [36]. PYTHIA describes reasonably well [37] the kinematics and cross sections (including the energy dependence) of D meson hadroproduction, as well as the semi-leptonic decays and the corresponding lepton distributions. Figure 17 shows that the dimuon mass spectra measured in p-A collisions are very well reproduced taking the high-mass region to normalize the Drell-Yan component and an open-charm cross-section in good agreement with direct measurements made by other experiments. The calculations do not include NLO QCD diagrams, particularly important for high-p_T Drell-Yan production.

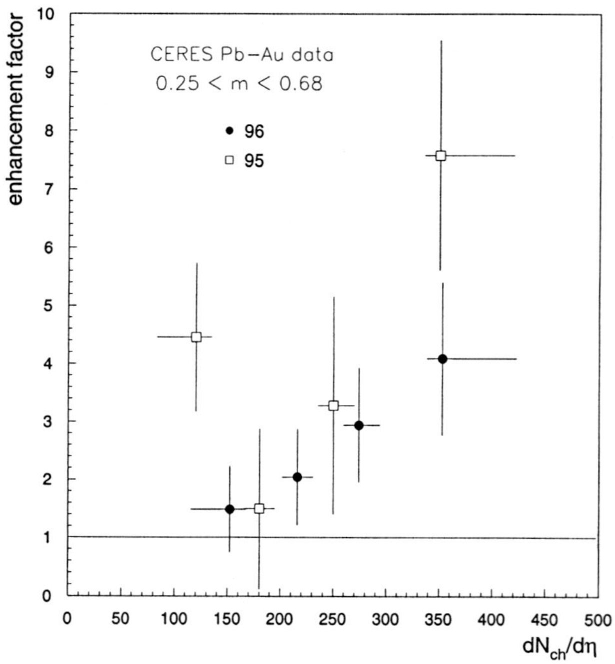

Fig. 16. Yield of excess dileptons versus the charged particle multiplicity. Is there an onset at $dN_{ch}/d\eta \sim 100$?

On the contrary, the superposition of Drell-Yan and open-charm contributions, with the nucleon-nucleon absolute cross sections scaled with the product of the mass numbers of the projectile and target nuclei (as expected for hard processes), fails to properly describe the dimuon yield measured in ion collisions.

Figure 18 shows, for central Pb-Pb collisions, how the sum of the expected sources underestimates the measured data. The same figure shows that the data can be reproduced by simply increasing the open-charm yield. The scaling factor by which the charm contribution should be multiplied to properly describe the measured spectra seems to grow linearly with the number of nucleons participating in the collision, as shown in Fig. 19. In this figure, the points "4-D analysis" are obtained with an improved deconvolution method to extract the physical kinematics from the measured values, affected by acceptance and finite resolution (smearing) effects. This analysis method accounts for physical correlations among kinematical variables and does not require any assumption on the specific shapes of their distributions [38].

The observed excess can be due to an overall enhancement of open-charm production in heavy-ion collisions [39]. An alternative explanation could be that the rescattering of the charm quarks or D mesons in the produced medium leads to a broader p_T distribution and would locally enhance the

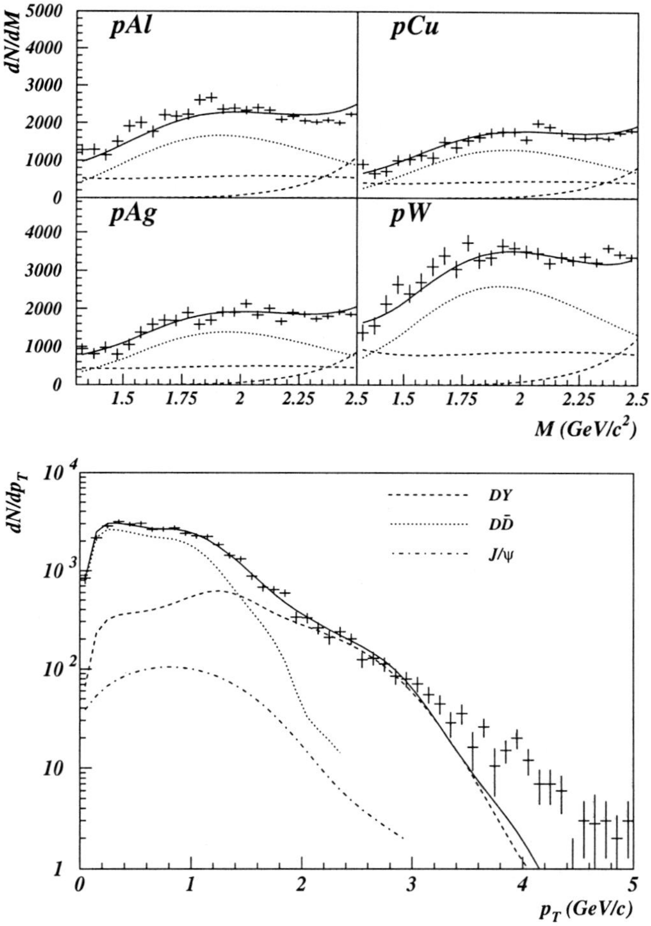

Fig. 17. Dimuon mass (top) and p_{T} (bottom) distributions measured in p-A collisions by NA50, compared to the corresponding expected sources

charm component in the limited phase-space domain covered by NA50 [40]. However, recent studies [41] have shown that the data cannot be accounted for by this last model.

The observed excess can also be due to the production of thermal dimuons, a signal that was the original motivation for the NA38 experiment and that has been recently revisited [42,43]. In particular, the intermediate-mass dimuons produced in the most central Pb-Pb collisions are well reproduced by adding thermal radiation [41], calculated according to the model of Ref. [42], to the Drell-Yan and charm contributions normally extrapolated from nucleon-nucleon collisions. This model explicitly includes a QGP phase transition with a critical temperature of 175 MeV. The best description of the data is obtained using ~ 250 MeV as the initial temperature of the QGP medium ra-

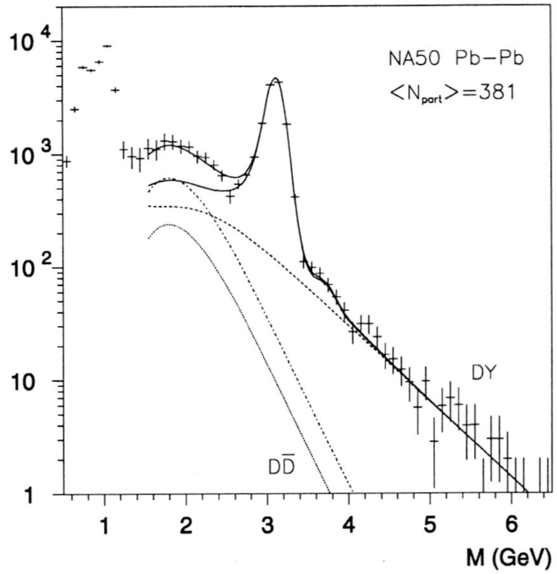

Fig. 18. Dimuon mass distribution measured in central Pb-Pb collisions, compared to the expected sources, with and without scaling up the charm contribution

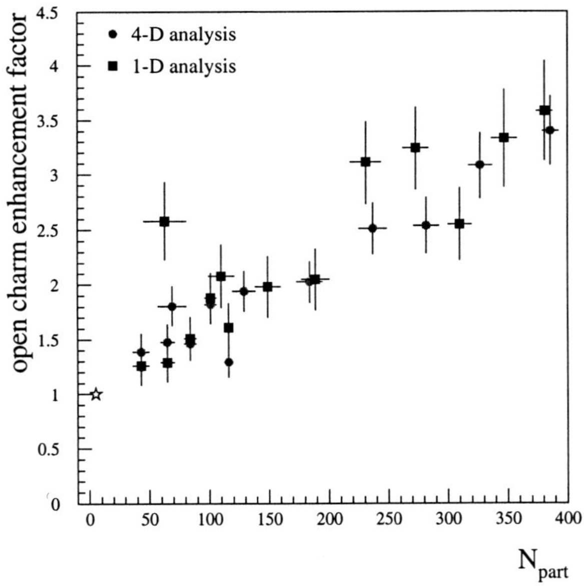

Fig. 19. Dependence on the number of participants of the scaling factor by which the charm contribution must be multiplied to properly describe the measured spectra

diating the virtual photons. The presently available data cannot distinguish between an absolute enhancement of charm production and the emission of thermal dilepton radiation. The clarification of the nature of the physical process behind the observed excess is also on the wish lists presented by E. Shuryak and B. Müller in their Quark Matter 1999 papers, and is the strongest physics motivation of the NA60 experiment.

7 Charmonia Production and Suppression

The formation of a deconfined medium should induce a considerable suppression of the charmonia production rate, due to the colour 'Debye' screening of the $c\bar{c}$ potential or to the breaking of the $c\bar{c}$ binding by scattering with energetic (deconfined) gluons [44]. However, even the relatively simple measurement of J/ψ production faces a big challenge when it comes to furnish a convincing logical case that proves, to the satisfaction of the experts in the field, that a deconfined state of matter has been formed. It is not enough to show that a certain observable changes from p-Pb to Pb-Pb collisions, for instance, or to argue that its value in the most central nucleus-nucleus collisions is different from what is calculated in a 'conventional physics' model. The best path to clearly establish a solid result and shed light on this complicated field is to build a robust set of measurements that establishes a precise reference baseline, relative to which the specific behaviour of heavy-ion collisions can be extracted. Such a baseline shows what is the 'normal' behaviour of the signal we are studying, with respect to which we look for changes due to QGP formation. Furthermore, we are in a much better position if nature provides us with a reference process, insensitive to the formation of a deconfined phase, specially if we can measure it with the same detector.

In the case of the J/ψ suppression topic, the baseline is built from the measurements done by NA38 and NA50 with pp, p-A and light-ion collisions [45]. Very peripheral Pb-Pb collisions have been successfully collected in the year 2000 and we will soon know how well they follow the "normal nuclear absorption" baseline. The best reference physics process, at SPS energies, is the rate of high-mass Drell-Yan dimuons, since the Drell-Yan process can be precisely calculated and depends on the collision system in a well known way. Figure 20 shows absolute cross sections of Drell-Yan production measured by the NA38, NA51 and NA50 experiments in pp, p-D, p-W, S-U and Pb-Pb collisions, at several energies, divided by the corresponding values calculated at leading order with the MRS A parton distribution functions.

From the measured J/ψ production yields we can derive the J/ψ cross section per nucleon, $B^{\psi}_{\mu\mu}\sigma^{\psi}/\mathrm{AB}$, displayed in Fig. 21 as a function of the product of the mass numbers of the two colliding nuclei. Contrary to what happens with the evolution of the Drell-Yan process, the Pb-Pb J/ψ point is completely different from the value indicated by the pattern established by the p-A and light-ion measurements.

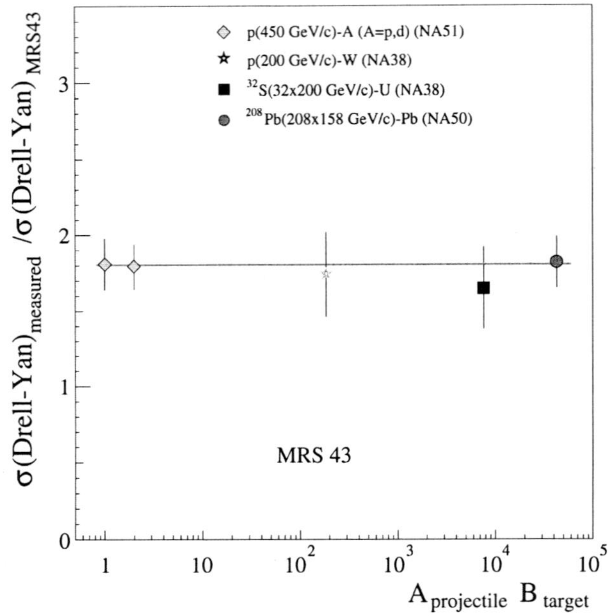

Fig. 20. The measured yield of Drell-Yan dimuons follows the expected values, from pp to Pb-Pb collisions

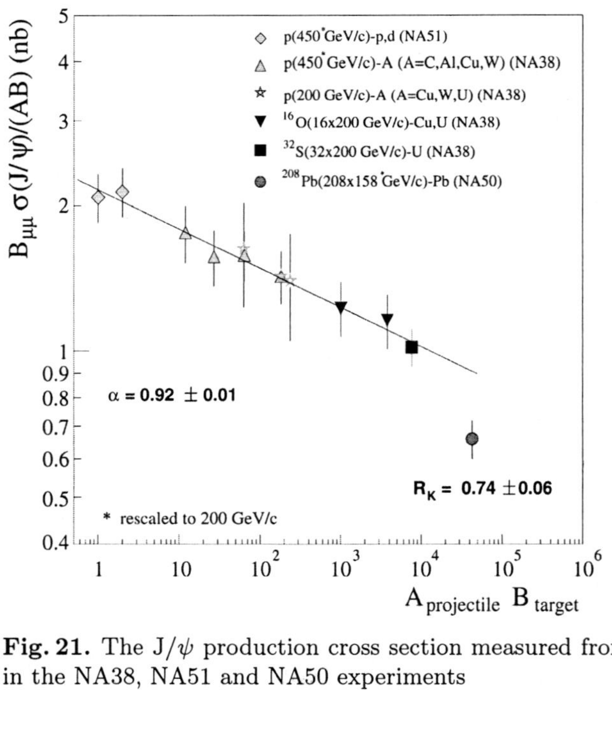

Fig. 21. The J/ψ production cross section measured from pp to Pb-Pb collisions, in the NA38, NA51 and NA50 experiments

The peculiar behaviour of the Pb-Pb data can be seen in much more detail by binning the collected event sample as a function of the centrality of the collisions. Figure 22 shows how the J/ψ production rate, with respect to the yield of Drell-Yan dimuons, decreases from peripheral to central Pb-Pb collisions, using either E_{ZDC} or E_{T} to estimate the centrality of the reactions. The same plots show the "normal J/ψ absorption line", determined by the reference data (from pp to S-U) shown in Fig. 21. The data points collected in the most central collisions show a very significant departure from the expected behaviour, while the most peripheral points seem to be in (rough) agreement with the absorption expected in normal nuclear matter.

The upper panel of Fig. 23 shows that the observed two-step suppression pattern is in clear disagreement with the predictions of the 'conventional' models [46–49] that attribute the disappearance of the J/ψ mesons to interactions with 'comoving' hadrons. A first departure from the conventional curves is seen for collisions that release around 40 GeV of neutral transverse energy in the electromagnetic calorimeter of NA50. A second substantial drop is seen for the most central Pb-Pb collisions, while the hadronic models systematically predict a smooth absorption trend. On the other hand, a two-step pattern, as seen in the data, is naturally expected if the charmonia states are dissolved in a deconfined medium, due to the different melting temperatures of the χ_c and J/ψ states (about 30–40 % of the observed J/ψ mesons result from the decay of χ_c states). The addition of the pp, p-A and S-U results makes the lower panel of Fig. 23 particularly illustrative of the unconventional nature of the observed J/ψ suppression pattern.

If we accept that this pattern indicates the production of a state of matter where colour is no longer confined, we must move on to the detailed understanding of how deconfinement sets in, and which physics variable governs the threshold behaviour of the (χ_c) suppression: (local) energy density, density of wounded nucleons, of percolation clusters, of produced gluons, etc. This requires collecting data with a smaller nuclear collision system like In-In. Indeed, it is possible to predict at which impact parameter, b, of In-In collisions the same threshold is reached in (local) energy density, or any other variable, as is reached in Pb-Pb collisions of $b \approx 8$ fm, where the χ_c state starts melting. According to the deconfinement model used in the calculation [50] shown in Fig. 24, the onset of anomalous J/ψ suppression should happen at an impact parameter $b = 2.0$–2.5 fm, in In-In collisions at 158 A GeV. A verification of such specific predictions would be the final element of proof that the deconfined quark-gluon phase sets in, and would provide fundamental information on the mechanisms behind the observed phenomena. In this context, it is also important to improve our knowledge of the nuclear dependence of χ_c production, in p-A collisions, at SPS energies.

The J/ψ data do not provide a direct measurement of the critical temperature. Finite-temperature lattice QCD tells us that the strongly bound J/ψ $c\bar{c}$ state should be screened when the medium reaches temperatures 30–40 %

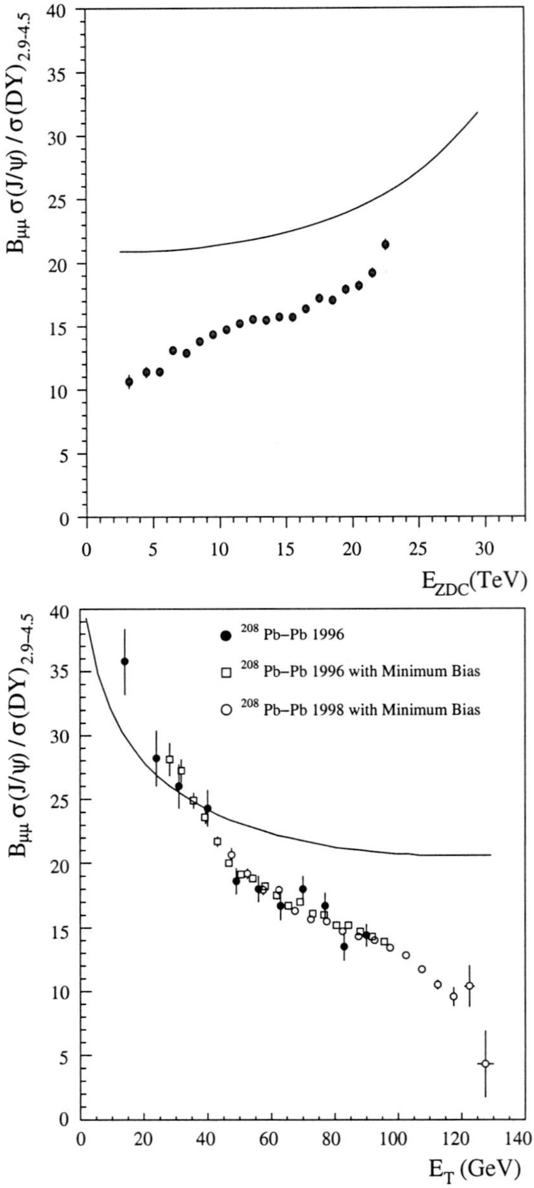

Fig. 22. The J/ψ suppression pattern as a function of the forward (top) and transverse (bottom) energy, in Pb-Pb collisions, as measured by NA50

higher than T_c, while the larger and more loosely bound ψ' state should melt near T_c. The ψ' is already significantly suppressed when going from p-U to peripheral S-U collisions but the presently existing results are not clear in

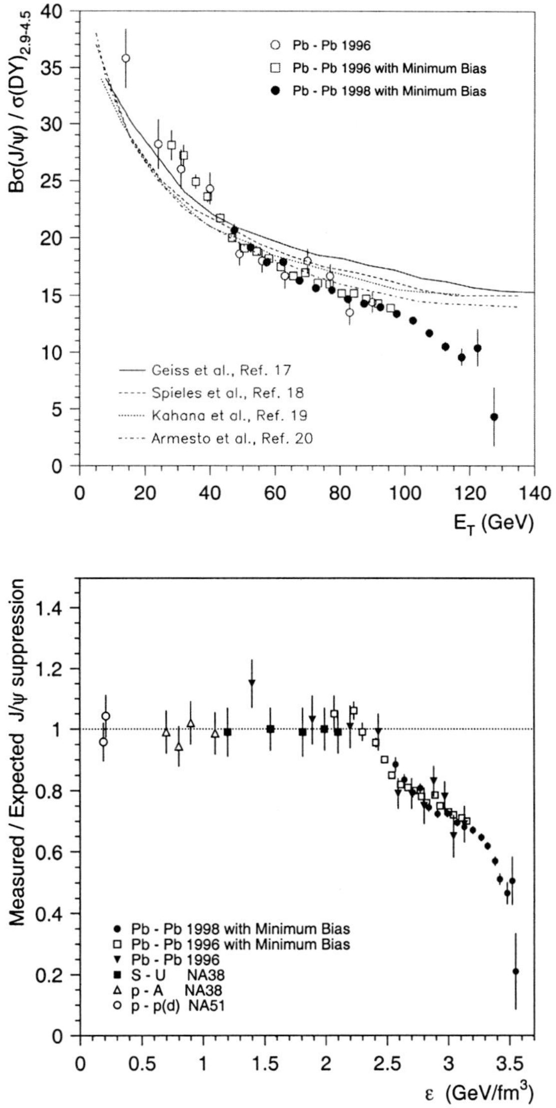

Fig. 23. The measured J/ψ suppression pattern rules out the presently available conventional models (top) and is in qualitative agreement with the two-step pattern expected in the deconfinement picture bottom)

what concerns the pattern of the ψ' suppression. Figure 25 shows the ratio between the ψ' and Drell-Yan production rates, as a function of L, the thickness of nuclear matter crossed by the charmonia states. The corresponding J/ψ (normal nuclear absorption) pattern is also shown, scaled down by the

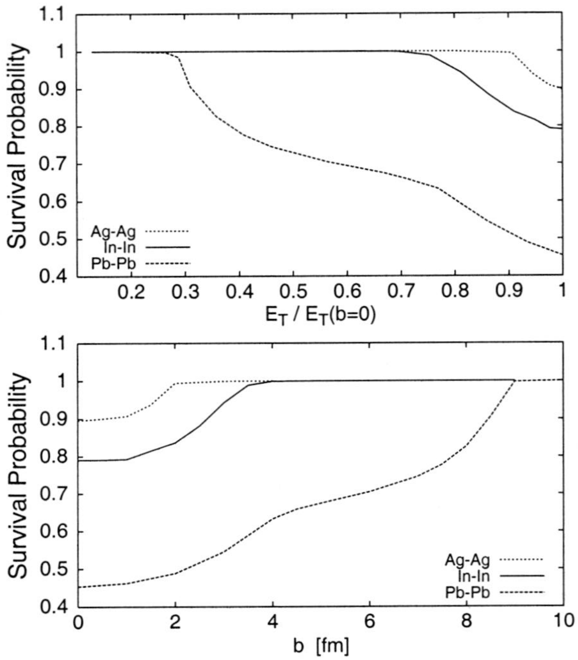

Fig. 24. The J/ψ suppression pattern predicted in a deconfinement model for different collision systems

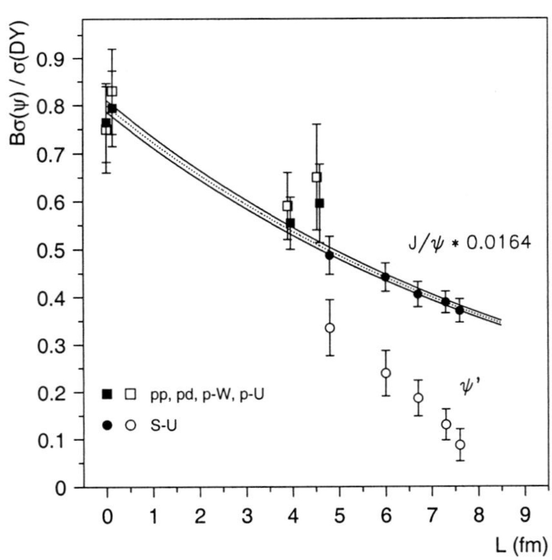

Fig. 25. The ψ' yield is strongly suppressed in S-U collisions. Hadronic absorption or Debye screening?

factor 1.64 %. Is the 'anomalous' ψ' suppression due to QGP melting or to hadronic absorption? If we see that this suppression happens more or less in an abrupt way, within a single-collision system rather than comparing p-U to S-U data, we would know that Debye screening is the mechanism responsible for the ψ' disappearance and we would have a clear measurement of T_c. This requires a new measurement, with improved mass resolution to have a cleaner separation between the ψ' and J/ψ peaks, and which scans an energy density region including the p-U and the S-U points.

Improved measurements of J/ψ and ψ' production, with intermediate mass nuclei, were also included in the wish lists of E. Shuryak and B. Müller, and are an important part of the physics program of the NA60 experiment, which will also measure χ_c production in p-A collisions.

8 Open-Charm Production

Knowing that the bound $c\bar{c}$ states are suppressed, it is natural to ask what happens to the *unbound* charm. Charm quarks are so heavy that they can only be produced at the earlier stages of the nuclear collision, before the eventual formation of the QGP state. Charm is the heaviest flavour that can be studied in heavy-ion collisions at the SPS energies. The production of charm quarks leads mainly to correlated pairs of D and $\bar{D}$ mesons. Only a few percent of the charmed quark pairs end up in the bound charmonia states presently studied by the NA50 experiment, and which exhibit a rather interesting "anomalous" behaviour. What happens to the vast majority of c quarks? Are they affected by energy loss while crossing the dense (partonic or hadronic) medium? Is charm production enhanced similarly to what has been seen in the strangeness sector? Finally, D meson production provides the natural reference with respect to which we should study the observed J/ψ suppression, since both production mechanisms depend on the same gluon distribution functions. If charm production is enhanced in nuclear collisions, it makes the J/ψ suppression even more anomalous. A direct observation of D meson production is clearly the most important new measurement that remains to be done at the SPS, and constitutes a basic reason for the construction and running of NA60.

9 Future Prospects

The results and open questions presented in the previous sections emphasize the importance of having a new experiment at the SPS that can significantly improve several existing observations and make a few new measurements, including a measurement of open-charm production in heavy-ion collisions. A dedicated experiment is needed that can cope with the very high particle multiplicities reached in the most central nuclear collisions (400 charged particles per unit rapidity at midrapidity) and with the rather small D production cross section.

The NA50 experiment has been using CERN's highest-intensity heavy-ion beam (more than 10^7 ions per second) and has a very selective dimuon trigger, quite appropriate to look for rare processes. The recently approved NA60 experiment [51] complements the muon spectrometer and zero-degree calorimeter already used in NA50 with two state-of-the-art silicon detectors, placed in the target region: a radiation-hard [52] beam tracker, consisting of four silicon microstrip detectors placed on the beam and operated at a temperature of 130 K, and a 10-plane silicon pixel tracking telescope, made with radiation tolerant [53] readout pixel chips, placed in a 2.5 T dipole magnetic field.

The NA60 experiment has been approved to run from 2001 to 2003, using proton, Pb and In beams. The following questions summarize the physics motivation of NA60.

- What is the origin of the dimuon excess seen in the intermediate-mass region? Thermal dimuon production?
- Is the open-charm yield enhanced in nucleus-nucleus collisions? How does it compare to the suppression pattern of bound charm states?
- What is the variable (local energy density, cluster density, etc.) that rules the onset of charmonia suppression?
- What is the physical origin of the ψ' suppression? If it is due to Debye screening, what is its melting temperature?
- Which fraction of the J/ψ yield comes from χ_c decays? Does it change from p-Be to p-Pb collisions?
- Are there medium-induced modifications in the ρ meson? Is there a threshold behaviour in the low-mass dilepton enhancement? What happens with the ω meson?
- Is the observed ϕ enhancement a specific feature of heavy-ion collisions? Is the ϕ sensitive to flow?

The high-granularity tracking telescope, placed in a powerful dipole field, gives access to the muon tracks at the vertex level and vastly improves the mass resolution of the dimuon measurement. This has been demonstrated in a very fast feasibility test done in 1998, using a small telescope (four half-planes) made of the previous generation of readout pixel chips. The results, shown in Fig. 26, confirm that the mass resolution improves from 70 to 20 MeV at the ω mass, as expected from the physics performance simulations illustrated on Fig. 27.

The NA60 beam tracker gives the transverse coordinates of the interaction point, on the targets, with enough accuracy (around 20 μm) to measure the impact parameter of the muon tracks, i.e. the minimum distance between the track and the collision vertex, in the transverse plane. Thanks to this information, NA60 will be able to separately study the production of prompt dimuons and the production of muons originating from the decay of charmed

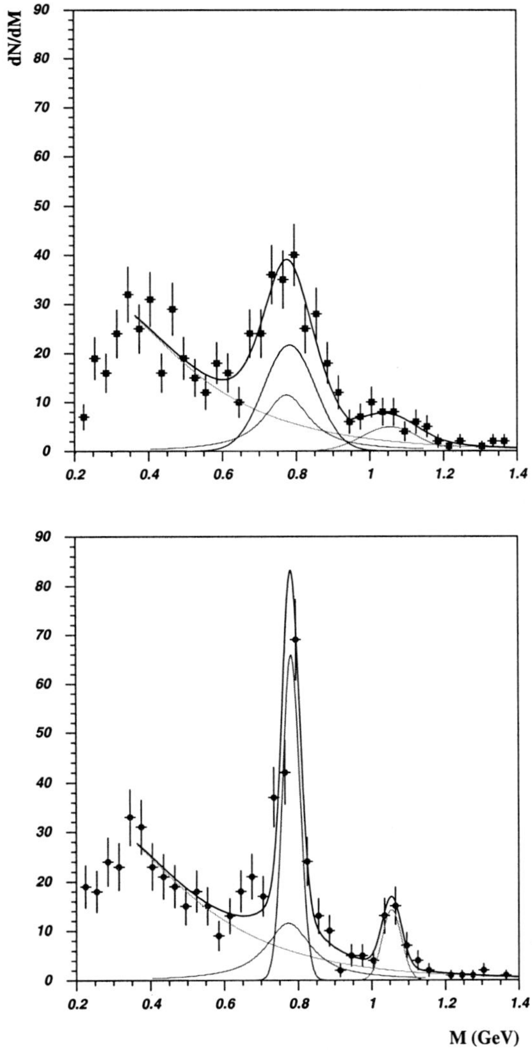

Fig. 26. Dimuon mass distributions measured in 1998, in p-Be collisions, before (top) and after (bottom) using the information of the test pixel telescope. The curves represent the low-mass vector meson resonances (ρ, ω and ϕ) on the top of a continuum. They are normalized to the same number of events in both figures. The collected statistics (600 events) correspond to a few minutes of NA60 running

mesons, in p-A and heavy-ion collisions. The prompt dimuon analysis will use events where both muons come from (very close to) the interaction vertex. The open-charm event sample is composed of those events where both muon tracks have a certain minimum offset with respect to the interaction point and a minimal distance between themselves at z_{vertex}. Figure 28 shows the

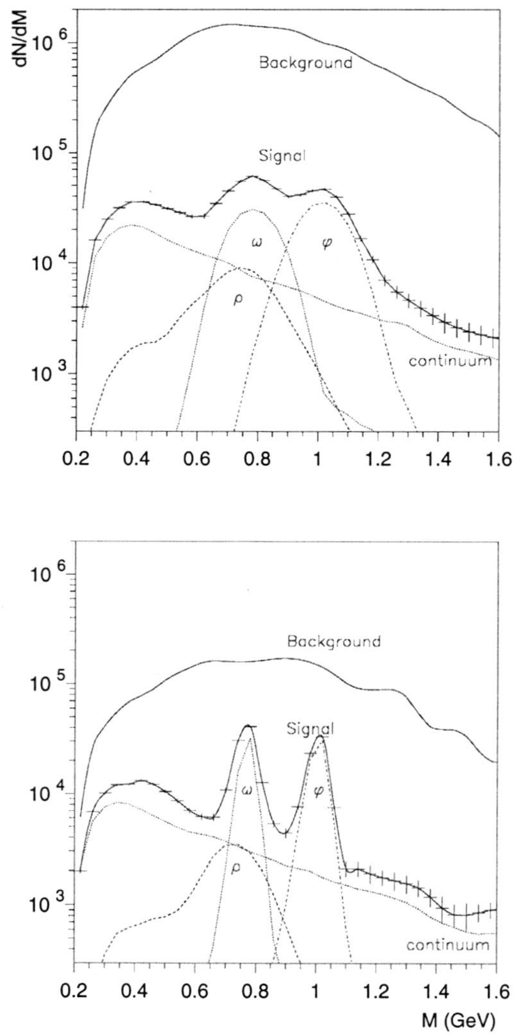

Fig. 27. Simulated dimuon mass distribution in Pb-Pb collisions, before (top) and after (bottom) using the NA60 pixel telescope information. The statistics corresponds to around one week of running

simulated mass spectra for both event samples. It should not be difficult to see which of these two event samples is enhanced by a factor of 2 or 3 in nuclear collisions of $N_{\text{part}} \sim 300$.

Figure 29 shows the accuracy of the determination of the interaction point, in the transverse plane. Along the beam axis the vertex is found with a precision of $\sim$ 100–150 μm. If the incident nucleus makes a peripheral collision

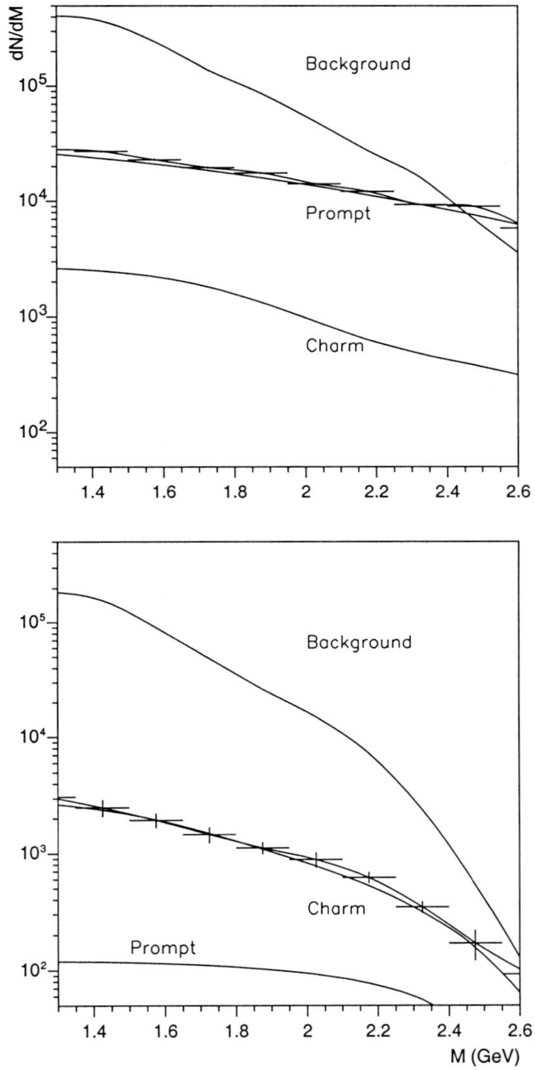

Fig. 28. Simulated dimuon mass distributions for the prompt (top) and charm (bottom) event selections. The background contribution is also shown, including pion/kaon decays and fake matches between the tracks in the muon and in the vertex spectrometers. The error bars in the signal points include the uncertainty from background subtraction

and the beam spectators fragment collides further down in the target, the double emission of particles should be distinguishable from what happens in a single (central) collision. This allows the use of a thicker target, with a corresponding gain in total effective luminosity. A high interaction rate is the

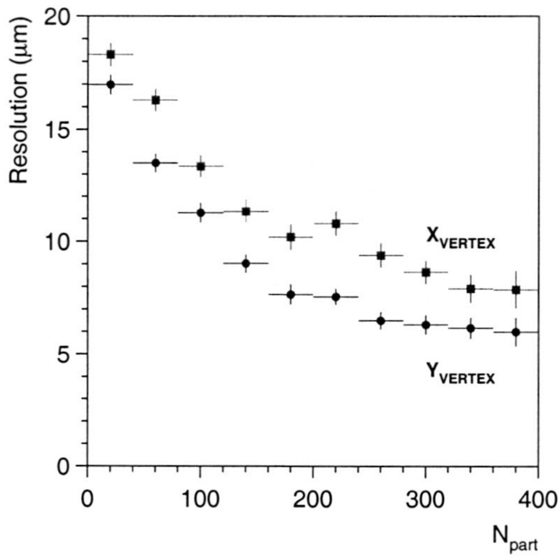

Fig. 29. Calculated resolution in the determination of the transverse coordinates of the interaction vertex

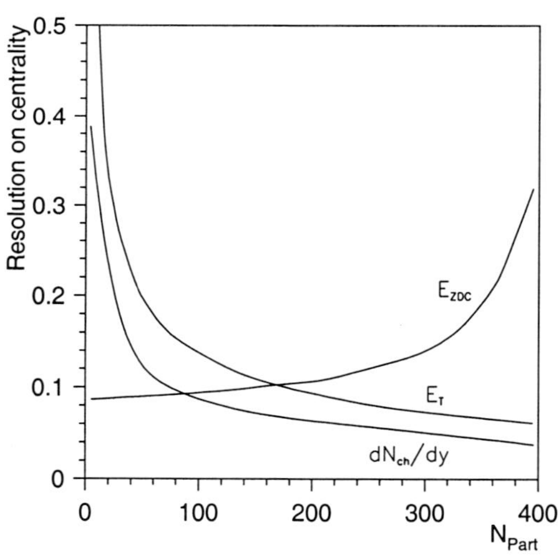

Fig. 30. Resolution on the measurement of the centrality of the collisions, using three different estimators

basis for enough statistics to study the charmonia production yield in many centrality bins, a necessary condition to accurately determine a step-wise suppression pattern. Besides statistics, it is also important to have a good accuracy in the measurement of the centrality of the collision. Figure 30 shows that a resolution between 5 % and 10 %, depending on the centrality, should be reached in NA60, using the forward energy, E_{ZDC}, and the charged particle multiplicity, $\mathrm{d}N_{\mathrm{ch}}/\mathrm{d}y$.

The studies of prompt dimuon and open-charm production in p-A collisions are important reference measurements, to understand the results obtained with nuclear collisions. In particular, the ratio between the open-charm and the Drell-Yan production cross sections will be determined with high accuracy in several p-A collision systems, revealing if these two hard processes have the same A-dependence or not. Figure 31 illustrates the foreseeable analysis of intermediate-mass dimuons production, in p-Be or p-Pb collisions, showing the p_{T} distribution of the prompt dimuons and the mass distribution of the open charm events.

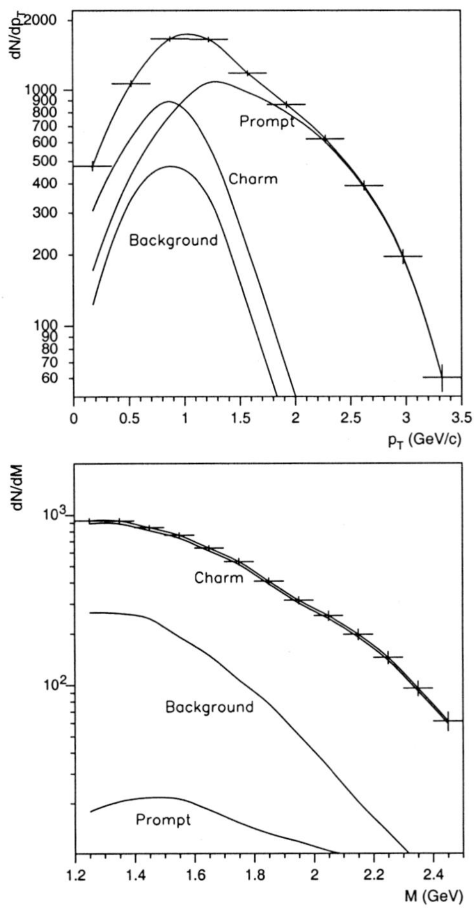

Fig. 31. Expected prompt (top) and charm (bottom) event samples selected in proton-induced collisions

More challenging will be the measurement of the dependence of χ_c production on the mass number of the target, in p-A collisions, by seeing how the ratio between χ_c and J/ψ yields changes from p-Be to p-Pb collisions. To minimize the systematical uncertainties due to the beam flux normalization, the measurement will be done using the Be and Pb targets simultaneously in the beam. The $\chi_c \to \psi\,\gamma \to \psi\; e^+e^-$ decays will be used for this study, with the photons converting in a Pb disk placed downstream of the targets, and the electron-positron pairs reconstructed in the silicon pixel telescope.

10 Summary and Conclusions

Starting from the questions and wishes expressed at the Quark Matter 1999 conference, concerning measurements that should be done at the CERN SPS before closing this facility, I have briefly mentioned some of the most recent developments and emphasized the issues that have imposed a continuation of the SPS heavy-ion physics program.

After 15 years of "learning curve", we can say that we have been unable to falsify the hypothesis of quark-gluon plasma formation at the CERN SPS. In fact, as predicted, strangeness is enhanced, the J/ψ is suppressed, the dimuon continuum looks as if thermal dileptons are produced, and there are modifications in the low-mass dilepton spectra, among other observations. These results provide extremely relevant information about the (predicted) formation of a deconfined state of matter in high-energy heavy-ion collisions. However, considerable homework remains to be done in view of converting "compelling evidence" into "conclusive evidence" that the quark-matter phase has indeed been formed at CERN. This is exactly the reason why the heavy-ion community must make a significant effort to further clarify the present results and reach a deeper understanding of the critical behaviour of QCD at SPS energies.

The renaissance of the heavy-ion physics program at the CERN SPS, with the extension of NA49 and the approval of the new NA60 experiment, represents an evolution from a broad physics program to a dedicated study of specific signals that already provided very interesting results. The new measurements of NA60 should give a significant contribution to the understanding of the presently existing results, and considerably help in building a convincing logical case that establishes beyond reasonable doubt the formation (or not) of a deconfined state of matter in heavy-ion collisions at the SPS.

Acknowledgements

Many people contributed, in one way or another, to the contents of this article. Special thanks are due to N. Carrer, A. Drees, U. Heinz, G. Paic and E. Scomparin, among others. H. Wöhri has kindly helped in preparing the manuscript. It is a pleasure to acknowledge very interesting discussions with J.-P. Blaizot, D. Kharzeev, L. McLerran, B. Müller and H. Satz.

References

1. See the article by F. Karsch in this volume
2. R. Stock: Nucl. Phys. A **661**, 282c (1999)
3. L. Kluberg: Nucl. Phys. A **661**, 300c (1999)
4. R. Stock: Phys. Lett. B **456**, 277 (1999)
   U. Heinz: Nucl. Phys. A **661**, 140c (1999)
5. M. Jacob, U. Heinz: CERN Press Release, Feb. 2000
6. A. Bialas et al.: Nucl. Phys. B **111**, 461 (1976)
7. F. Antinori et al. (NA57 Coll.): CERN-EP-2000-002
8. See the article by M. Guylassy in this volume
9. T. Alber et al. (NA49 Coll.): Phys. Rev. Lett. **75**, 3816 (1995)
10. J.D. Bjorken: Phys. Rev. D **27**, 140 (1983)
11. S. Margetis et al. (NA49 Coll.): J. Phys. G **25**, 189 (1999)
12. F. Sikler et al. (NA49 Coll.): Nucl. Phys. A **661**, 45c (1999)
13. K. Redlich: *Proceedings of Quark Matter 2001*
14. F. Antinori et al. (WA97 Coll.): Nucl. Phys. A **661**, 130c (1999)
15. N. Carrer et al. (NA57 Coll.): *Proceedings of Quark Matter 2001*
16. T. Susa et al. (NA49 Coll.): *Proceedings of Quark Matter 2001*
17. N. Willis et al. (NA50 Coll.): Nucl. Phys. A **661**, 534c (1999)
18. V. Friese et al. (NA49 Coll.): *Proceedings of Quark Matter 2001*
19. P. Braun-Munzinger et al.: Phys. Lett. B **465**, 15 (1999)
20. F. Antinori et al. (WA97 Coll.): Eur. Phys. J. C **14**, 633 (2000)
21. U. Heinz: Nucl. Phys. A **685**, 414c (2001)
22. H. van Hecke et al.: Phys. Rev. Lett. **81**, 5764 (1998)
23. H. Appelshauser et al. (NA49 Coll.): Eur. Phys. J. C **2**, 661 (1998)
24. B. Tomasik et al.: nucl-th/9907096
25. G. Agakichiev et al. (CERES Coll.): Eur. Phys. J. C **4**, 231 (1998)
26. G. Agakichiev et al. (CERES Coll.): Phys. Lett. B **422** 405, (1998)
    B. Lenkeit et al. (CERES Coll.): Nucl. Phys. A **661**, 23c (1999)
27. V. Koch: Int. J. Mod. Phys. E **6**, 203 (1997)
28. R. Rapp and J. Wambach: hep-ph/9909229
29. E. Shuryak: Nucl. Phys. A **661**, 119c (1999)
30. B. Müller: Nucl. Phys. A **661**, 272c (1999)
31. A. Marin et al. (CERES Coll.): Nucl. Phys. A **661**, 673c (1999)
32. H. Appelshauser et al. (CERES Coll.): *Proceedings of Quark Matter 2001*
33. M.C. Abreu et al. (NA38/50 Colls.): Eur. Phys. J. C **14**, 443 (2000)
34. L. Capelli et al. (NA50 Coll.): *Proceedings of Quark Matter 2001*
35. T. Sjostrand: Comp. Phys. Comm. **82**, 74 (1994)
36. A.D. Martin et al.: Phys. Rev. D **51**, 4756 (1995)
37. P. Braun-Munzinger et al.: Eur. Phys. J. C **1**, 123 (1998)
38. M.C. Abreu et al. (NA38 Coll.): Nucl. Instr. Meth. A **405**, 139 (1998)
39. C.Y. Wong, Z.Q. Wang: Phys. Lett. B **367**, 50 (1996)
40. Z. Lin, X.-N. Wang: Phys. Lett. B **444**, 245 (1998)

41. L. Capelli et al. (NA50 Coll.): *Proc. of the 4th Rencontres du Vietnam*, Hanoi, Vietnam, July 2000
42. R. Rapp, E. Shuryak: Phys. Lett. B **473**, 13 (2000)
43. K. Gallmeister, B. Kampfer, O.P. Pavlenko: Phys. Lett. B **473**, 20 (2000)
44. H. Satz: in *"International School of Subnuclear Physics"*, A. Zichichi (Ed.), Erice, Italy, 1997, BI-TP-97-47, hep-ph/9711289
45. E. Scomparin et al. (NA50 Coll.): *Proceedings of Quark Matter 2001*
46. J. Geiss et al.: Phys. Lett. B **447** 31, (1999); J. Geiss et al.: *Proc. Workshop on Quarkonium Production in Relativistic Nuclear Collisions*, Seattle, USA, May 1998 (nucl-th/9810059)
47. C. Spieles et al.: Phys. Rev. C **60**, 054901 (1999)
48. D.E. Kahana, S.H. Kahana: Prog. Part. Nucl. Phys. **42**, 269 (1999); D.E. Kahana, S.H. Kahana: nucl-th/9908063
49. N. Armesto, A. Capella, E.G. Ferreiro: Phys. Rev. C **59**, 395 (1999)
50. D. Kharzeev, M. Nardi, H. Satz: BI-TP-97-33, hep-ph/9707308
    M. Nardi, H. Satz: Phys. Lett. B **442**, 14 (1998); hep-ph/9805247
    M. Nardi: private communication, May 2000
51. A. Baldit et al. (NA60 Coll.): Proposal CERN/SPSC 2000-010, March 2000
52. K. Borer et al.: Nucl. Instr. Meth. A **440**, 5 (2000)
53. M. Campbell et al.: IEEE Trans. on Nucl. Sci. **46-3**, 156 (1998)

Theory of High-Energy A+A at RHIC

Miklos Gyulassy

Department of Physics, Pupin Lab
Columbia University, New York, NY 10027, USA

Summary. In this article I introduce aspects of current theory used to interpret the preliminary data on ultra-relativistic nuclear collisions at RHIC energies in terms of the physical properties of QCD matter at extreme densities. Topics covered include: What are the physics questions at SPS and RHIC? Geometrical vs. dynamical features of A+A. The interplay of computable hard perpurbative QCD vs. phenomenological soft dynamics. Baryon number transport and junctions. How can we compute and get experimental control over the initial conditions? How to reconcile apparent hydrodynamic behavior with partonic/hadronic transport theory? I use the preliminary RHIC data available up to June 1, 2001 to illustrate these topics. Most technical details are deferred to the literature. However, since the main new observable at RHIC relative to SPS is jet quenching, I elaborate more on this "tomographic" probe of ultra-dense matter. The possible discovery of jet quenching at RHIC by STAR and PHENIX is highlighted.

1 Introduction

Finally, after 20 years of preparation [1], a new chapter in nuclear/particle physics commenced on June 12, 2000 with the measurement of the first $Au + Au$ collisions at $\sqrt{s} = 56$ AGeV (GeV per nucleon pair) in the Relativistic Heavy Ion Collider (RHIC) at the Brookhaven National Lab (BNL). Soon thereafter collisions at $\sqrt{s} = 130$ AGeV were also measured. The first results were reported at Quark Matter 2001 [2] from the four major experiments, STAR [3], PHENIX [4], PHOBOS [5], and BRAHMS [6]. A small army of $\sim$ 1000 experimentalists measured the flavor, rapidity, and transverse momentum distributions of the approximately 4000 charged particles produced in each central (head on) collision at 130 AGeV. In the summer of 2001, it is anticipated that RHIC will reach its design energy, and $p + p$ and $Au + Au$ collisions at $\sqrt{s} = 200$ AGeV will come under experimental scrutiny.

This article provides a very condensed introduction to current theoretical work aimed to provide a consistent interpretation of observables measured in such reactions in terms of the properties of dense QCD matter. The color slides of the original lectures can be found on my WWW site [7]. This article is designed to supplement those slides and update them with the preliminary RHIC data available as of June 1, 2001.

The theoretical work on the new physics that may exist in QCD matter at extreme densities began in the mid 1970's with the realization that the asymptotic freedom property of QCD implies the existence of a new phase of strongly interacting matter called the Quark-Gluon Plasma (QGP) [8]-[13]. Unlike familiar nuclear or more generally hadronic matter consisting of composite "elementary" particles $(\pi, K, \rho, p, \Delta, \Lambda, \cdots)$ in which quarks and gluons are permanently confined, the QGP phase at very high temperature and/or baryon chemical potential ($T, \mu_B \gg \Lambda_{QCD} \sim 200$ MeV $= 1$/fm) is one where the interactions between quarks and gluons become relatively weak and short range

$$V(r) \sim \frac{g^2}{4\pi}\frac{e^{-\mu_D r}}{r} \;, \quad \alpha_s = \frac{g^2}{4\pi} \propto \frac{1}{\log(T \text{ or } \mu_B)} \to 0 \;.$$

The color electric (Debye) screening mass $\mu_D(T, \mu_B)$ increases linearly with T or μ_B modulated by a slowly varying factor of the running coupling, $g(T, \mu_B)$ (see the article of T. Rebhan in this book). The thermodynamic properties of this deconfined and chirally symmetric (~massless) phase of matter are thus expected in perturbation theory to reduce approximately to an ideal Stefan-Boltzmann gas of quarks and gluons. For the Standard Model with 3 colors and N_f flavors of "light" quarks relative to T, μ_B ($SU_c(3) \otimes SU_f(N_f)$), the Stefan Boltzmann constant for $\mu_B = 0$ is

$$K_{SB} = \frac{3P}{T^4} = \frac{\epsilon}{T^4} = (2_s \times 8_c + \frac{7}{8} \times 2_s \times 2_{q+\bar{q}} \times 3_c \times N_f)\frac{\pi^2}{30}(1 + O(g^2)) \sim 12$$

taking the helicity, color, flavor, and antiquark degrees of freedom into account. In reality the severe infrared singularities of perturbative QCD (pQCD) lead to large non-perturbative corrections to the ideal gas equation of state for all temperatures and chemical potentials accessible experimentally even beyond the future Large Hadronic Collider. Only numerical lattice QCD (lQCD) methods [14] (see the article of F. Karsch in this book) can provide reliable predictions for the thermodynamic properties of the QGP phase of matter. Effective models and resumed many-body techniques (see the articles of T. Alford, T. Rebhan, J.P. Blaizot, and E. Shuryak in this book) are, however, needed to interpret the lQCD "data" and provide physical insight especially at finite chemical potential. However, it is sobering to recall that pQCD thermodynamic expansion of the pressure in powers of g shows no sign of convergence [13] even before the Linde infrared catastrophe at $O(g^6)$, and non-perturbative corrections to the pQCD Debye mass, μ_D, remain about a factor of ~ 3 up to $T \sim 200$ GeV [16]. The full theoretical understanding of the structure of the non-Abelian plasma phase of QCD therefore remains a fundamental open problem in physics because it involves strongly correlated, nonperturbative and possibly turbulent dynamical features [15].

One of the essential and intriguing aspects of the QCD many-body problem is that the *physical* vacuum is an extraordinarily complex coherent many-body medium. The gluon and quark condensates lower the energy density of the physical vacuum by an amount $B \sim \Lambda_{QCD}^4 \sim 200$ MeV/fm^3. Drilling a perturbative vacuum bubble of volume V in this condensate costs an enormous energy BV. The QGP, if formed in V, must counteract the physical vacuum pressure B to prevent it from collapsing. This is only possible when the temperature exceeds $T_c \approx (3B/K_{SB})^{1/4} \sim 150$ MeV.

The driving force behind the experimental effort at CERN and BNL over the past 20 years has been to try to create the extreme conditions necessary to produce and diagnose this new form of matter in the laboratory. Over the past 15 years experiments at the AGS/BNL ($\sqrt{s} = 5$ AGeV) and the SPS/CERN ($\sqrt{s} = 20$ AGeV) have searched systematically via a very large set of observables for evidence of the QGP phase (see the article of C. Lourenco in this volume). In these lectures I will focus on the most recent developments in that search that has just begun at RHIC. I must emphasize from the onset that most of the data shown here are of *PRELIMINARY* nature and could change as better control over the systematic errors is achieved in the next few years. Nevertheless, the new data are so exciting that it is worth trying a first pass to give an overview and possible interpretation.

2 Geometry and Dynamics in A+A

The main obstacle in interpreting data on collisions of finite nuclei (at any energy) is that the matter created undergoes quantum (perhaps semi-classical) many-body *dynamics* that may be approximated by *thermodynamics* only over a limited (low-p_T) kinematic range. Experimentalists do not have the luxury of lattice or perturbative QCD theorists of tapping into the infinite gedanken volume or reservoir with a fixed temperature and pressure. Nuclear collisions produce dense matter in a highly dynamical environment, and the matter produced expands anisotropically near the speed of light. It is far from clear whether local thermal and chemical equilibrium concepts apply, and even so, over what domain of the 8-dimensional (x^μ, p^μ) phase space can they be used.

Before a collision, the partons of the two colliding nuclei are locked into a coherent field configuration. The dense virtual cloud of gluons and quarks may be described in the colinearly factorized QCD approximation by A times the known structure functions, $f_{a/p}(x,Q)$, of nucleons when the resolution scale is high enough $Q > 1 - 2$ GeV. However, many-body initial-state interactions could lead to strong modifications of this naive parton picture (see the article of L. McLerran in this volume). The nuclear QCD fields continue to interact after the nuclear valence quark pancakes pass through each other. The interaction spans a space-time hyperbola over a proper time $\sqrt{t^2 - z^2} \sim 30$ fm/c $= 10^{-22}$ s. Then a "miracle" happens! The field quanta hadronize in a way that is unfortunately not well understood. The dense final hadronic debris can further interact as it expands toward the detector elements. From CERES/SPS data [17] there is evidence that the in-medium mass-width (spectral function) of vector mesons may change drastically [18].

Hard probes (jets, leptons, photons, heavy quarks) are of special interest because they provide "tomographic" tools with which one can map out this evolution experimentally. Hard probes are effective "external" tomographic probes because they are produced with a pQCD computable initial distribution on a much shorter time scale, $\sim 1/m_\perp$, than the plasma formation time, $\sim 1/3T$. Modification of their known initial distributions therefore provides information on the medium through which they propagate in analogy to conventional X-ray or positron tomography used in medicine [19]. The primary advantage of RHIC over lower energy machines (AGS, SPS) is that hard pQCD probes are produced at RHIC orders of magnitude more abundantly over a significantly larger kinematic range. This greatly improves their tomographic resolution power.

Figure 1 shows the rapid growth of high-p_T $Au+Au \to \pi^0 + X$ predicted by pQCD from SPS to RHIC and LHC. As discussed for example in [20,21], it is useful to decompose the nuclear geometry dependence of invariant hadron distributions produced in $A+B \to h+X$ at impact parameter $\mathbf{b}$ into a phenomenological "soft" and pQCD calculable "hard" components as

$$E\frac{dN_{AB}(\mathbf{b})}{d^3p} = N_{part}(\mathbf{b})\,\frac{dN_{soft}(\mathbf{b})}{dyd^2\mathbf{p}_\mathrm{T}} + N_{coll}(\mathbf{b})\,\frac{1}{\sigma_{in}^{pp}}\,\frac{d\sigma_{hard}(\mathbf{b})}{dyd^2\mathbf{p}_\mathrm{T}}\;, \tag{1}$$

where $N_{coll}(\mathbf{b}) = \sigma_{in}^{pp} T_{AB}(\mathbf{b})$ is the number of binary NN collisions and $N_{part}(\mathbf{b})$ is the number of nucleon participants at impact parameter $\mathbf{b}$. The nuclear geometry of hard collisions is expressed in terms of the Glauber profile density per unity area $T_{AB}(\mathbf{b}) = \int d^2\mathbf{r}\; T_A(\mathbf{r})T_B(\mathbf{r}-\mathbf{b})$, where $T_A(\mathbf{r}) = \int dz\; \rho_A(\mathbf{r},z)$ (see Fig. 2). The hard part scales with the number of binary collisions $\propto A^{4/3}$ because their probability is small built up from all possible independent parton scattering processes. The soft part scales with only $N_{part} \propto A^1$ because their probability is large and therefore "shadowed".

The (textbook) computable lowest-order pQCD differential cross section for inclusive $p+p \to h+X$ invariant cross section is given by

$$E_h\frac{d\sigma_{hard}^{pp\to h}}{d^3p} = K\sum_{abcd}\int dx_a dx_b f_{a/p}(x_a,Q_a^2) f_{b/p}(x_b,Q_b^2) \tag{2}$$

$$\frac{d\sigma}{d\hat{t}}(ab\to cd)\frac{D_{h/c}(z_c,Q_c^2)}{\pi z_c}$$

where $x_a = p_a/P_A, x_b = p_b/P_B$ are the initial momentum fractions carried by the interacting partons, $z_c = p_h/p_c$ is the momentum fraction carried by the final observable hadron, $f_{\alpha/p}(x_\alpha,Q_\alpha^2)$ is the proton structure function for parton of flavor α, and $D_{h/c}(z_c,Q_c^2)$ is the fragmentation function for the parton of flavor c into h. The UA1 data on $p\bar{p}$ hadron production with $p_\mathrm{T} > 2$ GeV can be well reproduced with this pQCD model expression.

The soft ($p_T < p_0 \sim 2$ GeV/c) nonperturbative contribution to the hadron yields can only be modeled phenomenological. The Dual Parton Model [22,23]

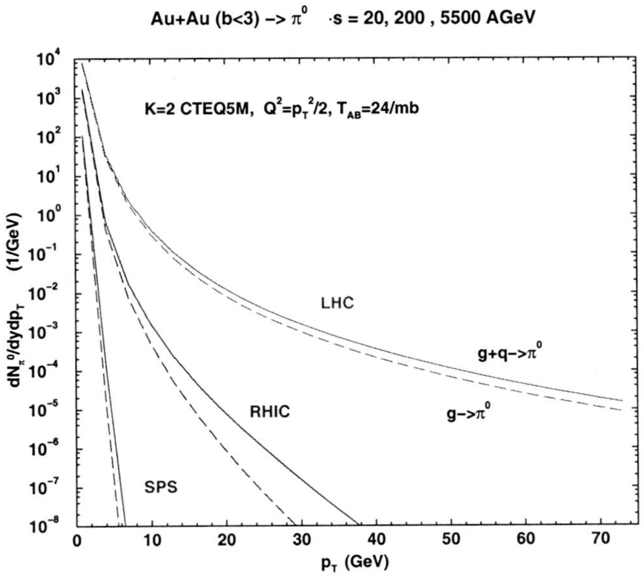

Fig. 1. Invariant distribution of hard pQCD produced π^0 in central $Au + Au$ collisions as a function of c.m. energy via Eqs. (1, 3) with CTEQ5M structure functions, K=2 factor, scale $Q = p_T/2$, and multiplied by nuclear overlap $T_{AB} = 24/mb$. The dashed curve shows the contribution from gluon jet fragmentation only

and the LUND string model [24,25] are the most extensive and successful low-p_T multiparticle phenomenologies. The basic pQCD matrix elements have been encoded into a Monte Carlo code, PYTHIA [26]. A variant of soft string phenomenology tuned to $pp, p\bar{p}$ data, with the hard part taken from PYTHIA, a hadronization scheme taken from the LUND JETSET hadronization, and an eikonal nuclear multiple collision geometry were combined into the Monte Carlo A+B collision generator in HIJING [27]. HIJING has been used over the past decade to predict many observables at RHIC [20,27,28]. The separate soft and hard components in HIJING with a fixed $A, \sqrt{s}$ independent scale $p_0 = 2$ GeV/c are illustrated in Fig. 3. Hard gluons in the LUND hadronization scheme are represented by kinks in the strings between valence quarks and diquarks of the $N_{part}(b)$ interacting baryons in $A + B$ collisions at impact parameter b. In this class of models no final-state interactions are taken into account.

The physics in $A + A$ reactions that must be understood in order to be able to interpret observables in terms of the properties of dense QCD matter requires extending the above class of event generator models to include

1. Initial Conditions: The formation physics responsible for creating an incoherent gas of gluons and quarks from the initial virtual nuclear fields;
2. Parton Transport: The (x^μ, p^μ) phase space evolution of that parton gas toward equilibrium;

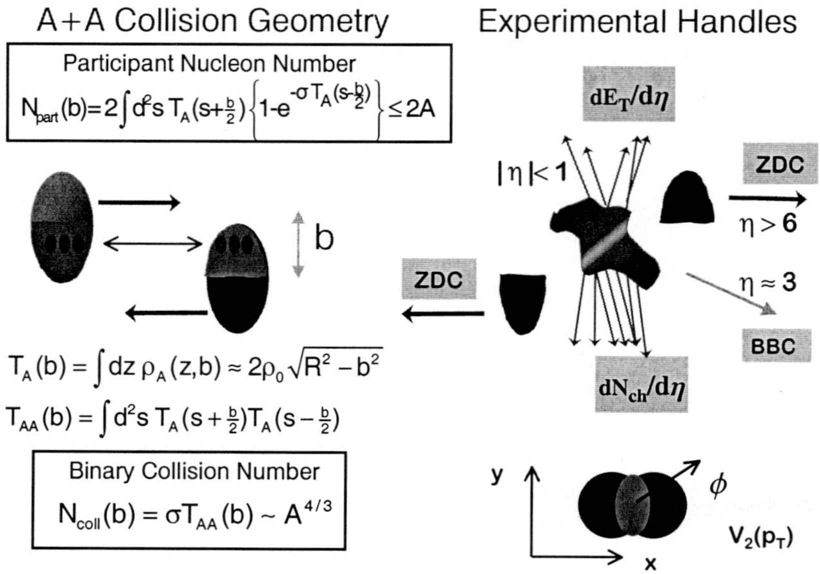

Fig. 2. Illustration of key aspects of the relation between the geometry of nuclear collisions and the participant and collision number at a fixed impact parameter. The observables (see [3] - [6]) used to constrain the geometry experimentally are also illustrated

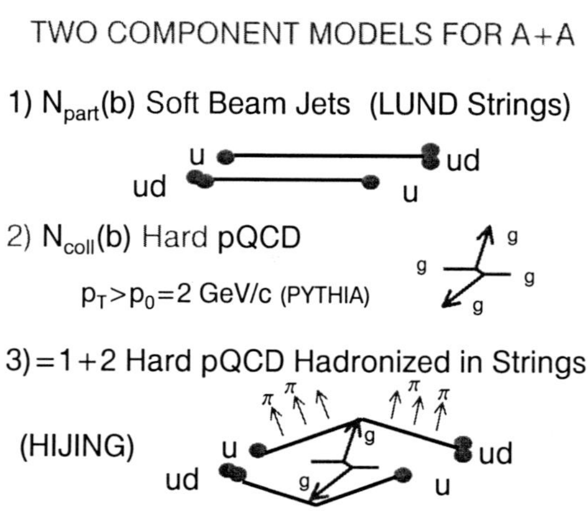

Fig. 3. Illustration of hard+soft two component models of A+A reactions like HIJING [27]

3. Hadronization: The dynamical mechanisms that convert the parton degrees of freedom in the confining physical vacuum into the observable hadronic ones;
4. Hadron Transport: The final-state interactions of the expanding dense hadronic matter prior to "freeze-out".

Each problem is fascinating in its own right but only bits and pieces are understood or phenomenologically mapped out up to now. There exists unfortunately no complete *computable* dynamical theory (like Magneto-Hydrodynamics for QED plasmas) that consistently takes into account all four elements. QCD is believed to be THE theory, but it is still not computable except at high p_T where perturbative or classical methods may apply. There exist instead several different dynamical "scenarios" to describe $A + A$ that attempt to patch together different approximation techniques and phenomenological models to address all the physics issues in turn.

Two generic approaches to $A + A$ can be classified by whether the Initial Conditions are (i) computed (via pQCD or classical Yang-Mills (cYM)) and subsequent evolution followed by a dynamical scenario for 2-4, or (ii) the initial conditions are fit by extrapolating final observables backwards via a suitable dynamical scenario. At lower energies (AGS, SPS) only the second approach is available since the momentum scales are simply too low to apply either pQCD or cYM. At collider energies RHIC and beyond, the copious production of mini-jets [20,27,34–36] with $p_T > p_0 \sim 2$ GeV shown in Fig. 1 makes it possible for the first time to pursue the first approach via pQCD Eq. (3). At very high energies classical Yang Mills theory [37–42] provides a general method to compute the Formation Physics which reduces to pQCD at high p_T. Whether RHIC or LHC energies are high enough is an open question.

The second approach, trying to "fit" the initial conditions by extrapolating the final distributions backwards with a suitable dynamical model has been traditionally based on relativistic hydrodynamics [43–46]. The approximate longitudinal boost invariant boundary conditions at ultra-relativistic energies simplify hydrodynamic equations greatly as pointed out by Bjorken [48]. For $\mu_B = 0$ the hydrodynamic equations are,

$$\partial_\mu T^{\mu\nu} = 0 \quad , \quad T^{\mu\nu}(x) = u^\mu u^\nu(\epsilon + P) - g^{\mu\nu} P \quad ,$$

where $\epsilon(x), P(x)$ are the proper energy density and pressure and $u^\mu(x)$ is the four velocity field of the fluid. The central assumption is that thermal and chemical equilibrium are maintained locally in spite of the possible large gradients in the fluid variables. The great advantage of hydrodynamics is that it provides a covariant dynamics depending only on the equation of state $P(T(x))$ that is directly related to the lQCD predictions. When a specific space-time freeze-out hypersurface is assumed together with the assumption, the Cooper-Frye prescription [49–52,54], the computed four-fluid velocity field can be used to predict the final anisotropic flow pattern of hadrons. Since this process is assumed to be reversible, the final distributions *together with an assumed freeze-out hypersurface* can be used to compute the initial conditions on any desired initial hypersurface. The disadvantage of this approach as emphasized in [55] is that both the initial and final freeze-out hypersurfaces must be guessed. Also finite mean-free-path physics is outside the scope of ideal hydrodynamics, and transport theory solutions [55] do not support "sharp" freeze-out hypersurfaces. Thus the inversion of data in this way to deduce the initial conditions is not unique. The neglect of dissipative effects such as viscosity also makes it impossible to relate central $A + A$ to peripheral and light-ion data, especially $p + p$. Finally, the assumption of homogeneous or slowly spatially varying initial conditions is questionable because of copious mini-jet production [56]. In spite of all the above theoretical problems, initial conditions for RHIC have been successfully constructed that lead via ideal hydrodynamics and idealized Cooper-Frye freeze-out to distributions that reproduce amazingly well many of the low-p_T observables at RHIC [57–60] (see the next section).

In order to bring the freeze-out assumption under better theoretical control covariant, nonequilibrium transport theory [61] must be solved. Until recently, only simplified 1+1D Bjorken transport theory was soluble in the linearized relaxation time approximation (see [62] and refs. therein). This is due to the great numerical complexity of the 3+1D nonlinear Boltzmann equations [55,63]:

$$p_1^\mu \partial_\mu f_1 = \iiint\limits_{234} (f_3 f_4 - f_1 f_2)\, W_{12\to 34} \delta^4(p_1 + p_2 - p_3 - p_4) + S(x, \mathbf{p}_1), \quad (3)$$

where W is the square of the $2 \to 2$ scattering matrix element, the integrals are shorthands for $\int\limits_i \equiv \int \frac{g\, d^3 p_i}{(2\pi)^3 E_i}$, where g is the number of internal degrees of freedom, while $f_j \equiv f(x, \mathbf{p}_j)$ is the parton phase space distribution. The initial conditions are specified by a source function $S(x, \mathbf{p})$ that corresponds to the assumed initial conditions.

Yang Pang's parton subdivision technique [63,64] and the speed of current workstations have finally made it possible to solve Eq. (3) numerically. (codes can be obtained from the OSCAR Web site [65]). The solutions [66,67] prove that elastic parton scattering with pQCD rates is insufficient at RHIC to keep the plasma in local equilibrium due to the extreme rapid longitudinal "Hubble" expansion of the system [62]. Unfortunately, there exists no *practical* algorithm at this time to solve the more nonlinear inelastic transport equations involving $gg \to ng$ processes. Therefore, if hydrodynamics applies to $A+A$ at RHIC, then most likely strong nonperturbative mechanisms must be assumed to exist on faith or hypothesis (see the article of E. Shuryak in this volume). This is an important open theoretical problem.

I would also like to call attention to a new class of hydrodynamic models [68] that side-step the final freeze-out problem by assuming that local equilibrium is maintained only up to an intermediate hyper-surface, just after hadronization on a $T = T_c - \epsilon$ isotherm. Using that intermediate freeze-out as the initial conditions of a hadronic transport theory, the subsequent evolution of the hadronic gas toward a dynamical freeze-out is then determined by known hadronic cross sections via URQMD [69,70].

3 Preliminary Results from RHIC

3.1 Global Constraints on Initial Conditions

The first results from RHIC, from PHOBOS [5], shown in Fig. 4, demonstrate that the energy dependence of the scaled charged particle (pseudo)rapidity density, $dN_{ch}/d\eta/N_{part}$, is different from $p+p$ and $p+\bar{p}$ systematics.

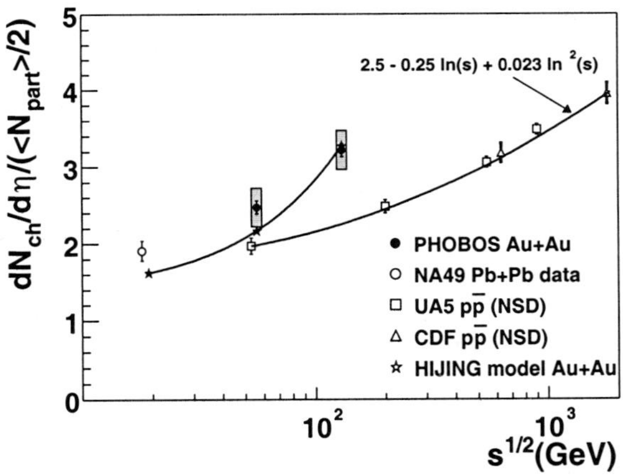

Fig. 4. Measured pseudorapidity density normalized per participant pair for central Au+Au collisions (PHOBOS [5,29]). Systematic errors are shown as shaded areas. Also shown are results of Pb+Pb data (CERN SPS), HIJING [20] simulations and a parameterization of $p\bar{p}$ data

Approximately 50% more particles are produced at mid rapidity per participating baryon in central $Au + Au$ collisions then in $p + p$ at the same energy per baryon. The curve shows that the two component HIJING model predicted well this result. However, as shown in Fig. 5, another model EKRT [36], was also found to predict the same multiplicity as HIJING for central collisions. In Ref. [20], we proposed that the centrality dependence of this observable could differentiate between these competing models of the initial conditions. The new data of PHENIX and PHOBOS [33] verified this prediction. While neither model accounts quantitatively for the data, but the two component HIJING model with its combined A^1 and $A^{4/3}$ dependence better describes the rate of increase of the scaled multiplicity with participant number. The observed increase of the scaled multiplicity with energy relative

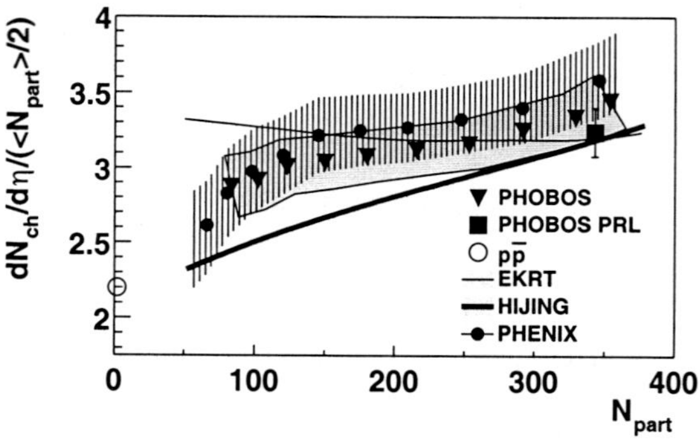

Fig. 5. Normalized pseudorapidity density (PHOBOS and PHENIX) [33] $dN_{ch}/d\eta|_{|\eta|<1}/(0.5\times N_{part})$ as a function of the number of participants. Predictions based on HIJING (thick solid) and EKRT [36] (thin) are shown

to $p+P$ and with participant number dependence supports the prediction of copious mini-jet production at RHIC. This is one of the necessary, though insufficient, conditions to form a dense gluon plasma in $A+A$.

The difference between HIJING and EKRT is that in the latter it is assumed that *all* the produced entropy (multiplicity) arises at RHIC energies from hard pQCD processes. EKRT assume that there is no significant soft component, i.e. $dN_{soft} \ll dN_{hard}$ in Eq. (1). However, the hard component is cutoff at scale p_0 that is allowed to vary with both A and hence N_{part} and with energy $\sqrt{s}$ based on the following assumption: independent and hence $T_{AB}(b)$ proportional numbers of gluons with $p_T > p_0$ are produced only in "resolvable" domains of finite area π/p_0^2. There are $p_0^2R^2$ such domains in the transverse plane in a central nuclear collision. This so called "final-state saturation" model is then specified by

$$\frac{dN_g}{dy} = \frac{N_{coll}(\mathbf{b}=0)1}{\sigma_{in}^{pp}} \int_{p_0}^{\infty} d^2\mathbf{p}_\mathrm{T} \frac{d\sigma_{hard}^{A+A\to g}}{dyd^2\mathbf{p}_\mathrm{T}} = \beta p_0^2 R^2 \ . \tag{4}$$

For $\beta = 1$ assumed in EKRT, the solution for the saturation scale is $p_0(\sqrt{s}, A) \equiv p_{sat} \approx 0.2A^{0.13}(\sqrt{s})^{0.19}$. This predicts $dN_g/dy \propto A^{0.93}$ in spite of the apparent proportionality of hard processes to $A^{4/2}$. The flat $(dN_g/dy)/N_{part} \sim A^{\sim 0}$ independence of the scaled multiplicity is a general feature of saturating QCD models of the initial conditions (see also the article of L.McLerran in this book). Such a flat behavior is, however, ruled out by the present data at RHIC.

An alternate (so-called initial-state saturation) model was proposed by Kharzeev and Nardi (KN [71]) based on the nonlinear QCD evolution equations of [72]. In this model of nuclear initial conditions, the number of liberated gluons is proportional to the number of virtual gluons participating in the reaction on a scale p_0. The produced number is then taken to

$fN_{part}xG(x,p_0)$ in terms of the nucleon gluon structure function, where $f \sim 1.2$ is a factor on the order of unity. Since the interaction probability is proportional to the running coupling $\alpha_s(p_0)$, the initial-state saturation condition is defined by

$$\frac{dN_g}{dy} = fN_{part}xG(x,p_0) = f\frac{2}{3\pi^2\alpha_s(p_0)}p_0^2R^2 \ . \tag{5}$$

The main difference between initial- and final-state saturation models is therefore due to the logarithmic dependence on p_0 introduced by the running coupling. In [71] a simple ansatz was assumed for $xG(x,Q) \propto logQ/\Lambda$ based on the linear (DGLAP) evolution equations. With this ansatz KN predicted a participant dependence surprisingly close to the observed data in Fig. 5.

However, the x independent ansatz of KN used for $xG(x,Q)$ for the scale $Q \sim 1$ GeV/c is a guess that cannot be supported by the available *ep* HERA data. At small $x \sim 0.01$ and low $Q \sim 1$ the pQCD factorization analysis of deep inelastic $e+p$ reactions breaks down and xG acquires a 100% systematic error bar as shown in Fig. 6. Initial-state saturation is a theoretically sound model only at very high energies or nuclei with $A \gg 200$, when $Q > 2$ GeV/c and the errors based on pQCD analysis become manageable.

While it is premature to conclude which approach is least wrong (see also [83]), in my opinion, it appears that the global multiplicity data and its centrality dependence can be used as indicators that the initial gluon rapidity density at RHIC is between HIJING's 200 and EKRT's 1000. The corresponding gluon density, $\rho_g(\tau) = dN_g/dy/(\tau\pi R^2)$ is thus $\sim 10 - 50/\mathrm{fm}^3$ at the corresponding formation time $1/p_0 = 0.1 - 0.2$ fm/c. Thus RHIC may have indeed created the densest gluon plasma ever in the laboratory. As I emphasize in a later section, fortunately there are many other observables, especially jet quenching that provide independent checks of this possibility.

It is important to emphasize that similar results for the multiplicity in central collisions in HIJING and EKRT are purely coincidental because the models differ by a factor of five on the initial gluon density. This is compensated for by the underlying very different hadronization schemes assumed. HIJING creates a large fraction of the observed hadrons at RHIC through its soft string fragmentation scheme, while EKRT assume that entropy conservation implies that $dN_\pi \approx dN_g$. The lack of a detailed hadronization theory can only be overcome phenomenologically by testing experimentally all the ramifications of any particular model.

Another observable that was suggested in [20] to help differentiate models of initial conditions is the shape and scaling of the whole rapidity distribution (see Fig. 7).

It is seen that HIJING predicts a somewhat narrower and stronger centrality dependence than observed by PHOBOS. This may be related to the baryon stopping power at RHIC. Unfortunately no predictions are available for either the initial or final saturation models on the predicted shape of the

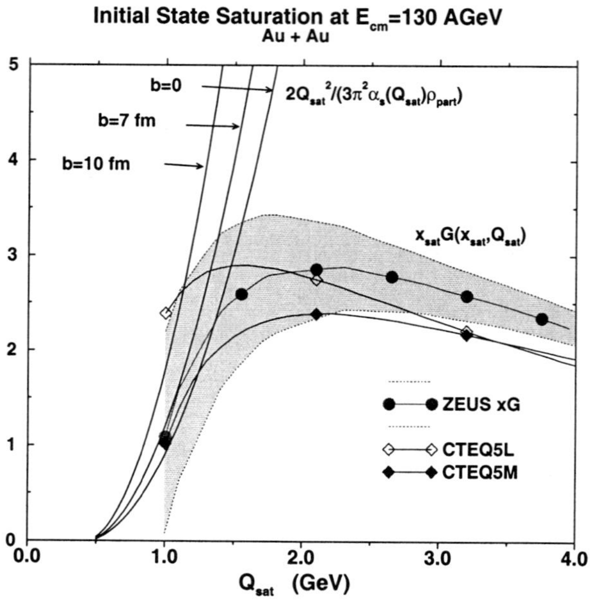

Fig. 6. Graphical solution to the initial-state saturation Eq. (5) for different impact parameters at 130 AGeV using the xG gluon structure functions from ZEUS and CTEQ. The energy dependence enters through the dependence on $x_{sat} = 2Q_{sat}/\sqrt{s}$. The participant number dependence of Q_{sat} follows from the intersection of the parabolic curves with xG. Unfortunately Au nuclei are too small, and the solutions in the $Q \sim 1$ GeV region are completely unreliable

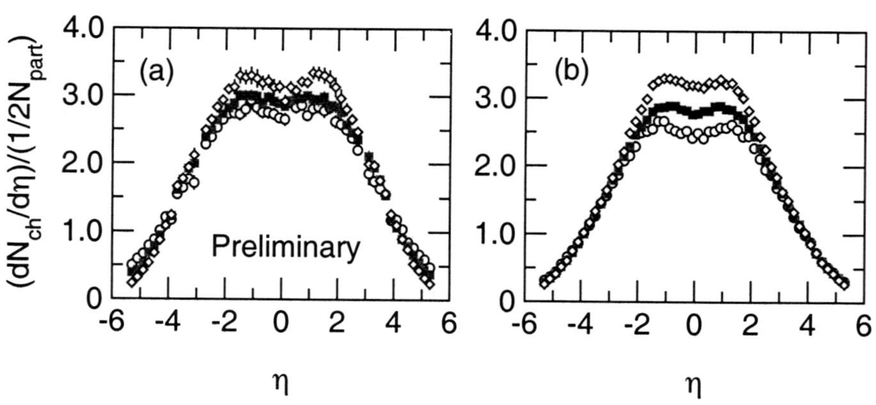

Fig. 7. (a). Measured $dN_{ch}/d\eta/(\langle N_{part}\rangle/2)$ for $\langle N_{part}\rangle$=102 (circles), 216 (squares) and 354 (diamonds) by PHOBOS [30]. (b). Same as (a) from HIJING

rapidity distribution. This observable is especially sensitive in those models to the x dependence of the saturation criteria.

The total integrated charge particle multiplicity is shown in Fig. 8. RHIC has produced about 4000 charged particles in $Au+Au$ at 130 AGeV. The non-linear enhancement near central collisions is interpreted in terms of HIJING as due to the onset of the mini-jet component.

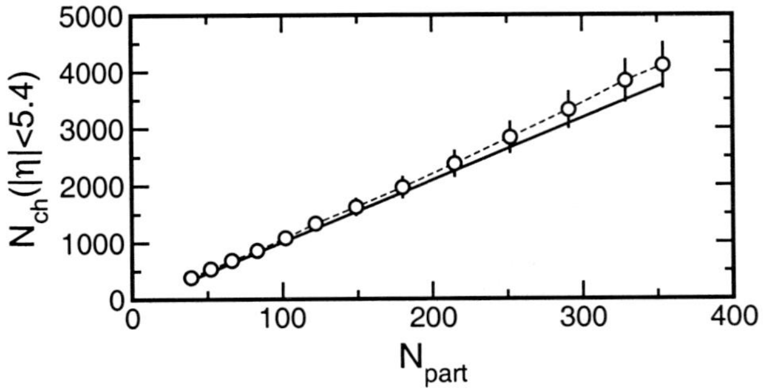

Fig. 8. PHOBOS total charged particle multiplicity vs. nucleon participant number [33]

3.2 Global Barometric Observable E_T/N_{ch}

An important global barometric measure of the internal pressure in the ultra-dense matter produced is the average transverse energy per charged particle. PHENIX data are shown in Fig. 9 compared to WA98 data from CERN. What is most amazing is that $E_T/N_{ch} \approx 0.8$ GeV almost independent of $\sqrt{s}$ from 20 to 130 AGeV and independent of centrality! HIJING predicts that it should rise from 0.8 to 0.9 GeV from CERN to SPS due to the enhanced mini-jet activity at RHIC. The EKRT initial state saturation model predicts a growth of this quantity in the initial state by about a factor of 3. The reason that EKRT remains viable after these data is that the assumed entropy conservation implies that a large amount of pdV work due to longitudinal expansion is performed by the plasma. In 1+1D hydrodynamics the energy per particle $\epsilon/\rho \approx 2.7\,T$ decrease as the system expands and cools $T \sim 1/\tau^{1/3}$. If the freeze-out is assumed to occur at all energies and impact parameters in $A + A$ on a fixed decoupling isotherm, then the energy per particle will always be the same. At RHIC this global transverse energy loss from the initial state is predicted to be about a factor 3. The theoretical problem of justifying hydrodynamics and the freeze-out prescription itself discussed in the previous section comes back to haunt us here [57]. The observed NULL effect in E_T/N_{ch} is very interesting because it is so difficult to obtain in any transport theory with finite pQCD relaxation rates.

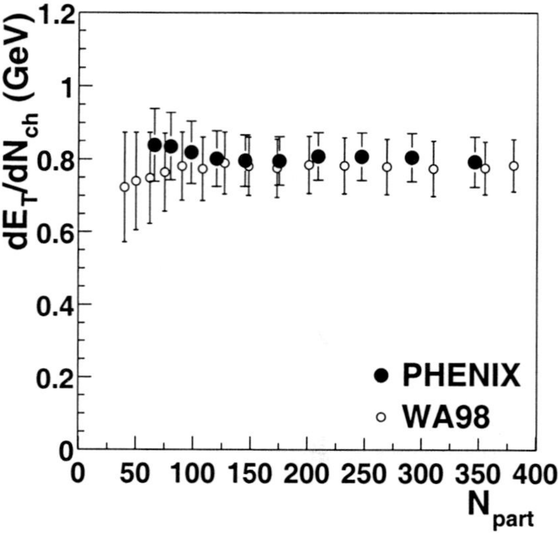

Fig. 9. Preliminary PHENIX data and WA98 data on dE_T/dN_{ch} as a function of participant number [33]. This barometric observable appears to be independent of $\sqrt{s} = 20 - 200$ and centrality!

3.3 Discovery of Jet Quenching

One of the predicted signatures [28,73,74] of dense matter formation is the suppression of jets and their high-p_T hadronic debris due to energy loss of the jet in the medium. However, the search for this effect at SPS by WA98 yielded the opposite result as shown in Fig. 10. Even a modest $dEdx = 0.2$ GeV/fm is completely ruled out by the data [81]. The problem is that at lower energies, multiple initial-state elastic scattering leads to a random walk in transverse momentum. This enhances the p_T of the scattered partons so that $\langle p_T^2 \rangle = p_0^2 + A^{1/3}\delta p_T^2$. This so-called Cronin effect has been well studied in $p + A$ reactions up to 800 GeV. At lower energies the very steep fall of the high-p_T tail makes the distribution extremely sensitive to this modest p_T enhancement. When convoluted through two nuclei, Wang predicted [81] that the Cronin enhancement at SPS in Pb+Pb should be a factor of two as verified in Fig. 11. What is plotted there is the ratio of the observed invariant cross section to the scaled binary collision number, $N_{coll}(b)$, scaled invariant cross section in p+p. Unity corresponds to naive superposition of N_{coll} independent elementary $p + p$ hard processes in the absence of any nuclear effects. The ratio starts below 1 since the low-p_T distribution grows only with the number of participants (divided by two) and $N_{part}(0)/2N_{coll}(0) \approx 0.15$.

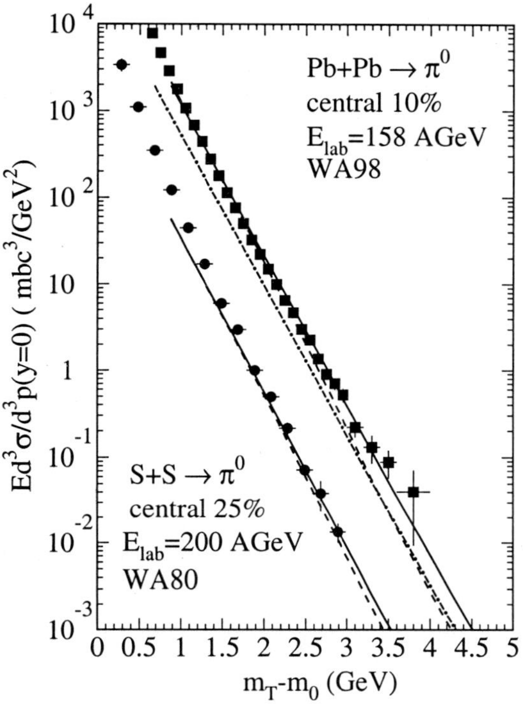

Fig. 10. Single-inclusive π^0 spectra in central $S+S$ at $E_{\text{lab}} = 200$ GeV and $Pb+Pb$ collisions at $E_{\text{lab}} = 158$ GeV. The solid lines are pQCD calculations (Wang [81]) with initial-k_T broadening and dashed lines are without. The $S+S$ data are from WA80 and $Pb+Pb$ data are from WA98 The dot-dashed line is obtained from the solid line for $Pb+Pb$ by shifting p_T by 0.2 GeV/c

In stark contrast to the SPS enhancement of high-p_T pions, a factor of two or more suppression of $p_T > 2$ GeV hadrons was reported by STAR [3,84] and PHENIX [85].

Fig. 12 shows that for $p_T < 2$ GeV a similar trend of increase due to the gradual transfer from participant to binary scaling is taking place as at SPS, but for $p_T > 2$ GeV the ratio for charged particles $\pi^\pm + K^\pm + p^\pm$ starts to drop again and reaches ~ 0.5 at 4 GeV/c.

The PHENIX data [85] show an even more dramatic quenching pattern for identified π^0 in Fig. 13. In this experiment, it was further verified that "peripheral" collisions are not quenched while central ones are. Fig. 14 shows that the suppression factor may reach a factor of three at 3 GeV/c. In this plot the ratio is not relative to pp data extrapolated to 130 GeV, but to "peripheral" collisions where the average number of participants and binary collisions is only ≈ 20. In contrast $N_{part} \approx 360$ and $N_{coll} \approx 857$ for the central collisions. It must be emphasized that current systematic errors are still much larger than statistical ones, but it is clear that the combined information from

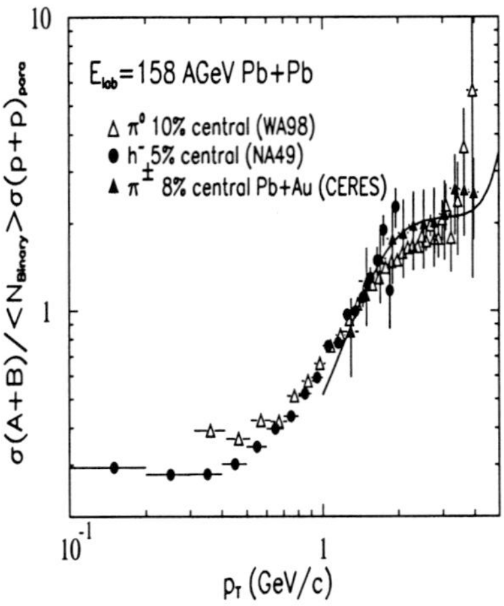

Fig. 11. The nuclear modification factor for hadron spectra in central $Pb + Pb$ collisions at the CERN-SPS exceeds unity at high-p_T due to the Cronin effect. The solid line is a pQCD calculation by Wang [83]

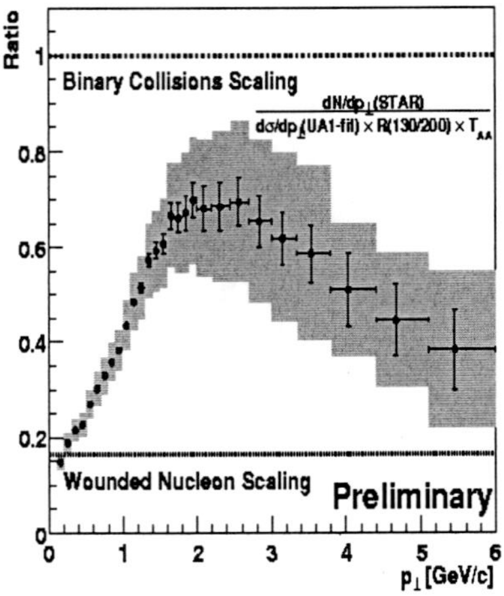

Fig. 12. The nuclear modification factor for charged hadrons in central $Au + Au$ RHIC from STAR [3,84]. In contrast to SPS, the high-p_T charged hadrons are suppressed

two independent experiments in Figs. 12 and 14 imply that something new has been discovered in $A + A$ collisions at RHIC. I believe that this is the predicted jet quenching as discussed in the next chapter.

The reason that this discovery is perhaps even more exciting than the famous J/ψ suppression effect discovered by NA50 [86,87] at the SPS is that J/ψ suppression was also seen in $p + A$. The cold nuclear suppression mechanism in $p+A$ is called "normal". The enhanced suppression in $Pb+Pb$ is "anomalous" because it is more than if the normal $p + A$ suppression pattern is extrapolated to $A+A$. That this is not theoretically fool proof was pointed out by Qiu et al. [88]. They showed that including radiative energy loss in cold nuclei could lead to non-linear enhancement of J/ψ suppression by decreasing their formation probability. Only a rather schematic model was presented, but it emphasizes the necessity of improving considerably the theory of the "normal" processes associated with heavy quark propagation through with nuclei. The situation is rather similar theoretically with regard to the Cronin effect. There also only rather schematic models are available to simulate the effect.

The big difference between the two cases is that for J/ψ the "normal" and "anomalous" components work in the *same* direction. The premium is thus high on developing an accurate theory "normal" nuclear suppression. In the jet quenching case, on the other hand, the "normal" Cronin effect

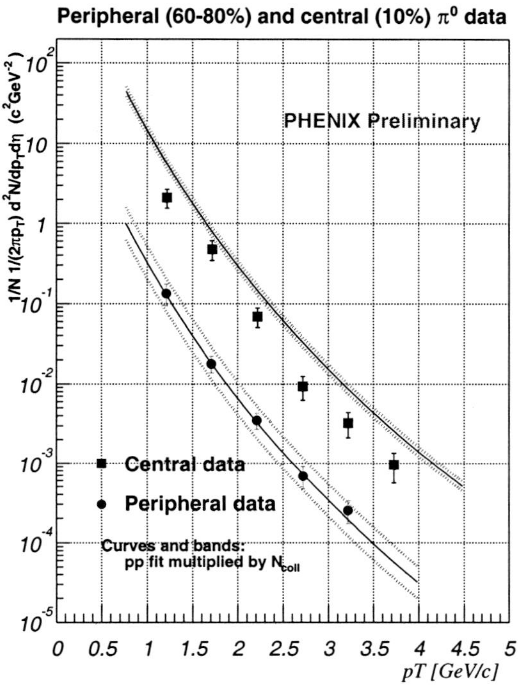

Fig. 13. Semi-inclusive π^0 p_T distribution $(1/N_{int})(dN_{\pi^0}/2\pi p_T dp_T dy))$ in the upper 60-80% peripheral events (solid circles) and the 10% most central events (solid squares) from PHENIX [85]. The lines are a parameterization of *pp* charged hadron spectra, scaled by the mean number of collisions $N_{coll} = 857, 19$, respectively. The bands indicate the possible range due to the systematic error on N_{coll}

works in the *opposite* direction to the "anomalous" new jet quenching mechanism. Of course, there are possibly other "normal" effects, such as gluon (anti?)shadowing, that may work in either direction at high p_T. To map out all the "normal" physics components will require detailed systematic measurements of $p + A$ at RHIC as done at the SPS. As a final remark, I want to emphasize the "normal" component of the dynamics is not dull run-of-the-mill background, but fundamentally interesting many-body QCD physics in its own right and deserves considerable more attention.

3.4 Where Have all the Baryons Gone?

One of the puzzling feature of Figs. 12 and 14 is that pions appear to be more quenched than the sum of charged particles. Usually we assume that pions are the most abundant hadron species at high p_T since both quark and gluon fragmentation functions prefer to make the lightest mesons [91,92].

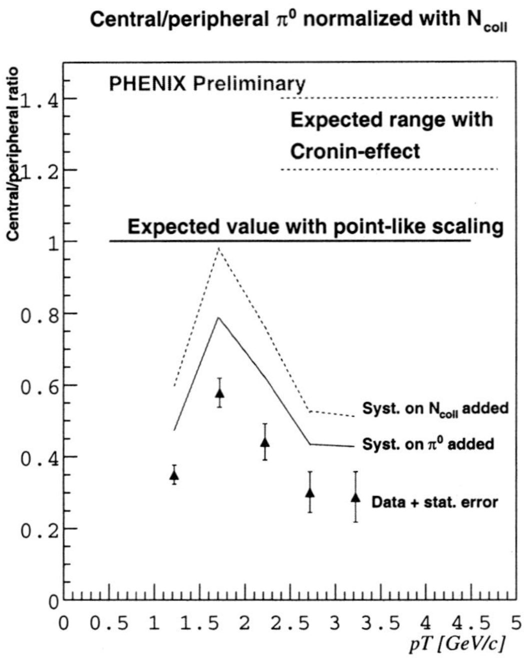

Fig. 14. Ratio of π^0 yields in central to peripheral collisions at RHIC reported by PHENIX [85] scaled by the number of binary collisions $N_{coll} = 857, 19$, respectively. The suppression of high-p_T pions at RHIC (in contrast to the Cronin enhancement at SPS in Fig. 11) is due to jet quenching in the ultra-dense matter formed at RHIC [21,27,73,28,75,83,89]

Surprisingly, the preliminary PHENIX data [90] on identified high-p_T hadron spectra suggest from Fig. 15 that baryons may be the most abundant species above $p_T > 2$ GeV/c. One possible source of such a non-pQCD like flavor distribution could be hydrodynamic transverse flow. For a longitudinal boost invariant (Bjorken) expansion with a transverse flow velocity field, $v_\perp = \tanh(\eta_r)$, the general formula [93] for the differential particle number is

$$E\frac{dN_s}{d^3\mathbf{p}} = \frac{d}{2\pi^2}\int_0^1 d\zeta\, r_f(\zeta)\tau_f(\zeta)\Bigg\{ -\frac{dr_f}{d\zeta} m_\mathrm{T} K_1\left(\frac{m_\mathrm{T} ch\eta_r}{T_f}\right) I_0\left(\frac{p_\mathrm{T} sh\eta_r}{T_f}\right) + \frac{d\tau_f}{d\zeta} p_\mathrm{T} K_0\left(\frac{m_\mathrm{T} ch\eta_r}{T_f}\right) I_1\left(\frac{p_\mathrm{T} sh\eta_r}{T_f}\right)\Bigg\}, \quad (6)$$

where $d = 2s+1$ is the degeneracy factor, $\eta_r = \mathrm{Artanh}(v_\perp(z=0))$ is the transverse fluid rapidity and $(r_f(\zeta), \tau_f(\zeta))$ is a parameterization (counter-clockwise) of the freeze-out surface (isotherm of temperature T_f).

Solutions for freeze-out surfaces with arbitrary transverse velocity fields $v_\perp(\xi)$ can be obtained by solving relativistic hydrodynamics. For the simplest

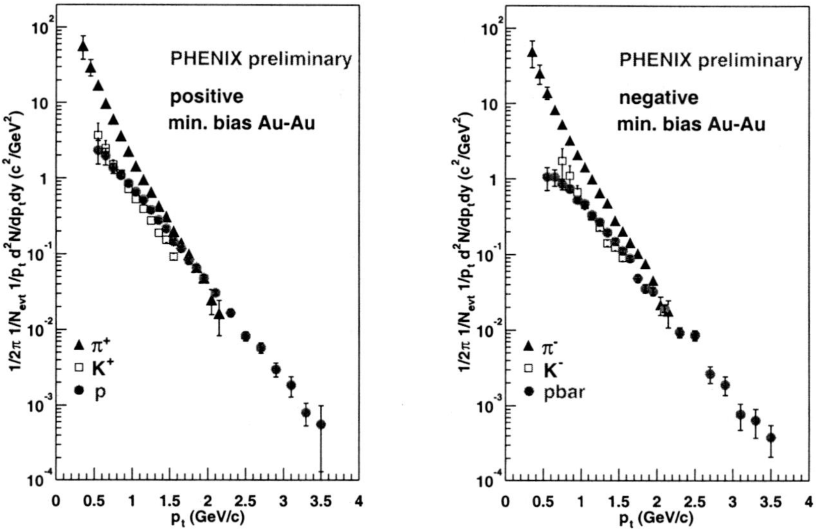

Fig. 15. Minimum bias transverse momentum distributions for positive (left) and negative (right) identified hadrons measured in PHENIX [90]. The error bars include statistical errors and systematic errors in the acceptance and decay corrections. Additional 20% systematic errors on the absolute normalization are not included

case with $v_\perp = \tanh \eta_r$ a constant and an isotherm freeze-out on a proper time hypersurface τ_f [94],

$$\begin{aligned}\frac{dN_s}{dyd^2\mathbf{p}_\mathrm{T}} &= \frac{d}{4\pi^2} R^2 \tau_f m_\mathrm{T} K_1 \left(\frac{m_\mathrm{T}\cosh\eta_r}{T_f}\right) I_0 \left(\frac{p_\mathrm{T}\sinh\eta_r}{T_f}\right) \\ &\overset{p_\mathrm{T}\gg m}{\longrightarrow} const \times d \, \exp\left(-\frac{p_\mathrm{T}}{T_f\exp(\eta_r)}\right), \qquad (7)\end{aligned}$$

which corresponds to a blue-shifted effective temperature $T_f e^{\eta_r}$. This is the uniform rapidity, transverse boosted Bjorken sausage parameterization of nuclear collision distributions.

Evidence for increased transverse flow phenomena at RHIC relative to SPS comes from low-p_T STAR data [99] shown in Figs. 16 and 17. The data can be fit up to $p_T < 1$ GeV/c with a rather radial flow velocity $v_\perp \sim 0.6$ c that is significantly larger than the radial flow ~ 0.4 c deduced from similar SPS spectra.

Another important experimental tool to search for collective flow effects is to study anisotropic multiparticle emission patterns [95–98]. A particularly useful measure of collective behavior in ultra-relativistic energies has turned out to be the differential second Fourier component [96] of the azimuthal

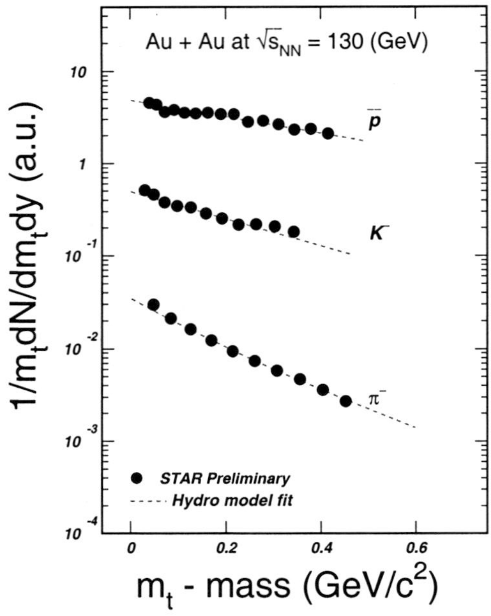

Fig. 16. Transverse momentum distribution for π, K, p measured by STAR [99] at RHIC. The dashed curves are fits with Eq. 7 leading to a freeze-out temperature $T_f = 100$ MeV boosted with a radial flow velocity field $v_\perp \sim 0.6$c

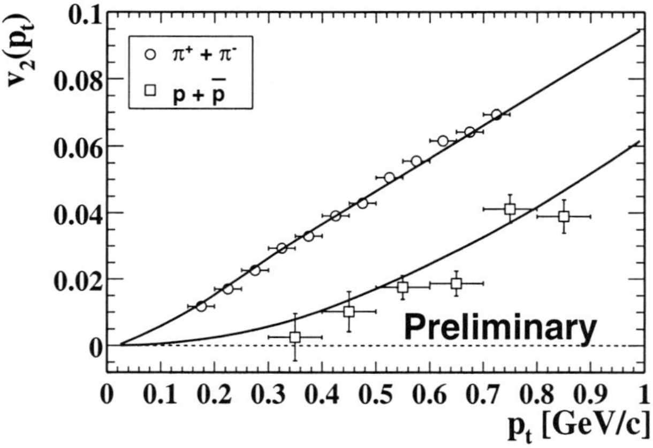

Fig. 17. Second azimuthal Fourier component of invariant pion and proton distributions from STAR [31] compared to ideal hydrodynamic flow pattern from [58]

distribution:

$$\frac{dN_h(\mathbf{b})}{dyd^2\mathbf{p}_\mathrm{T}} = \frac{dN_h(\mathbf{b})}{dydp_\mathrm{T}^2}\frac{1}{\pi}\left(1 + 2v_2^h(p_\mathrm{T})\cos(2\phi)\right) , \tag{8}$$

where ϕ is measured relative to a "reaction plane" event by event as determined in [98].

Azimuthal or "elliptic" flow results from the initial spatial anisotropy of the dense matter in semi-peripheral $A + A$ collisions. The hydrodynamic model predicts an elliptic flow pattern at RHIC [58,59] that can be approximately parameterized as

$$v_2^s(p_\mathrm{T}) \approx \tanh(p_\mathrm{T}/(10 \pm 2\ \mathrm{GeV})) . \tag{9}$$

Up to about $p_T < 1$ GeV, this agrees remarkably well with STAR data. At high p_T this hydrodynamic flow component breaks down because of the emergence of the hard pQCD hadrons.

The transverse boosted Bjorken sausage model Eq. (7) predicts that asymptotically the baryon/meson ratios $p/\pi+ = \bar{p}/\pi^- \to 2$ for any flow velocity because nucleons have two spin states. However, numerically this ratios exceed unity only at $p_T > 3 - 4$ GeV. Thus transverse flow alone cannot account for the anomalous baryon dominance of high-p_T spectra in Fig. 15 as emphasized in [92].

Another observation [99] that possibly provides a hint that the answer to the puzzling result may lie in novel baryon dynamics at RHIC can be seen in Figs. 18 and 19. As was shown by Kharzeev [101], the energy and rapidity dependence of the inclusive baryon production at mid-rapidity can be obtained using Mueller's generalized optical theorem in the double Regge limit. Here, the exchanges of a Pomeron and a M_0^J baryon-anti-baryon "junction" pair lead to the following form for single mid-rapidity baryon production

$$E_B\frac{d^3\sigma^{(1)}}{d^3p_B} = C_B f_B(m_t^2)\left(\frac{s_0}{s}\right)^{1/4}\cosh(y/2) , \tag{10}$$

where C_B is a constant that reflects the couplings of the Reggeon and Pomeron to the proton, $f_B(m_t^2)$ is an unknown function of m_t and $s_0 \simeq 1$ GeV is a Regge energy scale. The $\cosh(y/2)$ rapidity dependence and the $1/\sqrt[4]{s}$ energy dependence follow from the assumed intercept [102], $\alpha_{M_0^J}(0) \approx 1/2$. In contrast to simpler diquark breaking models as in the Dual Parton Model, the multiplicity of junction also enhanced events is enhanced by a factor of 5/4 in $p+p$, and the strangeness content is also enhanced by a large factor. The junction mechanism for baryon number (vs. valence quark number) transport predicts for the unique possibility of producing $S = -3$ Ω^- baryons at midrapidity, as were observed at the SPS in WA97. In the Monte Carlo event generator HIJING/$B\bar{B}$ [65,100], baryon junctions are implemented in terms of Y shaped strings spanning valence quarks.

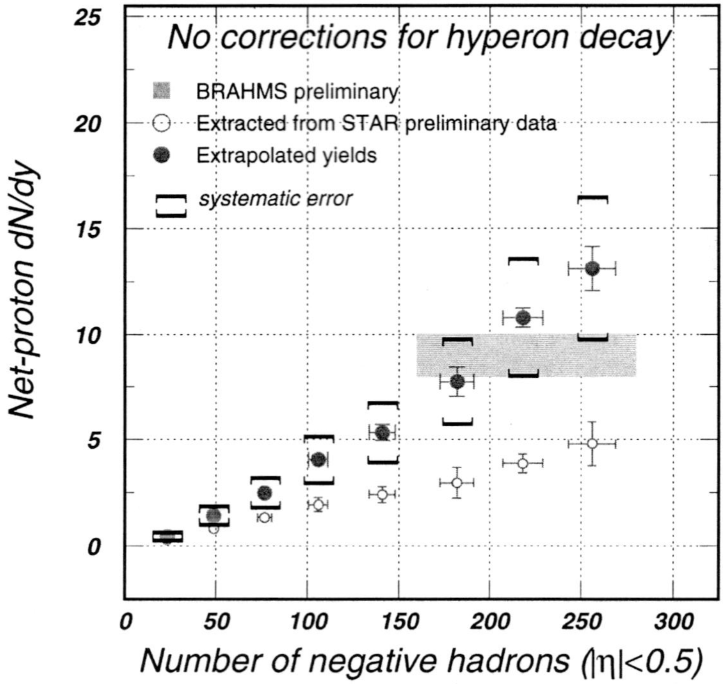

Fig. 18. The valence proton rapidity density measured by STAR [99] at RHIC as a function of N_{part}. Preliminary BRAHMS data are also indicated

The junction is a topological knot in the gluon field connecting the color flux from three quarks into a color singlet state [102]. The intriguing aspect of junctions is that the conserved baryon number resides in the gluon knot and not in the valence quarks [101]. In a nuclear collision some or all of the valence quarks may fragment into mesons. However, the gluonic junctions insure that baryon number is conserved. The understanding of the dynamics of junction exchange and pair production is still rather primitive, but the consistency of the baryon stopping power at SPS and now RHIC with HIJING/$B\bar{B}$ predictions suggest that baryon dynamics at central rapidities may be especially interesting at RHIC. See Ref. [103] for a discussion of possible novel junction network physics that may lead to femto-scale buckyball and even CP odd junction network production in $A + A$.

3.5 Quenching of Elliptic Flow

As seen in Figs. 17 and 20 strong elliptic flow was discovered at RHIC consistent with hydrodynamic predictions at low transverse momentum $p_T < 1$ GeV. However, the preliminary data from STAR [104] shows that above $p_T > 2$ GeV the elliptic flow saturates and the azimuthal asymmetry deviates more and more from hydrodynamic behavior as seen in Fig. 21. This

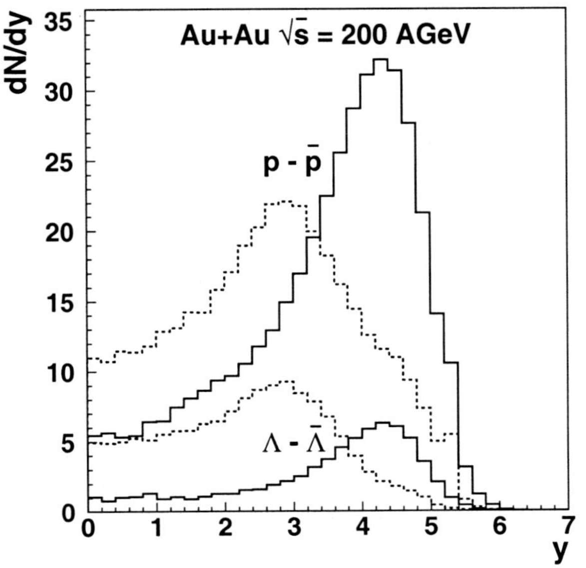

Fig. 19. Predicted valence proton rapidity density at RHIC from [100] showing a factor of two enhancement expected if baryon junction exchange is included in HIJING/B. The dashed curves are result of HIJING including only standard LUND diquark fragmentation

information provides insight into how hydrodynamic behavior breaks down at high p_T due to the finite energy loss of partons in the plasma. As shown in detail in [21] the saturation pattern at high p_T depends on the energy dependence of the gluon energy loss as well as on the geometry of the plasma density at finite impact parameters. It therefore provides tomographic information about the density profile and its evolution in $A + A$. See section 4.4 for more details.

3.6 Where Did the Slowly Burning Plasma Log Vanish?

The last major RHIC result that I highlight here is on pion interferometry. Relativistic combustion theory [105,106,93] predicts that if there were a sufficiently rapid cross over between the QGP and hadronic phases of ultra-dense matter, then a deflagration burn front may appear between two phases. The main characteristic of that burn front is its very small velocity in case the entropy density jump across it is sufficiently large and no high degree of non-equilibrium supercooling arises. Even with a smooth cross-over transition, such slowly burning plasma solutions were shown to exist as long as the width of the transition region is ($\Delta T_c/T_c < 0.08$). The lifetime of a Bjorken plasma log is therefore significantly enhanced $\tau \sim R/v_d$, where $v_d \sim 1/25$ is the small deflagration velocity in the static 1+1D case.

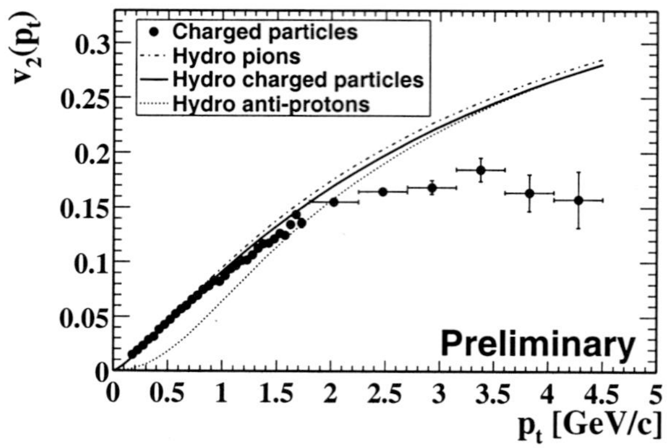

Fig. 20. Saturation of elliptic flow as measured by STAR [104]. Curves are the extrapolations of the hydrodynamic model predictions from [58,59] to high p_T

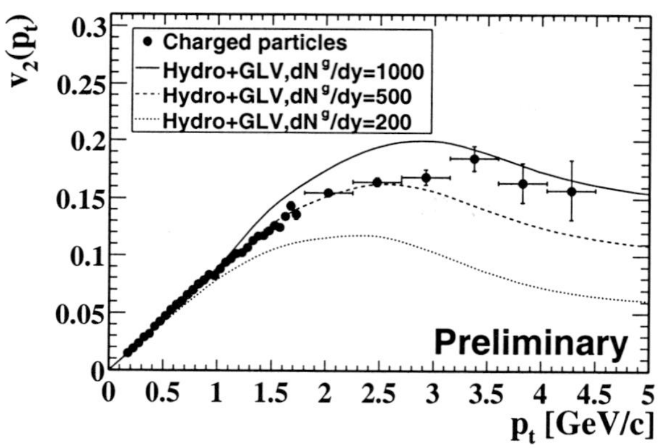

Fig. 21. Curves show saturation of elliptic flow due to finite energy loss of partons in a gluon plasma with rapidity density $dN_g/dy = 200, 500, 1000$ from [21]

This characteristic time delay of the hadronization from a QGP state was suggested in [107,108] to be testable via pion interferometry. In Ref. [93] the 3+1D hydrodynamic equations were solved to study this plasma "stall" phenomenon in detail.

The two-pion correlation function measures the coincidence probability $P(\mathbf{p}_1, \mathbf{p}_2)$ of two (identical) bosons with momenta $\mathbf{p}_1$, $\mathbf{p}_2$ relative to the probability of detecting uncorrelated particles from different events,

$$C_2(\mathbf{p}_1, \mathbf{p}_2) = \frac{P(\mathbf{p}_1, \mathbf{p}_2)}{P(\mathbf{p}_1)P(\mathbf{p}_2)} \,. \tag{11}$$

If the average 4–momentum is denoted as $K^\mu = (p_1^\mu + p_2^\mu)/2$ and the relative 4–momentum as $q^\mu = p_1^\mu - p_2^\mu$, then under the assumption that the particle source is chaotic and sufficiently large,

$$C_2(\mathbf{p}_1, \mathbf{p}_2) = 1 + \frac{\left| \frac{1}{(2\pi)^3} \int_\Sigma \mathrm{d}\Sigma \cdot K \; \exp\left[i\, \Sigma \cdot q\right] \, f\left(\frac{u \cdot K}{T}\right) \right|^2}{E_1 \frac{\mathrm{d}N}{\mathrm{d}^3\mathbf{p}_1} \; E_2 \frac{\mathrm{d}N}{\mathrm{d}^3\mathbf{p}_2}} , \tag{12}$$

where [47]

$$E \frac{\mathrm{d}N}{\mathrm{d}^3\mathbf{p}} = \frac{1}{(2\pi)^3} \int_\Sigma \mathrm{d}\Sigma \cdot p \; f\left(\frac{u \cdot p}{T}\right) \tag{13}$$

is the single inclusive momentum distribution, $f(x) = (e^x - 1)^{-1}$, and u^μ the fluid 4–velocity. The integrals run over the assumed freeze–out hypersurface. In general, that hypersurface is represented by a 3–parametric (4–vector) function $\Sigma^\mu(\zeta, \eta, \phi)$, and the normal vector on the hypersurface is determined by

$$\mathrm{d}\Sigma_\mu = \epsilon_{\mu\alpha\beta\gamma} \frac{\partial \Sigma^\alpha}{\partial \zeta} \frac{\partial \Sigma^\beta}{\partial \eta} \frac{\partial \Sigma^\gamma}{\partial \phi} \, \mathrm{d}\zeta \, \mathrm{d}\eta \, \mathrm{d}\phi , \tag{14}$$

where $\epsilon_{\mu\alpha\beta\gamma}$ is the completely antisymmetric 4-tensor. For the common isotherm freeze-out temperature T_f hypersurface, the fluid velocity generally varies $u^\mu = u^\mu(\Sigma)$.

For the Bjorken cylinder geometry, it is useful to restrict consideration to particles emitted at midrapidity, $K^z = q^z = 0$. Rotational symmetry around the z–axis in central collisions makes it possible to choose the average transverse momentum as $\mathbf{K}_\perp = (K, 0, 0)$, and consequently, $C_2(K, q_{\mathrm{out}}, q_{\mathrm{side}})$ is a function of three independent variables only. The so called out and side projections of the relative momenta are $\mathbf{q}_{\mathrm{out}} = (q_{\mathrm{out}}, 0, 0)$, $\mathbf{q}_{\mathrm{side}} = (0, q_{\mathrm{side}}, 0)$. As shown in [107,108] the width, $1/R_{side}$, of the correlation function in q_{side} is a measure of the transverse decoupling or freeze-out radius, while the width $1/R_{out}$ of the q_{out} correlation function is also sensitive to the duration of hadronization, $\Delta\tau$,

$$R_{out}^2 \approx R_{side}^2 + v^2 \Delta\tau^2 \; .$$

Thus a QGP stall would manifest experimentally in $R_{out} \gg R_{side}$. In [93] it was found that for possibly realistic parameters, $R_{out}/R_{side} \sim 2 - 3$, could be observed if a QGP stall occurred.

With this "warm-up" review of pion interferometry, we are now ready for the rude awakening from ideal gedanken considerations with the first splash of "cold" RHIC pion interferometry data shown in Fig. 22. The preliminary PHENIX data show that $R_{out} \sim R_{side}$ and even more disturbing all the deduced interferometry parameters are virtually identical to values seen at the AGS and SPS. To add insult to injury, it appears that $R_{out} < R_{side}$ for $p_T > 0.4$ GeV. Preliminary STAR data [3] show the same tendency.

Of course, scenarios may be invented to "explain" the data a-posteriori, but if these data are confirmed by further measurements, then they are indeed surprising and call into question our picture of the space-time evolution

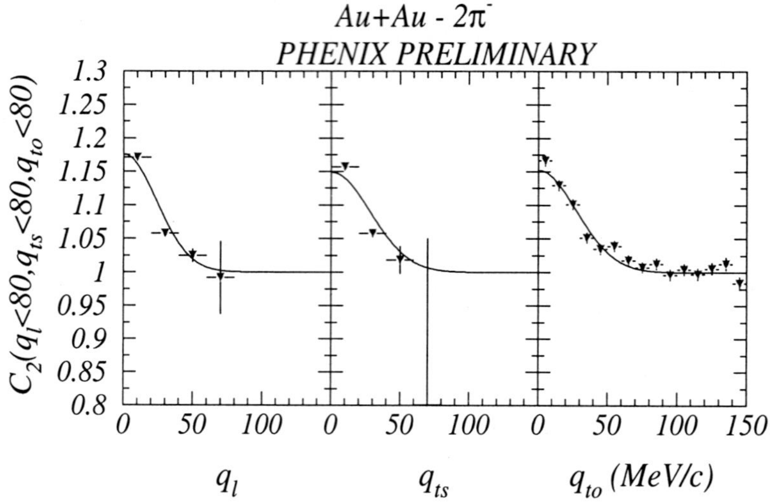

Fig. 22. Preliminary PHENIX pion interferometry data [109] vs. different projections of the relative momenta. Similar preliminary STAR data [3] were also shown at QM01. Unlike predictions [93] there is no hint of the expected stall or time delay of the QGP transition

of $A+A$. That this problem is not restricted to idealized hydrodynamics can be seen from the results of Ref. [110]. It was shown that $R_{out} > R_{side}$ is also predicted in a calculation where the entropy jump is small and pion decoupling is dynamically handled via URQMD. Among the theoretical questions that should now be further investigated is whether the pion interferometry theory based on *chaotic* ensembles [111] is in fact applicable to $A+A$. Another question that needs further study is whether the assumed ensemble of initial conditions was too restrictive and whether highly inhomogeneous and turbulent initial conditions apply [56].

4 Jet Quenching: Theory

Having had a brief tour of some of the interesting new data harvested from RHIC during the first round of experiments, I turn next to the more specific theoretical problem of computing the energy loss per unit length of a fast parton penetrating a finite, expanding quark-gluon plasma. As I emphasized above, high-p_T many-body pQCD physics is a new frontier at RHIC and higher energies. This requires the development of the non-Abelian analogue of the radiative energy loss theory familiar from classical E&M. The interesting new twist is that we have no external beams of quarks or gluons and the medium is very thin due to the fact that nuclei are tiny. Also the formation time physics of Landau-Pomeronchuk-Migdal (LPM) results in major destructive interference effects that must be taken into account. Work on this problem over the past five years has advanced considerably but many open problems remain.

I will highlight only one of those direction, namely the opacity expansion reaction operator method that we developed in Refs. [75]. The reader is referred to BDMS [77], Z [78], and U [79] for alternative methods and approximations.

In Ref. [74] we proposed a simple model to study induced gluon radiation due to multiple elastic scattering of a high-energy jet propagating in a locally color neutral amorphous plasma.

4.1 GLV Formalism

In [76] we developed a systematic graphical method to compute medium induced gluon radiation amplitudes as shown in Fig. 23. The exponential growth of the number of graphs with the number of interactions makes it very tedious to go beyond order three. In GLV [75] we overcame the combinatorial problem by developing a new algebraic operator technique to solve the inclusive radiated gluon distributions recursively. The first step is to compute the three direct (single Born) and four surviving virtual (contact double Born) diagrams shown in Fig. 24.

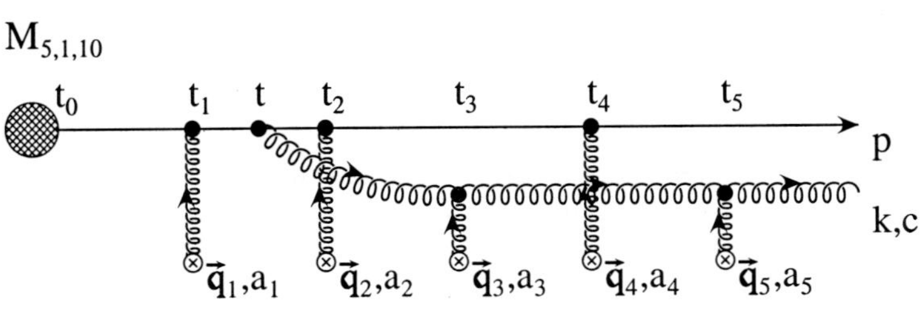

Fig. 23. Induced radiation amplitude [76] contributing to fifth order and higher order in the opacity expansion of QCD energy loss in the GW model [74]. The crosses denote static color screened Yukawa interactions on a scale μ. The blob is the initial hard jet amplitude without final-state interactions

For scattering off n scattering centers located at depths z_i in a transverse homogeneous medium of large area $((\mu R)^2 \gg 1)$, we can write the inclusive radiated gluon spectrum, $P_n(\mathbf{k}, c)$, as a sum over products of partial sums of amplitudes and complementary complex conjugate amplitudes. Every term in the sum contributes to the same $O(g^{2n})$. The average value of n is referred to as the opacity of the medium. The partial sums of diagrams at order n in such and opacity expansion can be conveniently expressed in a tensor notation

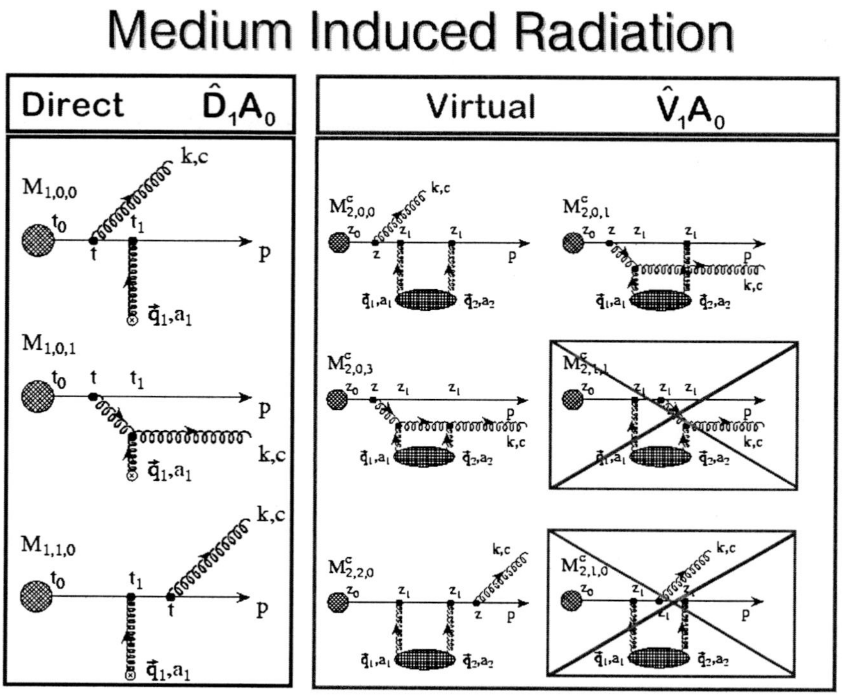

Fig. 24. Three first-order (single Born) direct and four surviving (double Born) virtual or contact amplitudes [77] from which the $\hat{D}_n$ and $\hat{V}_n$ components of the reaction operator in Eq. (16) are derived in [75]

and constructed by repeated operations of $\hat{\mathbf{1}}, \hat{D}_i$, or $\hat{V}_i$ corresponding to no, direct, or virtual interactions at scattering center i

$$\mathcal{A}_{i_1 \cdots i_n}(x, \mathbf{k}, c) = \prod_{m=1}^{n} \left(\delta_{0,i_m} + \delta_{1,i_m} \hat{D}_m + \delta_{2,i_m} \hat{V}_m \right) G_0(x, \mathbf{k}, c) \ .$$

Here G_0 is the initial hard $q + g$ color matrix amplitude. In the inclusive probability each class contracts with a unique complementary class

$$P_n(x, \mathbf{k}) = \bar{\mathcal{A}}^{i_1 \cdots i_n}(c) \mathcal{A}_{i_1 \cdots i_n}(c)$$

with the complementary class constructed as

$$\bar{\mathcal{A}}^{i_1 \cdots i_n}(x, \mathbf{k}, c) \equiv G_0^\dagger(x, \mathbf{k}, c) \prod_{m=1}^{n} \left(\delta_{0,i_m} \hat{V}_m^\dagger + \delta_{1,i_m} \hat{D}_m^\dagger + \delta_{2,i_m} \right) .$$

Fig. 25 shows an example of how this formalism works at 4th order in opacity for elastic and inelastic inclusive distributions.

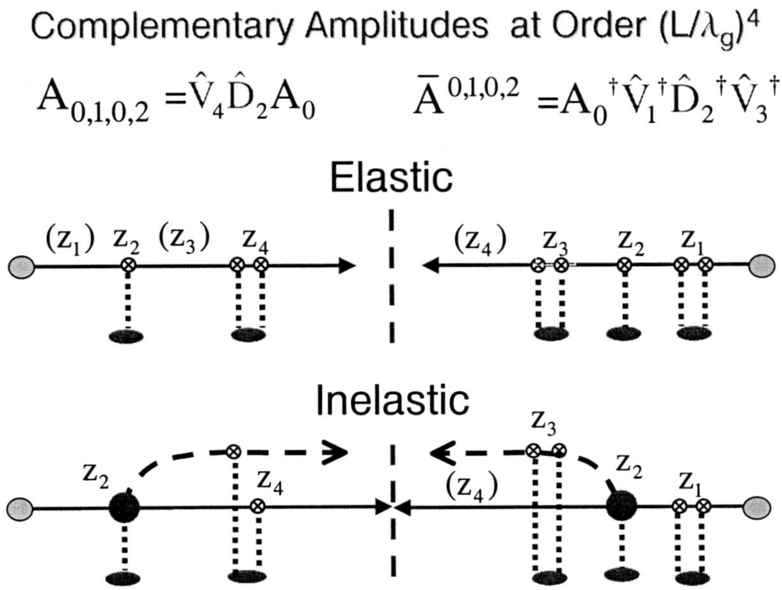

Fig. 25. Example of graphs constructed via $\hat{D}_i, \hat{V}_i$ that contribute to the 4th order in opacity in elastic and inclusive inelastic final-state interactions. The longitudinal depth of active scattering centers are denoted by z_i and inactive (created with $\hat{1}_i$) by (z_i). The form of $\hat{D}_i, \hat{V}_i$ depend on the process type but the tensorial bookkeeping of partial sums of amplitudes is the same

Direct interactions enlarge rank $n-1$ class elements as follows:

$$\hat{D}_n \mathcal{A}_{i_1\cdots i_{n-1}}(x, \mathbf{k}, c) \equiv (a_n + \hat{S}_n + \hat{B}_n)\mathcal{A}_{i_1\cdots i_{n-1}}(x, \mathbf{k}, c)$$

$$= a_n \mathcal{A}_{i_1\cdots i_{n-1}}(x, \mathbf{k}, c) + e^{i(\omega_0-\omega_n)z_n} \mathcal{A}_{i_1\cdots i_{n-1}}(x, \mathbf{k}-\mathbf{q}_n, [c, a_n]) -$$
$$\left(-\frac{1}{2}\right)^{N_v(\mathcal{A}_{i_1\cdots i_{n-1}})} \mathbf{B}_n \, e^{i\omega_0 z_n}[c, a_n] T_{el}(\mathcal{A}_{i_1\cdots i_{n-1}}),$$

where $\mathbf{B}_n = \mathbf{H} - \mathbf{C}_n = \mathbf{k}/\mathbf{k}^2 - (\mathbf{k}-\mathbf{q}_n)/(\mathbf{k}-\mathbf{q}_n)^2)$ is the so-called Bertsch-Gunion amplitude for producing a gluon with transverse momentum $\mathbf{k}$ in an isolated single collision with scattering center n. The momentum transfer to the jet is $\mathbf{q}_n$. The notation $\omega_n = (\mathbf{k}-\mathbf{q}_n)^2/2\omega$, for a gluon with energy ω and a_n is the color matrix in the d_R dimensional representation of the jet with color Casimir C_R. $N_v = \sum_{m=1}^{n-1} \delta_{i_m,2}$ counts the number of virtual interactions in $\mathcal{A}_{i_1\cdots i_{n-1}}$.

Unitarity (virtual forward scattering) corrections to the direct processes involve the sum of four double born contact diagrams in Fig. 24 that enlarge rank $n-1$ classes as follows:

$$\hat{V}_n = -\frac{1}{2}(C_A + C_R) - a_n \hat{S}_n - a_n \hat{B}_n = -a_n \hat{D}_n - \frac{1}{2}(C_A - C_R). \qquad (15)$$

This *key* operator relation between direct and virtual insertions that we discovered in [75] makes it possible to solve the problem algebraically.

The tensor classification of classes of diagrams makes it possible to construct the distribution of radiated gluons in the case of n interactions, P_n, recursively from lower rank (opacity) classes via a "reaction" operator

$$P_n = \bar{\mathcal{A}}^{i_1\cdots i_{n-1}} \hat{R}_n \mathcal{A}_{i_1\cdots i_{n-1}} \,, \qquad \hat{R}_n = \hat{D}_n^\dagger \hat{D}_n + \hat{V}_n + \hat{V}_n^\dagger. \qquad (16)$$

Using the key identity (15), the reaction matrix simplifies to

$$\hat{R}_n = (\hat{D}_n - a_n)^\dagger (\hat{D}_n - a_n) - C_A = (\hat{S}_n + \hat{B}_n)^\dagger (\hat{S}_n + \hat{B}_n) - C_A.$$

The next major simplification occurs because both $\hat{S}$ and $\hat{B}$ involve the same gluon color rotation through if^{ca_nd}. This reduces the color algebra to multiplicative Casimir factors

$$\bar{\mathcal{A}}^{i_1\cdots i_{n-1}} (\hat{S}_n^\dagger \hat{S}_n - C_A) \mathcal{A}_{i_1\cdots i_{n-1}}$$

$$= C_A \left(P_{n-1}(\mathbf{k} - \mathbf{q}_n) - P_{n-1}(\mathbf{k})\right) = C_A \left(e^{i\mathbf{q}_n\cdot\hat{\mathbf{b}}} - 1\right) P_{n-1}(\mathbf{k})$$

$$\bar{\mathcal{A}}^{i_1\cdots i_{n-1}} \hat{B}_n^\dagger \hat{B}_n \mathcal{A}_{i_1\cdots i_{n-1}} = 0$$

$$2\mathbf{Re}\, \bar{\mathcal{A}}^{i_1\cdots i_{n-1}} \hat{B}_n^\dagger \hat{S}_n \mathcal{A}_{i_1\cdots i_{n-1}} = -2C_A\, \mathbf{B}_n \cdot \left(\mathrm{Re}\; e^{-i\omega_n z_n} e^{i\mathbf{q}_n\cdot\hat{\mathbf{b}}} \mathbf{I}_{n-1}\right).$$

$\mathbf{I}_n$ obeys a recursion relation from which the inclusive radiation probability is found to obey the soluble recursion relation

$$P_n(\mathbf{k}) = C_A (P_{n-1}(\mathbf{k} - \mathbf{q}_n) - P_{n-1}(\mathbf{k})) - 2C_A\, \mathbf{B}_n \cdot \left(\mathbf{Re}\; e^{-i\omega_n z_n} e^{i\mathbf{q}_n\cdot\hat{\mathbf{b}}} \mathbf{I}_{n-1}\right)$$

$$+ \delta_{n,1} C_A C_R |\mathbf{B}_1|^2,$$

where $\hat{\mathbf{b}} = i\nabla_{\mathbf{k}}$ is the transverse momentum shift operator. The initial condition for this recursion relation is the initial hard vertex radiation amplitude without final-state interactions that is given by $P_0 = C_R\, \mathbf{H}^2 = C_R/\mathbf{k}_\perp^2$.

The complete solution to the problem can therefore be expressed in closed form as

$$P_n(\mathbf{k}) = -2 C_R C_A^n\, \mathbf{Re} \sum_{i=1}^{n} \left\{ \prod_{j=i+1}^{n} (e^{i\mathbf{q}_j\cdot\hat{\mathbf{b}}} - 1) \right\} \otimes \mathbf{B}_i \cdot\; e^{i\mathbf{q}_i\cdot\hat{\mathbf{b}}} e^{-i\omega_0 z_i} \times$$

$$\left\{\prod_{m=1}^{i-1}(e^{i(\omega_0-\omega_m)z_m}e^{i\mathbf{q}_m\cdot\hat{\mathbf{b}}}-1)\right\}\otimes\mathbf{H}(e^{i\omega_0 z_1}-e^{i\omega_0 z_0}).$$

This expression can be averaged over any spatial distribution of interaction centers, z_i as well as any z_i dependent momentum transfers q_n. This form is thus ideally suitable for Monte Carlo implementation for arbitrary $\mathbf{q}_i, z_i$ medium ensemble averages.

4.2 Non-Abelian Energy Loss at Finite Opacity

The first application [75] of our general solution to the energy loss problem was to calculate numerically the total radiated energy loss as a function of jet energy E, plasma depth L, and infrared screening scale μ. In the absence of a medium, the gluon radiation associated with a spin $\frac{1}{2}$ parton jet is distributed as

$$x\frac{dN^{(0)}}{dx\,d\mathbf{k}_\perp^2}=\frac{C_R\alpha_s}{\pi}\left(1-x+\frac{x^2}{2}\right)\frac{1}{\mathbf{k}_\perp^2}\ , \tag{17}$$

where $x=k^+/E^+\approx\omega/E$, and C_R is the Casimir of the (spin 1/2) jet in the d_R dimensional color representation. The differential energy distribution outside a cone defined by $\mathbf{k}_\perp^2>\mu^2$ is given by

$$\frac{dI^{(0)}}{dx}=\frac{2C_R\alpha_s}{\pi}\left(1-x+\frac{x^2}{2}\right)E\log\frac{|\mathbf{k}_\perp|_{\max}}{\mu}\ , \tag{18}$$

where the upper kinematic limit is $\mathbf{k}_{\perp\,\max}^2=\min\left[4E^2x^2,4E^2x(1-x)\right]$. The energy loss outside the cone in the vacuum is then given by

$$\Delta E^{(0)}=\frac{4C_R\alpha_s}{3\pi}E\log\frac{E}{\mu}. \tag{19}$$

While this overestimates the radiative energy loss in the vacuum (self-quenching), it is important to note that $\Delta E^{(0)}/E\sim 50\%$ is typically much larger than the medium induced energy loss.

Averaging over the momentum transfer $\mathbf{q}_{1\perp}$ via the color Yukawa potential leads to a very simple first-order opacity result for the $x\ll 1$ gluon double differential distribution

$$x\frac{dN^{(1)}}{dx\,d\mathbf{k}_\perp^2}=x\frac{dN^{(0)}}{dx\,d\mathbf{k}_\perp^2}\frac{L}{\lambda_g}\int_0^{q^2_{\max}}d^2\mathbf{q}_{1\perp}\frac{\mu^2_{eff}}{\pi(\mathbf{q}_{1\perp}^2+\mu^2)^2}$$
$$\frac{2\,\mathbf{k}_\perp\cdot\mathbf{q}_{1\perp}(\mathbf{k}-\mathbf{q}_1)_\perp^2L^2}{16x^2E^2+(\mathbf{k}-\mathbf{q}_1)_\perp^4L^2}, \tag{20}$$

where the opacity factor $L/\lambda_g=N\sigma_{el}^{(g)}/A_\perp$ arises from the sum over the N distinct targets. Note that the radiated gluon mean free path $\lambda_g=(C_A/C_R)\lambda$ appears rather than the jet mean free path. The upper kinematic bound on the momentum transfer $q^2_{\max} = s/4 \simeq 3E\mu$, $(1/\mu^2_{eff} = 1/\mu^2 - 1/(\mu^2 + q^2_{\max}))$. For SPS and RHIC energies, this finite limit cannot be ignored as we show below.

The second-order contribution in opacity involving the sum of 7^2 direct and 2×86 virtual and results in [75]

$$\begin{aligned} P^{(2)} \propto C_R C_A^2 d_R \, [& 2\,\mathbf{C}_1 \cdot \mathbf{B}_1 \, (1 - \cos(\omega_1 \Delta z_1)) \\ & + 2\,\mathbf{C}_2 \cdot \mathbf{B}_2 \, (\cos(\omega_2 \Delta z_2) - \cos(\omega_2(\Delta z_1 + \Delta z_2)) \\ & - 2\,\mathbf{C}_{(12)} \cdot \mathbf{B}_2 \, \big(\cos(\omega_2 \Delta z_2) - \cos(\omega_{(12)} \Delta z_1 + \omega_2 \Delta z_2)\big) \\ & - 2\,\mathbf{C}_{(12)} \cdot \mathbf{B}_{2(12)} \, \big(1 - \cos(\omega_{(12)} \Delta z_1)\big) \,] \, , \end{aligned} \tag{21}$$

where with $\mathbf{C}_{(mn)}$ and $\omega_{(mn)}$ obtained from $\mathbf{H}$ and ω_0 through the substitution $\mathbf{k}_\perp \Rightarrow \mathbf{k}_\perp - \mathbf{q}_{\perp m} - \mathbf{q}_{\perp n}$ and $\mathbf{B}_{m(nl)} \equiv \mathbf{C}_m - \mathbf{C}_{(nl)}$ [75].

Numerical results comparing the first three orders in opacity corrections to the hard distribution Eq. (17) were presented in [75]. To illustrate the result consider a quark jet in a medium with $\lambda_g = 1$ fm, a screening scale $\mu = 0.5$ GeV and $\alpha_s = 0.3$. The total radiative energy loss could be expressed as

$$\Delta E^{(1)} = \frac{C_R \alpha_s}{N(E)} \frac{L^2 \mu^2}{\lambda_g} \log \frac{E}{\mu} \, , \tag{22}$$

with $N(\infty) = 4 \log(E/\mu)/\tilde{v}$ if the kinematic bounds were ignored as in the approximations of Ref. [77]. We found that finite kinematic constraints and the form of the first-order result cause $N(E)$ to deviate considerably from the asymptotic value for all energies accessible in the RHIC range. Together with the logarithmic dependence on energy, these kinematic effects suppress greatly the energy loss at lower (SPS) energies as seen in Fig. 26. This is in sharp contrast to the approximately energy independent result in BDMS-ZW where the finite kinematic bounds were neglected because only the asymptotic limits were considered. Another remarkable result demonstrated numerically is that the second- and third-order contributions to the integrated energy loss remains surprisingly small in the physical range of nuclear opacities $L/\lambda_g \sim 5$. The rapid convergence of the opacity expansion even for realistic opacities results from the fact that the effective expansion parameter is actually the product of the opacity and the gluon formation probability $L\mu^2/2xE$. The leading quadratic dependence of the energy loss on nuclear thickness discovered in BDMS [77] therefore already emerges from the dominant first-order term in the opacity expansion.

At SPS energies kinematic effects suppress greatly the energy loss relative to BDMS. Our estimates provide a natural explanation for the absence of jet quenching in $Pb + Pb$ at 160 AGeV observed by WA98. At RHIC energies, on the other hand, a significant nonlinear (in A) pattern of suppression of high-$p_\perp$ hadrons relative to scaled pp data is predicted.

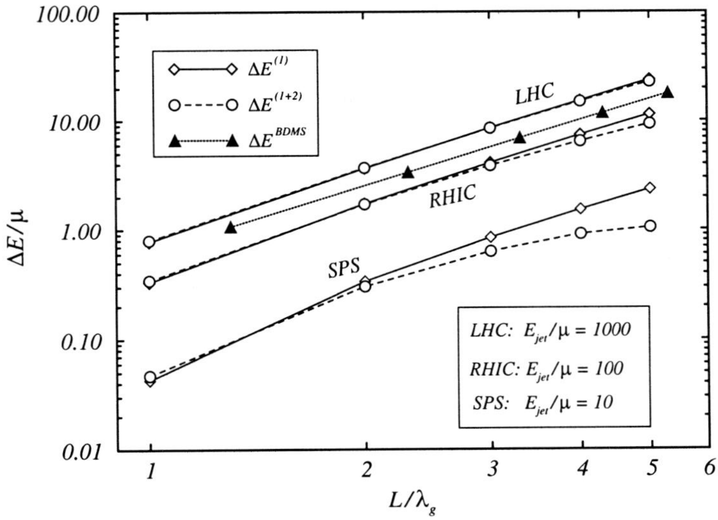

Fig. 26. The GLV radiated energy loss [75] of a quark jet with energy $E_{jet} = 5, 50, 500$ GeV (at SPS, RHIC, LHC) is plotted as a function of the opacity L/λ_g. ($\lambda_g = 1$ fm, $\mu = 0.5$ GeV). Solid curves show first order, while dashed curves show results up to second order in opacity. The asymptotic energy loss (solid triangles) of BDMS [77] is shown for comparison. The energy dependence of GLV suppressing radiative energy loss of low energy jets explains why no jet quenching was observed at the SPS (see Fig. 10)

4.3 The Opacity of the QGP at RHIC

As a second application of the GLV energy loss, in Ref. [89] we computed the quenched pQCD distribution of high-p_T hadrons as a function of the effective static plasma opacity, L/λ_g. In Figs. 27 and 28, the jet energy dependence of the GLV energy loss for gluons is shown. The most important feature to note is that $\Delta E_{GLV}/E$ is approximately constant in the energy range accessible at RHIC.

In order to compute the pion spectrum, note that jet quenching reduces the energy of the jet before fragmentation. We concentrate on mid-rapidity ($y_{\rm cm} = 0$), where the jet transverse momentum before fragmentation is shifted by the energy loss as in [112], $p_{\rm c}^*(L/\lambda) = p_{\rm c} - \Delta E(E, L)$. This shifts the $z_{\rm c}$ parameter in the fragmentation function of the integrand (23) to $z_{\rm c}^* = z_{\rm c}/(1 - \Delta E/p_{\rm c})$.

The invariant cross section of hadron production in central $A+A$ collision is then given by [91]

$$E_{\rm h}\frac{{\rm d}\sigma_{\rm h}^{\rm AA}}{{\rm d}^3p} = \int {\rm d}^2b\ {\rm d}^2r\ t_{\rm A}(\mathbf{b})t_{\rm B}(\mathbf{b}-\mathbf{r}) \sum_{\rm abcd}\int {\rm d}x_{\rm a}{\rm d}x_{\rm b}{\rm d}z_{\rm c}{\rm d}^2k_{\perp,{\rm a}}{\rm d}^2k_{\perp,{\rm b}} \cdot$$

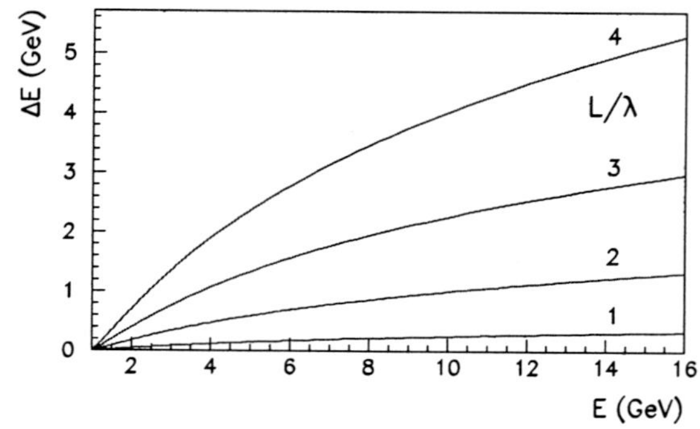

Fig. 27. Non-Abelian energy loss of a gluon jet calculated in the GLV picture [75]

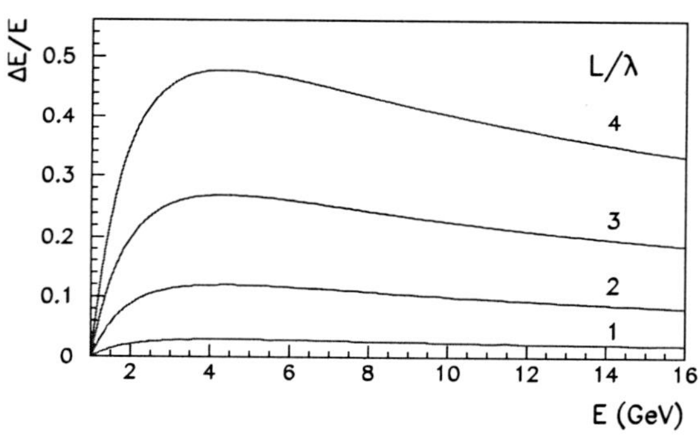

Fig. 28. The relative energy loss ($\Delta E/E$) is approximately constant at medium energy, $2 \leq E \leq 10$ GeV

$$f_{\mathrm{a/A}}(x_{\mathrm{a}}, k_{\perp,\mathrm{a}}(\mathbf{b}), Q^2) f_{\mathrm{b/A}}(x_{\mathrm{b}}, k_{\perp,\mathrm{b}}(\mathbf{b}-\mathbf{r}), Q^2)\, \frac{\mathrm{d}\sigma}{\mathrm{d}\hat{t}}$$

$$\frac{z_{\mathrm{c}}^*}{z_{\mathrm{c}}} \frac{D_{\mathrm{h/c}}(z_{\mathrm{c}}^*, \widehat{Q}^2)}{\pi z_{\mathrm{c}}^2}\, \hat{s}\delta(\hat{s}+\hat{t}+\hat{u}) \ , \tag{23}$$

where the upper limit of the impact parameter integral is $b_{\mathrm{max}} = 4.7$ fm for 10 % central Au+Au collisions. Here $t_{\mathrm{A}}(b)$ is the usual (Glauber) thickness function. The factor $z_{\mathrm{c}}^*/z_{\mathrm{c}}$ appears because of the in-medium modification of the fragmentation function [112]. Thus, the invariant cross section (23) depends on the average opacity or collision number, $\bar{n} = L/\lambda_{\mathrm{g}}$. The calculated spectra for pions are displayed for $\bar{n} = 0, 1, 2, 3, 4$ in Fig. 29. Fig. 30 shows their ratios to the non-quenched spectra at $\bar{n} = 0$. We note that in contrast to previous energy independent estimates for the energy loss, the GLV energy-dependent energy loss leads to constant suppression of the high-p_T domain in agreement with the preliminary data. The peripheral collisions are consistent with a rather small opacity in contrast to central collisions, as expected.

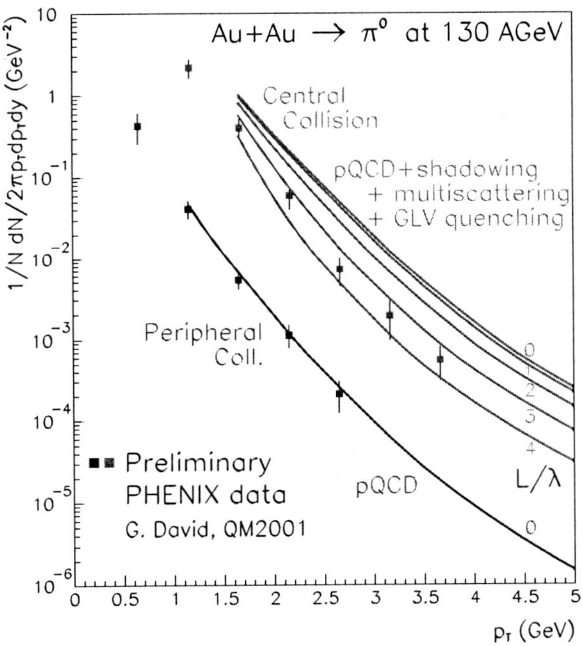

Fig. 29. Pion production in *Au+Au* collision including jet quenching with opacity $L/\lambda = 1, 2, 3, 4$. Preliminary QM01 PHENIX data shown (see updated data from [85] in Fig.13)

The ratio of central to peripheral PHENIX [85] data from QM01 shown in Fig. 30 clearly reveals that jet quenching at RHIC overcomes the Cronin enhancement at zero (final state) opacity. This is in stark contrast to data at SPS energies, where WA98 found no evidence for quenching in $Pb + Pb$ at 160 AGeV but a factor of two Cronin enhancement as discussed before.

Figs. 29 and 30 indicate that an effective static plasma opacity $L/\lambda = 3-4$ is sufficient to reproduce the preliminary jet quenching pattern observed at RHIC. In Ref. [83] it was shown that a rather small constant $dE/dx \approx 0.25$ GeV/fm was also found to be consistent with the data. However, it is important to emphasize that these effective *static* plasma opacities and parameters hide the underlying rapid dilution of the plasma due to expansion. The GLV formalism including the kinematic constraints at first order has been further

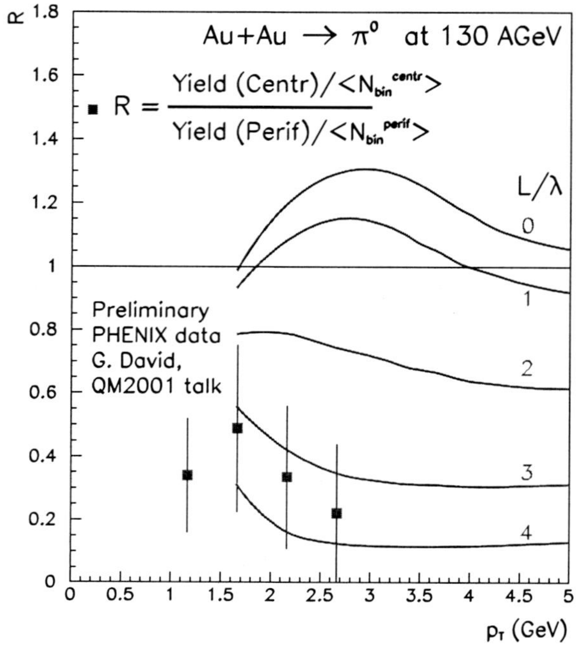

Fig. 30. The ratio of the central to the peripheral pion yields (normalized by the number of binary collisions, 857 and 5.5). (Note that updated data are shown in Fig. 14)

generalized to include effects of expansion in [21]. It was found in [21] that the inclusion of longitudinal expansion modifies the static plasma results in such a way that the moderate static plasma opacity actually implies that the produced mini-jet plasma rapidity density may have reached $dN_g/dy \sim 500$.

4.4 Jet Tomography from Quenched Elliptic Flow

So far we have not included the dilution effect of expansion on the energy loss. The generalization of GLV to the case of expanding plasmas is [21]

$$\frac{dI_{GLV}}{dx} = \frac{9C_R E}{\pi^2} \int_{z_0}^{\infty} dz\, \rho(z) \int^{|\mathbf{k}|_{\max}} d^2\mathbf{k}\, \alpha_s \int^{|\mathbf{q}|_{\max}} \frac{d^2\mathbf{q}\, \alpha_s^2}{(\mathbf{q}^2 + \mu(z)^2)^2}$$
$$\frac{\mathbf{k}\cdot\mathbf{q}}{\mathbf{k}^2(\mathbf{k}-\mathbf{q})^2} \left[1 - \cos\left(\frac{(\mathbf{k}-\mathbf{q})^2}{2xE}(z - z_0)\right)\right] , \qquad (24)$$

where $\rho(z)$ is the plasma density at time z along the jet path at position z from the production point and where the screening scale $\mu(z)$ may also depend on time.

Consider a density evolution of the form as in [77],

$$\rho(z) = \rho_0 \left(\frac{z_0}{z}\right)^\alpha \theta(L - z) , \tag{25}$$

where $\alpha = 0$ corresponds to a static uniform medium of thickness L, while $\alpha = 1$ to Bjorken 1+1D longitudinal expansion transverse to the jet propagation axis.

Analytic expressions can only be obtained again for asymptotic jet energies when the kinematic boundaries can be ignored. In that case, all but the path integral can be done giving

$$\Delta E \approx \frac{C_R \alpha_s}{2} \int_{z_0}^{\infty} dz \, \frac{\mu^2(z)}{\lambda(z)} \, (z - z_0) \log \frac{2E}{L\mu^2(z)} , \tag{26}$$

which is a linear weighed line integral over the local transport coefficient [77] $(\mu^2(z)/\lambda(z)) \approx \frac{9}{2}\pi\alpha_s^2 \rho(z)$, however, enhanced by a $\log 2E/L\mu^2(z)$ factor that results from the structure of the GLV integral missing in the BDMS asymptotic limit. For an expanding plasma as in (25)

$$\Delta E_\alpha(L, z_0) \approx \frac{C_R \alpha_s}{2} \frac{\mu^2(L) L^\alpha}{\lambda(L)} \frac{L^{2-\alpha}}{2 - \alpha} \tilde{v} . \tag{27}$$

Here $\tilde{v} = \log 2E/L\mu^2(L)$ and we used that $\mu^2(L)L^\alpha/\lambda(L)$ is a constant independent of L for this type of expansion. We also took the $z_0 \to 0$ limit. We therefore recover the asymptotic BDMSZ energy loss for both static and expanding media modulated by a $\log E/\omega(L)$ factor that is important at RHIC energies. Using the Bjorken relation between the gluon density and the rapidity density then gives

$$\Delta E_{\alpha=1}(L) = \frac{9 C_R \pi \alpha_s^3}{4} \left(\frac{1}{\pi R^2} \frac{dN^g}{dy} \right) L \log \frac{2E}{L\mu^2(L)} . \tag{28}$$

In practice, it is straight forward to integrate GLV numerically including the finite kinematic constraints.

For non-central collisions the GLV line integral depends, of course, on the azimuthal direction ϕ of the jet. The variation of the azimuthal energy loss with respect to ϕ at a given impact parameter b can be expressed in terms of

$$R(\mathbf{b}, \phi) = \frac{\Delta E(\mathbf{b}, \phi)}{\Delta E(0)}$$

with results shown in Fig. 31 The effect of this azimuthal variation of the energy loss is to induce an apparent elliptic flow at high p_T not related to hydrodynamic phenomena of low p_T. In [21] we proposed a simple interpolation between the hydrodynamic and jet quenched p_T eikonal regimes

$$v_2(p_{\rm T}) \approx \frac{v_{2s}(p_{\rm T})dN_s + v_{2h}(p_{\rm T})dN_h}{dN_s + dN_h} \, . \tag{29}$$

This interpolates between the hydrodynamic and the pQCD regimes because at high $p_{\rm T}$, $dN_h \gg dN_s$. For our numerical estimates the low-$p_{\rm T}$ interpolation is achieved by turning off the pQCD curves with a switch function $f_h(p_{\rm T}) = 0.5[1 + \tanh(3(p_{\rm T} - 1.5\,{\rm GeV}))]$.

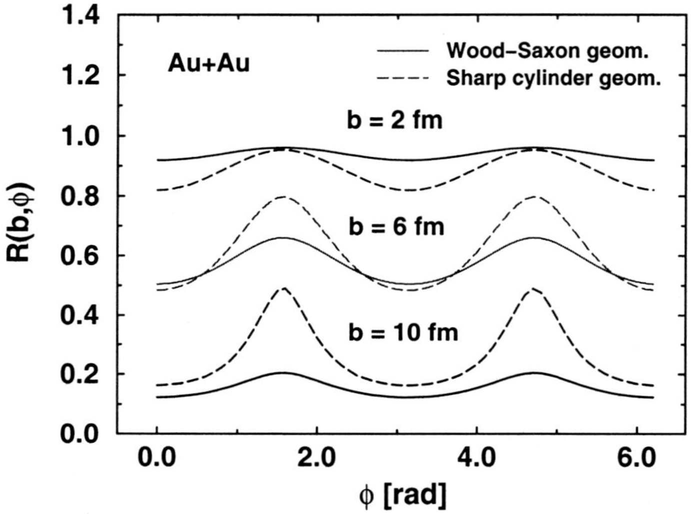

Fig. 31. The modulation function $R(\mathbf{b}, \phi)$ is plotted vs. ϕ for impact parameters $\mathbf{b} = 2, 6, 10$ fm. Diffuse Woods-Saxon and uniform sharp cylinder geometries are compared. The most drastic difference between these geometries occurs at high impact parameters

We see in Fig. 32 that the magnitude and shape of the high-p_T elliptic flow provides a complementary probe of the initial gluon density and is also sensitive to the geometrical distribution of the plasma. The saturated v_2 increases systematically with increasing initial plasma density and thus provides an important complementary constraint on the maximum initial parton density produced in central $\mathbf{b} = 0$ collisions. The consistency of the quenched elliptic flow in non-central with the central quench pattern will be very important to test when the final data become available.

5 Summary

If confirmed by further measurements and theoretical refinements, jet quenching may have already provided the first evidence that initial parton densities

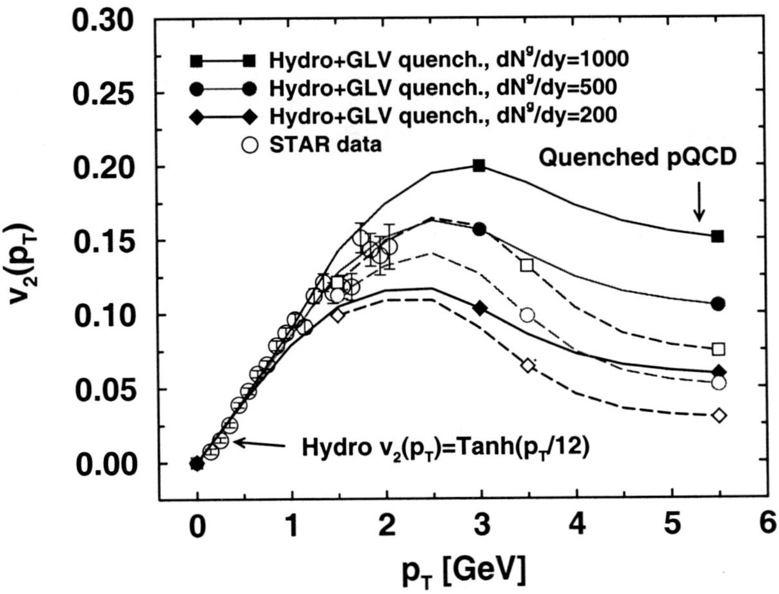

Fig. 32. The interpolation of $v_2(p_T)$ between the soft hydrodynamic [58] and hard pQCD regimes [21] is shown for different gluon rapidity densities in central $b = 0$ collisions. The gluon rapidity density at $b \neq 0$ is assumed to scale with the binary collision number. Solid (dashed) curves correspond to sharp cylindrical (diffuse Woods-Saxon) geometry

on the order of 100 times nuclear matter density may have been produced at RHIC. The full analysis of flavor composition, shape, and azimuthal moments of the high-p_T spectra appears to be a promising diagnostic probe of the evolution of the gluon plasma produced at RHIC. However, it is too early to tell what the preliminary say about the properties of that extremely dense form of matter. There are too many pieces of the puzzle that simply do not fit well into any scenario. The beam energy and centrality independence of the transverse energy per charged particle is one of them. The anomalous baryon number transport to high transverse momenta and central rapidities is another. Finally, the puzzling beam energy independence of the preliminary pion interferometry results is a mystery. As the tera-bytes of RHIC data continue to stream in during the next few years, they will certainly pose many interesting new QCD many-body problems. The new chapter on the physics of ultra-dense matter and the dynamics of ultra-relativistic nuclei is now unfolding at RHIC.

Acknowledgments

I especially thank T. Csorgo, A. Dumitru, P. Levai, L. McLerran, D. Molnar, D. Rischke, and I. Vitev for extensive collaborative work related to the topics of these lectures. This work was supported by the Director, Office of Energy Research, Office of High Energy and Nuclear Physics, Division of Nuclear Physics, and by the Office of Basic Energy Science, Division of Nuclear Science, of the U.S. Department of Energy under Contract DE-FG02-93ER40764. I also thank the Collegium Budapest for hospitality and partial support.

References

1. G. Baym: hep-ph/0104138
2. QM01: Proceedings of Quark Matter 2001;
   Slides: http://www.rhic.bnl.gov/qm2001/program.html
3. J. Harris: "Results from the STAR Experiment", QM01
4. W. Zajc: "Results from the PHENIX Experiment", QM01
5. C. Roland et al.: [PHOBOS Collaboration], hep-ex/0105043
6. F. Videbaek: "Results from the BRAHMS Experiment", QM01; J. Phys. **G27**, 671 (2001)
7. M. Gyulassy: Lectures delivered at the 40th Schladming Winter School on Dense Matter, March 3-10, 2001, Austria. Slides of the lectures are posted on the site:
   http://nt3.phys.columbia.edu/people/gyulassy/Talks/Schladming3_6_01/
8. J.C. Collins, M.J. Perry: Phys. Rev. Lett. **34**, 1353 (1975)
9. E.V. Shuryak: Phys. Lett. **B78**, 150 (1978)
10. J.I. Kapusta: Nucl. Phys. **B148**, 461 (1979)
11. D.J. Gross, R.D. Pisarski, L.G. Yaffe: Rev. Mod. Phys. **53**, 43 (1981)
12. J. Blaizot, E. Iancu: hep-ph/0101103
13. E. Braaten, A. Nieto: Phys. Rev. **D53**, 3421 (1996) [hep-ph/9510408]
14. F. Karsch: hep-ph/0103314; Nucl. Phys. Proc. Suppl. **83**, 14 (2000)
15. Z. Zheng, J. Li: Phys. Lett. **B409**, 45 (1997) [hep-th/9707052]
16. K. Kajantie, M. Laine, J. Peisa, A. Rajantie, K. Rummukainen, M. Shaposhnikov: Phys. Rev. Lett. **79**, 3130 (1997) [hep-ph/9708207]
17. G. Agakichiev et al [CERES Collab]: Phys. Lett. **B422**, 405 (1998)
18. R. Rapp, J. Wambach: Adv. Nucl. Phys. **25**, 1 (2000) [hep-ph/9909229]
19. I thank Istvan Lovas for suggesting the tomography analogy of hard observables (publication in preparation)
20. X. Wang, M. Gyulassy: Phys. Rev. Lett. **86**, 3496 (2001) [nucl-th/0008014]
21. M. Gyulassy, I. Vitev, X.N. Wang: Phys. Rev. Lett. **86**, 2537 (2001) [nucl-th/0012092]
22. A. Capella, U. Sukhatme, C. Tan, J. Tran Thanh Van: Phys. Rep. **236**, 225 (1994); A. Capella, C. Pajares, A.V. Ramallo: Nucl. Phys. **B241**, 75 (1984)
23. A. Capella, J. Tran Thanh Van, J. Kwiecinski: Phys. Rev. Lett. **58**, 2015 (1987)
24. B. Andersson, G. Gustafson, G. Ingelman, T. Sjostrand: Phys. Rep. **97**, 31 (1983)

25. B. Andersson, G. Gustafson, B. Nilsson-Almqvist: Nucl. Phys. **B281**, 289 (1987)
26. H. Bengtsson and T. Sjostrand: Comp. Phys. Comm. **46**, 43 (1987)
27. X. Wang, M. Gyulassy: Phys. Rev. **D44**, 3501 (1991); **D45**, 844 (1992); Comp. Phys. Comm. **83**, 307 (1994)
28. X. Wang, M. Gyulassy: Phys. Rev. Lett. **68**, 1480 (1992)
29. B.B. Back et al. [PHOBOS Collaboration]: Phys. Rev. Lett. **85**, 3100 (2000) [hep-ex/0007036]
30. B.B. Back et al.: PHOBOS collaboration nucl-ex/0105011.
31. K.H. Ackermann et al. [STAR Collaboration]: Phys. Rev. Lett. **86**, 402 (2001) [nucl-ex/0009011]
32. K. Adcox et al. [PHENIX Collaboration]: Phys. Rev. Lett. **86**, 3500 (2001) [nucl-ex/0012008]
33. P. Steinberg: QM01, nucl-ex/0105013
34. T.K. Gaisser, F. Halzen: Phys. Rev. Lett. **54**, 1754 (1985)
35. K. Kajantie, P. Landshoff, J. Lindfors: Phys. Rev. Lett. **59**, 2527 (1987); K. Eskola, K. Kajnatie, J. Lindfors: Nucl. Phys. **B323**, 327 (1989)
36. K.J. Eskola, K. Kajantie, P.V. Ruuskanen, K.Tuominen: Phys. Lett. **B497**, 39 (2001); Nucl. Phys. **B570**, 379 (2000)
37. L. McLerran, R. Venugopalan: Phys. Rev. **D49**, 2233 (1994); ibid. **D49**, 3352 (1994); Phys. Rev. **D50**, 2225 (1994); Phys. Rev. **D59**, 094002 (1999)
38. A. Kovner, L. McLerran, H. Weigert: Phys. Rev. **D52**, 3809 (1995); ibid. **D52**, 6231 (1995)
39. J. Jalilian-Marian, A. Kovner, L. McLerran, H. Weigert: Phys. Rev. **D55**, 5414 (1997)
40. M. Gyulassy, L. McLerran: Phys. Rev. **C56**, 2219 (1997)
41. A. Krasnitz, R. Venugopalan: Phys. Rev. Lett. **84**, 4309 (2000); Nucl. Phys. **B557**, 237 (1999)
42. Y.V. Kovchegov: hep-ph/0011252
43. L.D. Landau: Izv. Akad. Nauk SSSR **17**, 51 (1953)
44. H. Stocker, W. Greiner: Phys. Rep. **137**, 277 (1986)
45. L.P. Csernai: *Introduction to Relativistic Heavy Ion Collisions* (Wiley, New York 1994)
46. *Proc. Int. Work. on Applicability of Relativistic Hydrodynamic Models in Heavy Ion Physics*, ed. L.P. Csernai: Heavy Ion Phys. **5**, 321 (1997)
47. F. Cooper, G. Frye, E. Schonberg: Phys. Rev. **D11**, 192 (1975)
48. J.D. Bjorken: Phys. Rev. **D27**, 140 (1983)
49. F. Cooper, G. Frye: Phys. Rev. **D10**, 186 (1974)
50. L.P. Csernai: Z. Lázár, D. Molnár: Heavy Ion Phys. **5**, 467 (1997)
51. K.A. Bugaev: Nucl. Phys. **A606**, 559 (1996)
52. Cs. Anderlik, Zs.I. Lázár, V.K. Magas, L.P. Csernai, H. Stöcker, W. Greiner: Phys. Rev. **C59**, 388 (1999); ibid. **C59**, 3309 (1999)
53. F. Grassi, Y. Hama, T. Kodama: Z. Phys. **C73**, 153 (1996); Phys. Lett. **B355**, 9 (1995)
54. V.K. Magas et al.: Heavy Ion Phys. **9**, 193 (1999) [nucl-th/9903045]
55. D. Molnar, M. Gyulassy: Phys. Rev. **C62**, 054907 (2000) [nucl-th/0005051]
56. M. Gyulassy, D. Rischke, B. Zhang: Nucl. Phys. **A613**, 397 (1997)
57. D.H. Rischke, S. Bernard, J.A. Maruhn: Nucl. Phys. **A595**, 346 (1995); A. Dumitru, D.H. Rischke: Phys. Rev. **C59**, 354 (1999)

58. P.F. Kolb, P. Huovinen, U. Heinz, H. Heiselberg: Phys. Lett. **B500**, 232 (2001); Phys. Lett. **B503**, 58 (2001)
59. P.F. Kolb, U. Heinz, P. Huovinen, K.J. Eskola, K. Tuominen: hep-ph/0103234
60. D. Teaney, J. Lauret, E.V. Shuryak: nucl-th/0104041
61. S.R. de Groot: *Relativistic Kinetic Theory: Principles and Applications*, (North Holland, Amsterdam 1980)
62. A. Dumitru, M. Gyulassy: Phys. Lett. **B494**, 215 (2000) [hep-ph/0006257]
63. B. Zhang: Comp. Phys. Comm. **109**, 193 (1998) [nucl-th/9709009]
64. D. Molnár: MPC 0.1.2 Parton Cascade, http://www-cunuke.phys.columbia.edu/people/molnard
65. *Proceedings of Open Standards for Cascade Models for RHIC (OSCAR)*, BNL-64912, June 23-27, 1997, eds. M. Gyulassy, Y. Pang; Source codes and documentation for transport models under the OSCAR standard can be downloaded from the OSCAR WWW site http://www-cunuke.phys.columbia.edu/OSCAR/
66. M. Gyulassy, Y. Pang, B. Zhang: Nucl. Phys. **A626**, 999 (1997)
67. D. Molnar, M. Gyulassy: nucl-th/0104073
68. S.A. Bass, A. Dumitru: nucl-th/0001033
69. S.A. Bass et al.: Prog. Part. Nucl. Phys. **41**, 225 (1998) [nucl-th/9803035]
70. M. Bleicher, E. Zabrodin, C. Spieles, S.A. Bass, C. Ernst, S. Soff, L. Bravina, M. Belkacem, H. Weber, H. Stöcker, W. Greiner: J. Phys. **G25**, 1859 (1999) [hep-ph/9909407]
71. D. Kharzeev, M. Nardi: nucl-th/0012025.
72. A. H. Mueller: Nucl. Phys. **B558**, 285 (1999); A. H. Mueller, Jian-wei Qiu: Nucl. Phys. **B268**, 427 (1986)
73. M. Gyulassy, M. Plumer: Phys. Lett. **B243**, 432 (1990)
74. M. Gyulassy, X. Wang: Nucl. Phys. **B420**, 583 (1994) [nucl-th/9306003]; M. Pluemer, M. Gyulassy, X.N. Wang: Nucl. Phys. **A590**, 511C (1995)
75. M. Gyulassy, P. Lévai, I. Vitev: Phys. Rev. Lett. **85** 5535 (2000) 5535; Nucl. Phys. **B594**, 371 (2001)
76. M. Gyulassy, P. Levai, I. Vitev: Nucl. Phys. **B571**, 197 (2000)
77. R. Baier et al.: Nucl. Phys. **B483**, 291 (1997); ibid. **B484**, 265 (1997); ibid. **B531**, 403 (1998)
78. B.G. Zakharov: hep-ph/0012360
79. U.A. Wiedemann, Nucl. Phys. **B582**, 409 (2000); **B588**, 309 (2000) [hep-ph/0103332]
80. M.M. Aggarwal et al. [WA98 Collaboration]: Phys. Rev. Lett. **81**, 4087 (1998) [Erratum: ibid. **84**, 578 (1998)] [nucl-ex/9806004]
81. X. Wang: Phys. Rev. Lett. **81**, 2655 (1998) [hep-ph/9804384]
82. G. Agakishiev et al. [CERES Collaboration]: hep-ex/0003012
83. X. Wang: nucl-th/0105053
84. J.C. Dunlop for the STAR Collab: QM01
85. G. David [PHENIX Collaboration]: nucl-ex/0105014
86. M.C. Abreu et al. [NA50 Collaboration]: Phys. Lett. **B477**, 28 (2000)
87. C. Lourenco: hep-ex/0105053
88. J. Qiu, J.P. Vary, X. Zhang: hep-ph/9809442
89. P. Levai, G. Papp, G. Fai, M. Gyulassy, G.G. Barnafoldi, I. Vitev, Y. Zhang: nucl-th/0104035
90. J. Velkovska: nucl-ex/0105012

91. P. Lévai, G. Papp, G. Fai, M. Gyulassy: nucl-th/0012017
92. I. Vitev, M. Gyulassy: nucl-th/0104066
93. D.H. Rischke, M. Gyulassy: Nucl. Phys. **A608**, 479 (1996); Nucl. Phys. **A597**, 701 (1996)
94. E. Schnedermann, J. Sollfrank, U. Heinz: Phys. Rev. **C48**, 2462 (1993) [nucl-th/9307020]
95. H. Stocker et al.: Phys. Rev. **C25**, 1873 (1982)
96. J.-Y. Ollitrault: Phys. Rev. **D46**, 229 (1992)
97. W. Reisdorf, H.G. Ritter: Ann. Rev. Nucl. Part. Sci. **47**, 663 (1997)
98. S.A. Voloshin, A.M. Poskanzer: Phys. Lett. **B474**, 27 (2000); A.M. Poskanzer, S.A. Voloshin for the NA49 Collaboration: Nucl. Phys. **A661**, 341c (1999); Phys. Rev. **C58**, 1671 (1998)
99. N. Xu, M. Kaneta: nucl-ex/0104021
100. S.E. Vance, M. Gyulassy, X.N. Wang: Phys. Lett. **B443**, 45 (1998) [nucl-th/9806008]; Phys. Rev. Lett. **83**, 1735 (1999); S.E. Vance: "The role of the baryon junction in relativistic heavy ion collisions," Columbia Univ. Thesis UMI-99-49054 (1999)
101. D. Kharzeev: Phys. Lett. **B378**, 238 (1996) [nucl-th/9602027]
102. G.C. Rossi, G. Veneziano: Nucl. Phys. **B123**, 507; Phys. Rep. **63**, 153 (1980)
103. T. Csorgo, M. Gyulassy, D. Kharzeev: hep-ph/0102282
104. R.J. Snellings: [STAR Collaboration]: QM01 [nucl-ex/0104006]
105. L. Van Hove: Z. Phys. **C21**, 93 (1983); **27**, 135 (1985)
106. M. Gyulassy, K. Kajantie, H. Kurki-Suonio, L. Mclerran: Nucl. Phys. **B237**, 477 (1984)
107. S. Pratt: Phys. Rev. **C49**, 2722 (1994); Phys. Rev. **D33**, 1314 (1986)
108. G. Bertsch, M. Gong, M. Tohyama: Phys. Rev. **C37**, 1896 (1988)
109. S.C. Johnson [The PHENIX Collaboration]: nucl-ex/010402
110. S. Soff, S.A. Bass, A. Dumitru: Phys. Rev. Lett. **86**, 3981 (2001) [nucl-th/0012085]
111. M. Gyulassy, S.K. Kauffmann, L.W. Wilson: Phys. Rev. **C20**, 2267 (1979)
112. X.-N. Wang, Z. Huang: Phys. Rev. **C55**, 3047 (1997)

Dense Quark Matter in Compact Stars

Mark Alford

Department of Physics and Astronomy, Glasgow University
Glasgow G12 8QQ, United Kingdom

1 Introduction

QCD is firmly established as the theory underlying all of strong-interaction physics, and a pillar of the standard model. Perturbative QCD has been verified in deep inelastic scattering, and the spectrum and structural properties of the hadrons are gradually being calculated by the nonperturbative lattice formulation of QCD.

Even so, there remain tantalizing questions. As well as predicting the behaviour of small numbers of particles, QCD should also be able to tell us about the thermodynamics of matter in the realm of extraordinarily high temperatures ($\gtrsim$ 100 MeV) and densities at which it comes to dominate the physics. These regions are of more than academic interest: neutron stars are believed to consist of matter squeezed beyond nuclear density by gravitational forces, and the whole universe was hotter than 100 MeV for the first crucial microseconds of its history. However, only in the last few years have these regions begun to be probed experimentally in heavy-ion collisions and astrophysical observations of neutron stars, and our theoretical understanding of them remains elementary.

High densities have proven particularly difficult to study, in part because lattice gauge theory has been blocked by the complexity of the fermion determinant. We are still trying to establish the symmetries of the ground state, and find effective theories for its lowest excitations. These questions are of direct physical relevance: an understanding of the symmetry properties of dense matter can be expected to inform our understanding of neutron star astrophysics and perhaps also heavy-ion collisions which achieve high baryon densities without reaching very high temperatures.

In this article I will review the progress that has been made in the last few years in understanding the possible phases of QCD at low temperatures and high densities, and go on to discuss the possible observable signatures in compact-stars phenomenology. Other reviews, with different explanations and emphasis, will also prove useful to the reader [1].

1.1 The Fermi Surface and Cooper Instability

One of the most striking features of QCD is asymptotic freedom: the force between quarks becomes arbitrarily weak as the characteristic momentum scale of their interaction grows larger. This immediately suggests that at sufficiently high densities and low temperatures matter will consist of a Fermi sea of essentially free quarks, whose behavior is dominated by the freest of them all: the high-momentum quarks that live at the Fermi surface.

Actually, things are quite different. It was shown by Bardeen, Cooper, and Schrieffer (BCS) [2] that in the presence of attractive interactions a Fermi surface is unstable. If there is *any* channel in which the quark-quark interaction is attractive, then there is a state of lower free energy that consists of a complicated coherent superposition of particle and hole pairs – "Cooper pairs".

This can be seen intuitively as follows. Consider a system of free particles. The Helmholtz free energy is $F = E - \mu N$, where E is the total energy of the system, μ is the chemical potential, and N is the number of particles. The Fermi surface is defined by a Fermi energy $E_F = \mu$, at which the free energy is minimized, so adding or subtracting a single particle costs zero free energy. Now, suppose a weak attractive interaction is switched on. BCS showed that this favors a complete rearrangement of the states near the Fermi surface, because it costs no free energy to make a pair of particles (or holes), and the attractive interaction makes it favorable to do so. Many such pairs will therefore be created, in all the modes near the Fermi surface, and these pairs, being bosonic, will form a condensate. The ground state will be a superposition of states with all numbers of pairs, breaking the fermion number symmetry. An arbitrarily weak interaction has lead to spontaneous symmetry breaking.

In condensed-matter systems, where the relevant fermions are electrons, the necessary attractive interaction has been hard to find. The dominant interaction between electrons is the repulsive electrostatic force, but in the right kind of crystal there are attractive phonon-mediated interactions that can overcome it. In these materials the BCS mechanism leads to superconductivity, since it causes Cooper pairing of electrons, which breaks the electromagnetic gauge symmetry, giving mass to the photon and producing the Meissner effect (exclusion of magnetic fields from a superconducting region). The Cooper-paired state is rare and delicate, easily disrupted by thermal fluctuations, so superconductivity only survives at low temperatures.

In QCD, by contrast, the dominant gauge-boson-mediated interaction between quarks is itself attractive [3–7]. The relevant degrees of freedom are those which involve quarks with momenta near the Fermi surface. These interact via gluons, in a manner described by QCD. The quark-quark interaction has two color channels available, the antisymmetric $\bar{\mathbf{3}}$, and the symmetric $\mathbf{6}$. It is attractive in the $\bar{\mathbf{3}}_A$: this can be seen in single-gluon exchange or by counting of strings.

Since pairs of quarks cannot be color singlets, the resulting condensate will break the local color symmetry $SU(3)_{\text{color}}$. We call this "color superconductivity". Note that the quark pairs play the same role here as the Higgs particle does in the standard model: the color-superconducting phase can be thought of as the Higgsed (as opposed to confined) phase of QCD.

It is important to remember that the breaking of a gauge symmetry cannot be characterized by a gauge-invariant local order parameter which vanishes on one side of a phase boundary. The superconducting phase can be characterized rigorously only by its global symmetries. In electromagnetism there is a non-local order parameter, the mass of the magnetic photons, that corresponds physically to the Meissner effect and distinguishes the free phase from the superconducting one. In QCD there is no free phase: even without pairing the gluons are not states in the spectrum. No order parameter distinguishes the Higgsed phase from a confined phase or a plasma, so we have to look at the global symmetries.

In most of this article we will take an approach similar to that used in analyzing the symmetry breaking of the standard model, and discuss the phases of dense QCD in terms of a gauge-variant observable, the diquark condensate, which is analogous to the vacuum expectation value (VEV) of the Higgs field. However, this is only a convenience, and we will be careful to label different phases by their unbroken global symmetries, so that they can always be distinguished by gauge-invariant order parameters.

1.2 The Gap Equation

To decide whether or not fermions condense in the ground state, one can explicitly construct a wave functional with the appropriate pairing, and use a many-body variational approach. But the field-theoretical approach, though less concrete, is more general, and I will briefly describe it here.

The important quantity is the quark self-energy, i.e. the one-particle irreducible (1PI) Green function of two quark fields. Its poles will give the gauge-invariant masses of the quasiquarks, the lowest energy fermionic excitations around the quark Fermi surface. To see if condensation (chiral condensation, flavor-singlet quark pairing, or whatever) occurs in some channel, one writes down a self-consistency equation, the gap equation, for a self-energy with that structure, and solves it to find the actual self-energy (the gap). If it is zero, there is no condensation in that channel. If not, there can be condensation, but it may just be a local minimum of the free energy. There may be other solutions to the gap equation, and the one with the lowest free energy is the true ground state.

There are several possible choices for the interaction to be used in the gap equation. At asymptotically high densities QCD is weakly coupled, so one-gluon exchange is appropriate. Such calculations [8–17] are extremely important, since they demonstrate from first principles that color superconductivity occurs in QCD. However, the density regime of physical interest for neutron stars or heavy-ion collisions is up to a few times nuclear density ($\mu \lesssim 500$ MeV) and weak-coupling calculations are unlikely to be trustworthy

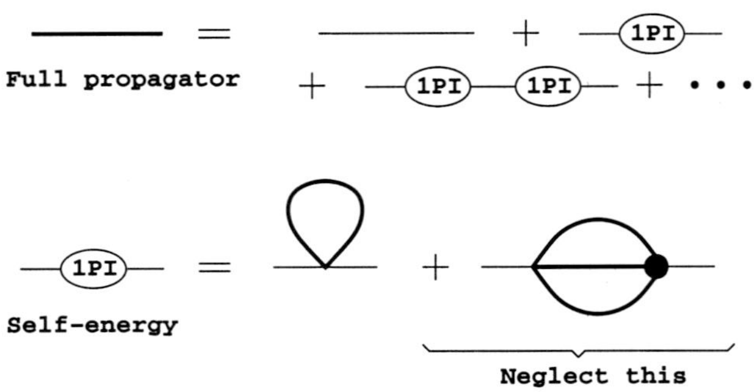

Fig. 1. Mean-field Schwinger-Dyson (gap) equations

in that regime. In fact, current weak-coupling calculations cannot be extrapolated below about 10^8 MeV because of gauge dependence arising from the neglect of vertex corrections [18]. There have also been some preliminary investigations of confinement-related physics such as a gluon condensate [87,88].

The alternative is to use some phenomenological interaction that can be argued to capture the essential physics of QCD in the regime of interest. The interaction can be normalized, to reproduce known low-density physics such as the chiral condensate, and then extrapolated to the desired chemical potential. In two-flavor theories, the instanton vertex is a natural choice [6,7,19,20], since it is a four-fermion interaction. With more flavors, the one-gluon exchange vertex without a gluon propagator [5,21,22] is more convenient. It has been found that these both give the same results, to within a factor of about 2. This is well within the inherent uncertainties of such phenomenological approaches. In the rest of this article we will therefore not always be specific about the exact interaction used to obtain a given result. One caveat to bear in mind is that the single-gluon exchange interaction is symmetric under $U(1)_A$, and so it sees no distinction between condensates of the form $\langle qCq\rangle$ and $\langle qC\gamma_5 q\rangle$ (C is the Dirac charge-conjugation matrix). However, once instantons are included the Lorentz scalar $\langle qC\gamma_5 q\rangle$ is favored [6,7], so in single-gluon exchange calculations the parity-violating condensate is usually ignored.

The mean-field approximation to the Schwinger-Dyson equations is shown diagramatically in Fig. 1, relating the full propagator to the self-energy. In the mean-field approximation, only daisy-type diagrams are included in the resummation, vertex corrections are excluded. Algebraically, the equation takes the form

$$\Sigma(k) = -\frac{1}{(2\pi)^4}\int d^4q\, M^{-1}(q)D(k-q), \qquad (1)$$

where $\Sigma(k)$ is the self-energy, M is the full fermion matrix (inverse full propagator), and $D(k-q)$ is the vertex, which in NJL models will be momentum-independent, but in a weak-coupling QCD calculation will include the gluon propagator and couplings. Since we want to study quark-quark condensation, we have to write propagators in a form that allows for this possibility, just as to study chiral symmetry breaking it is necessary to use 4-component Dirac spinors rather than 2-component Weyl spinors, even if there is no mass term in the action. We therefore use Nambu-Gorkov 8-component spinors, $\Psi = (\psi, \bar{\psi}^T)$, so the self-energy Σ can include a quark-quark pairing term Δ. The fermion matrix M then takes the form

$$M(q) = M_{\text{free}} + \Sigma = \begin{pmatrix} \not{q} + \mu\gamma_0 & \gamma_0 \Delta \gamma_0 \\ \Delta & (\not{q} - \mu\gamma_0)^T \end{pmatrix} . \tag{2}$$

Equations (1) and (2) can be combined to give a self-consistency condition for Δ, the gap equation. If the interaction is a point-like four-fermion NJL interaction then the gap parameter Δ will be a color-flavor-spin matrix, independent of momentum. If the gluon propagator is included, Δ will be momentum-dependent, complicating the analysis considerably.

In NJL models, the simplicity of the model has allowed renormalization-group analyses [23,24] that include a large class of four-fermion interactions, and follow their running couplings as modes are integrated out. This confirms that in QCD with two and three massless quarks the most attractive channels for condensation are those corresponding to the two-flavor superconducting (2SC) and color-flavor locked (CFL) phases studied below. Calculations using random matrices, which represent very generic systems, also show that diquark condensation is favored at high density [25].

Following through the analysis outlined above, one typically finds gap equations of the form

$$1 = K \int_0^{\Lambda} k^2 dk \, \frac{1}{\sqrt{(k-\mu)^2 + \Delta^2}}, \tag{3}$$

where K is the NJL four-fermion coupling. In the limit of small gap, the integral can be evaluated, giving

$$\Delta \sim \Lambda \exp\Big(\frac{\text{const}}{K\mu^2}\Big). \tag{4}$$

This shows the non-analytic dependence of the gap on the coupling K. Condensation is a nonperturbative effect that cannot be seen to any order in perturbation theory. The reason it can be seen in the diagrammatic Schwinger-Dyson approach is that there is an additional ingredient: an ansatz for the form of the self-energy. This corresponds to guessing the form of the ground-state wave function in a many-body variational approach. All solutions to gap equations therefore represent possible stable ground states, but to find the favored ground state their free energies must be compared, and even then one can never be sure that the true ground state has been found, since there is always the possibility that there is another vacuum that solves its own gap equation and has an even lower free energy.

In weak-coupling QCD calculations, where the full single-gluon exchange vertex complete with gluon propagator is used, the gap equation takes the form [3,4,8,12]

$$\Delta \sim \mu \frac{1}{g^5} \exp\Bigl(-\frac{3\pi^2}{\sqrt{2}}\frac{1}{g}\Bigr) , \tag{5}$$

or, making the weak-coupling expansion in the QCD gauge coupling g more explicit,

$$\ln\Bigl(\frac{\Delta}{\mu}\Bigr) = -\frac{3\pi^2}{\sqrt{2}}\frac{1}{g} - 5\ln g + \text{const} + \mathcal{O}(g). \tag{6}$$

This gap equation has two interesting features. Firstly, it does not correspond to what you would naively expect from the NJL model of single-gluon exchange, in which the gluon propagator is discarded and $K \propto g^2$, yielding $\Delta \sim \exp(-1/g^2)$. The reason [4,8] is that at high density the gluon propagator has an infrared divergence at very small angle scattering, since magnetic gluons are only Landau damped, not screened. This divergence is regulated by the gap itself, weakening its dependence on the coupling.

Secondly, in (5) we have left unspecified the energy scale at which the coupling g is to be evaluated. Natural guesses would be μ or Δ. If we use $g(\mu)$ and assume it runs according to the one-loop formula $1/g^2 \sim \ln\mu$ then the exponential factor in (5) gives very weak suppression, and is in fact overwhelmed by the initial factor μ, so that the gap rises without limit at asymptotically high density, although Δ/μ shrinks to zero so that weak-coupling methods are still self-consistent. This means that color superconductivity will inevitably dominate the physics at high enough densities.

2 Two Massless Quark Flavors

In the real world there are two light quark flavors, the up (u) and down (d), with masses $\lesssim$ 10 MeV, and a medium-weight flavor, the strange (s) quark, with mass $\sim$ 100 MeV. A first approximation is to ignore the strange, and set $m_{u,d} = 0$.

The gap equation for this scenario has been solved using various interactions: pointlike four-fermion interactions based on the instanton vertex [3,6,7,19], a full instanton vertex including all the form factors [20], and a weakly coupled gluon propagator [9,12,10,13–15]. All agree that the quarks will pair in the color $\bar{\mathbf{3}}$ flavor singlet channel, a pattern that we call the two-flavor superconducting (2SC) phase,

$$\text{2SC phase:}\quad \Delta^{\alpha\beta}_{ij} = \langle q^\alpha_i q^\beta_j \rangle_{1PI} \propto C\gamma_5 \varepsilon_{ij}\varepsilon^{\alpha\beta 3} \tag{7}$$

(color indices α,β run from 1 to 3, flavor indices i,j run from 1 to 2, Dirac indices are suppressed, and C is the Dirac charge-conjugation matrix). The four-fermion interaction calculations also agree on the magnitude of Δ:

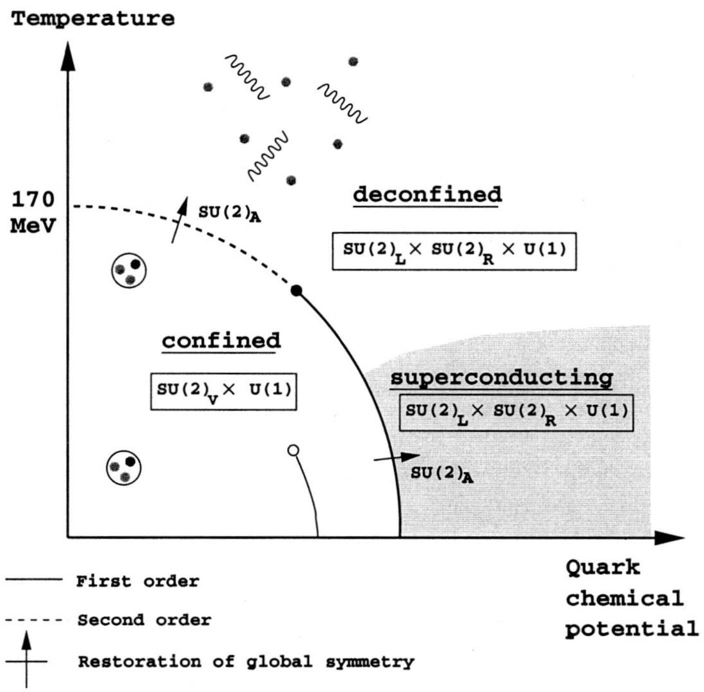

Fig. 2. Two massless flavor phase diagram

around 100 MeV. This is found to be roughly independent of the cutoff, although the chemical potential at which it is attained is not. Such calculations are based on calibrating the coupling to give a chiral condensate of around 400 MeV at zero density, and turning μ up to look for the maximum gap.

As with any spontaneous symmetry breaking, one of the degenerate ground states is arbitrarily selected. In this case, quarks of the first two colors (red and green) participate in pairing, while the third color (blue) does not. The ground state is invariant under an $SU(2)$ subgroup of the color rotations that mixes red and green, but the blue quarks are singled out as different. The pattern of symmetry breaking is therefore (with gauge symmetries in square brackets)

$$\begin{aligned} & [SU(3)_{\text{color}}] \times [U(1)_Q] \times SU(2)_L \times SU(2)_R \\ \longrightarrow\ & [SU(2)_{\text{color}}] \times [U(1)_{\tilde{Q}}] \times SU(2)_L \times SU(2)_R \end{aligned} \tag{8}$$

The expected phase diagram in the μ-T plane is shown in Fig. 2. The features of this pattern of condensation are

- The color gauge group is broken down to $SU(2)$, so five of the gluons will become massive, with masses of order the gap (since the coupling is of order 1). The remaining three gluons are associated with an unbroken

$SU(2)$ red-green gauge symmetry, whose confinement distance scale rises exponentially with density [26].

- The red and green quark modes acquire a gap Δ, which is the mass of the physical excitations around the Fermi surface (quasiquarks). There is no gap for the blue quarks in this ansatz, and it is an interesting question whether they find some other channel in which to pair. The available attractive channels appear to be weak so the gap will be much smaller, perhaps in the keV range [6,27]. It has even been suggested that 'tHooft anomaly matching may prevent any condensation [28,29].
- Electromagnetism is broken, but this does not mean that the 2SC phase is an electromagnetic as well as a color superconductor. Just as in the standard model the Higgs VEV leaves unbroken a linear combination Q of the weak W_3 and hypercharge Y bosons, so here a linear combination $\tilde{Q}$ of the eighth gluon T_8 and the electric charge Q is left unbroken. This plays the role of a "rotated" electromagnetism. We will discuss some of its physical effects in a later section.
- No global symmetries are broken (although additional condensates that break chirality have been suggested [30]) so the 2SC phase has the same symmetries as the quark-gluon plasma (QGP), so there need not be any phase transition between them. Again, this is in close analogy to the physics of the standard model, where the Higgs VEV breaks no global symmetries: the phase transition line between the unbroken and broken phases ends at some critical Higgs mass, and the two regimes are analytically connected. The reader may wonder why one cannot construct an order parameter to distinguish the 2SC phase using the fact that the quark pair condensate blatantly breaks baryon number, which is a global symmetry. However, in the two-flavor case baryon number is a linear combination of electric charge and isospin, $B = 2Q - 2I_3$, so baryon number is already included in the symmetry groups of (8). Just as an admixture of gluon and photon survives unbroken as a rotated electromagnetism, so an admixture of B and T_8 survives unbroken as a rotated baryon number.

3 Three Massless Quark Flavors

In QCD with three flavors of massless quarks the Cooper pairs *cannot* be flavor singlets, and both color and flavor symmetries are necessarily broken [22] (see also [31] for zero density). The gap equation has been solved for pointlike 4-fermion interactions with the index structure of single-gluon exchange [22,32,33] as well as a weakly coupled gluon propagator [16,17]. They agree that the attractive channel exhibits a pattern called color-flavor locking (CFL),

$$\text{CFL phase:} \quad \begin{aligned} \Delta^{\alpha\beta}_{ij} = \langle q^\alpha_i q^\beta_j \rangle_{1PI} &\propto C\gamma_5 [\varepsilon^{\alpha\beta X} \varepsilon_{ijX} + \kappa(\delta^\alpha_i \delta^\beta_j + \delta^\alpha_j \delta^\beta_i)] \\ &\propto C\gamma_5 [(\kappa+1)\delta^\alpha_i \delta^\beta_j + (\kappa-1)\delta^\alpha_j \delta^\beta_i] \end{aligned} \qquad (9)$$

(color indices α, β and flavor indices i, j all run from 1 to 3). The first line shows the connection between this and the 2SC pairing pattern. When $\kappa = 0$, the pairing is in the $(\bar{\mathbf{3}}_A, \bar{\mathbf{3}}_A)$ channel of color and flavor, which corresponds to three orthogonal copies of the 2SC pairing: the red and green u and d pair as in 2SC, in addition the red and blue u and s pair, and finally the green and blue d and s pair. The term multiplied by κ corresponds to pairing in the $(\mathbf{6}_S, \mathbf{6}_S)$. It turns out that this additional condensate, although not highly favored energetically (the color $\mathbf{6}_S$ is not attractive in single-gluon exchange, instanton vertex, or strong coupling) breaks no additional symmetries and so κ is in general small but not zero [22,34]. A weak-coupling calculation [16] shows that κ is suppressed by one power of the coupling, $\kappa = g\sqrt{2}\log(2)/(36\pi)$.

The second line of (9) exhibits the color-flavor locking property of this ground state. The Kronecker deltas dot color indices with flavor indices, so that the VEV is not invariant under color rotations, nor under flavor rotations, but only under simultaneous, equal and opposite, color and flavor rotations. Since color is only a vector symmetry, this VEV is only invariant under vector flavor rotations, and breaks chiral symmetry.

The pattern of symmetry breaking is therefore

$$[SU(3)_{\text{color}}] \times \underbrace{SU(3)_L \times SU(3)_R}_{\supset [U(1)_Q]} \times U(1)_B \longrightarrow \underbrace{SU(3)_{C+L+R}}_{\supset [U(1)_{\tilde{Q}}]} \times \mathbb{Z}_2 . \qquad (10)$$

The expected phase diagram in the μ-T plane is shown in Fig. 3. The features of this pattern of condensation are:

- The color gauge group is completely broken. All eight gluons become massive. This ensures that there are no infrared divergences associated with gluon propagators.
- All the quark modes are gapped. The nine quasiquarks (three colors times three flavors) fall into an $\mathbf{8} \oplus \mathbf{1}$ of the unbroken global $SU(3)$, so there are two gap parameters. The singlet has a larger gap than the octet.
- Electromagnetism is no longer a separate symmetry, but corresponds to gauging one of the flavor generators. A rotated electromagnetism ("$\tilde{Q}$") survives unbroken. Just as in the 2SC case it is a combination of the original photon and one of the gluons, although the relative coefficients are different.
- Two global symmetries are broken, the chiral symmetry and baryon number, so there are two gauge-invariant order parameters that distinguish the CFL phase from the QGP, and corresponding Goldstone bosons which are long-wavelength disturbances of the order parameter. The order parameter for the chiral symmetry is $\langle \bar{\psi}_L \gamma_\mu \lambda^A \psi_L \bar{\psi}_R \gamma_\mu \lambda^A \psi_R \rangle$ where λ^A are the flavor generators [16] (which only gets a vacuum expectation value beyond the mean-field approximation). The chiral Goldstone bosons form a pseudoscalar octet, like the zero-density $SU(3)_{\text{flavor}}$ pion

octet. The breaking of the baryon number symmetry has order parameter $\langle udsuds \rangle = \langle \Lambda\Lambda \rangle$, and a singlet scalar Goldstone boson which makes the CFL phase a superfluid.
If a quark mass were introduced, then it would explicitly break the chiral symmetry and give a mass to the chiral Goldstone octet, but the CFL phase would still be a superfluid, distinguished by its baryon number breaking.

- Quark-hadron continuity. It is striking that the symmetries of the 3-flavor CFL phase are the same as those one might expect for 3-flavor hypernuclear matter [32]. In hypernuclear matter one would expect the hyperons to pair in an $SU(3)_{\text{flavor}}$ singlet ($\langle \Lambda\Lambda \rangle, \langle \Sigma\Sigma \rangle, \langle N\Xi \rangle$), breaking baryon number but leaving flavor and electromagnetism unbroken. Chiral symmetry would be broken by the chiral condensate. This means that one might be able to follow the spectrum continuously from hypernuclear matter to the CFL phase of quark matter – there need be no phase transition. The pions would evolve into the pseudoscalar octet mentioned above. The vector mesons would evolve into the massive gauge bosons. This will be discussed in more detail below for the 2+1 flavor case.

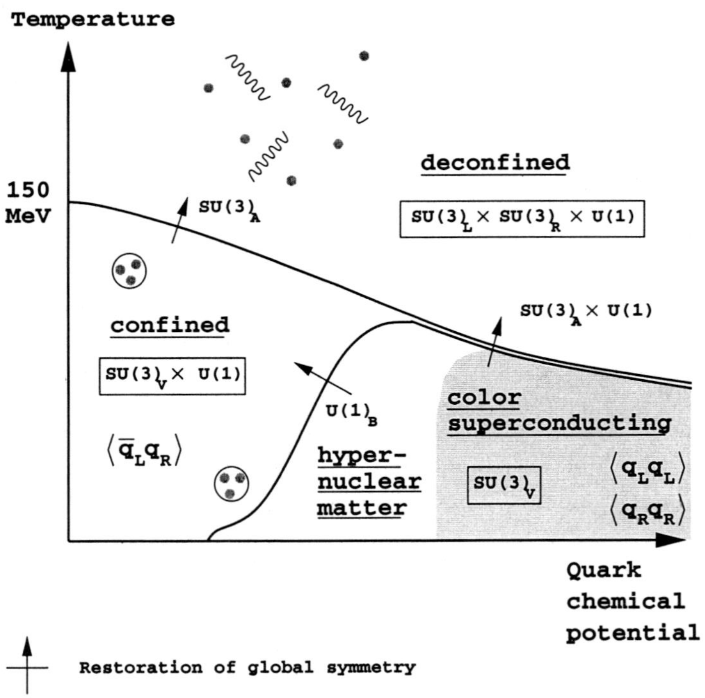

Fig. 3. Three massless flavor phase diagram

Table 1. Symmetries of phases of QCD

phase	electromagnetism	chiral symmetry	baryon number
QGP	Q	unbroken	B
2 flavor nuclear matter	broken	broken	broken
2 flavor quark pairing (2SC)	$\tilde{Q} = Q - \frac{1}{2\sqrt{3}} T_8$	unbroken	$\tilde{B} = \tilde{Q} + I_3$
3 flavor nuclear matter	Q	broken	broken
3 flavor quark pairing (CFL)	$\tilde{Q} = Q + \frac{1}{\sqrt{3}} T_8$	broken	broken

We can now draw a hypothetical phase diagram for 3-flavor QCD (Fig. 3). Comparing with the 2-flavor case, we see that the 2SC quark-paired phase is easy to distinguish from nuclear matter, since it has restored chiral symmetry, but hard to distinguish from the QGP. The CFL phase is easy to distinguish from the QGP, but hard to distinguish from hypernuclear matter.

We conclude that dense quark matter has rather different global symmetries for $m_s = 0$ than for $m_s = \infty$. Since the real world has a strange quark of middling mass, it is very interesting to see what happens as one interpolates between these extremes.

4 Two Massless + One Massive Quark Flavors

A nonzero strange quark mass explicitly breaks the flavor $SU(3)_L \times SU(3)_R$ symmetry down to $SU(2)_L \times SU(2)_R$. If the strange quark is heavy enough then it will decouple, and 2SC pairing will occur. For a sufficiently small strange quark mass we expect a reduced form of color-flavor locking in which an $SU(2)$ subgroup of $SU(3)_{\text{color}}$ locks to isospin, causing chiral symmetry breaking and leaving a global $SU(2)_{\text{color}+V}$ group unbroken.

As m_s is increased from zero to infinity, there has to be some critical value at which the strange quark decouples, color and flavor rotations are unlocked, and the full $SU(2)_L \times SU(2)_R$ symmetry is restored. It can be argued on general grounds (see below) that a simple unlocking phase transition must be first order, although there are strong indications that there is a crystalline intermediate phase (see Sect. 5).

An analysis of the unlocking transition, using a NJL model with interaction based on single-gluon exchange [35,32] confirms this expectation. Although the quantitative results from NJL models can only be regarded as rough approximations, it is interesting that the calculations indicate that for realistic values of the strange quark mass chiral symmetry breaking may be present for densities all the way down to those characteristic of baryonic matter. This raises the possibility that quark matter and baryonic matter may be continuously connected in nature, as Schäfer and Wilczek have conjectured for QCD with three massless quarks [32]. The gaps due to pairing at the quark Fermi surfaces map onto gaps due to pairing at the baryon Fermi surfaces in superfluid baryonic matter consisting of nucleons, Λ's, Σ's, and Ξ's (see below).

Based on the NJL calculations, the zero-temperature phase diagram as a function of chemical potential and strange quark mass has been conjectured [35] to be as shown in Fig. 4. Electromagnetism was ignored in this calculation, and it was assumed that wherever a baryon Fermi surface is present, baryons always pair at zero temperature. To simplify our analysis, we assume that baryons always pair in channels which preserve rotational invariance, breaking internal symmetries such as isospin if necessary. So the real phase diagram may well be even more complicated.

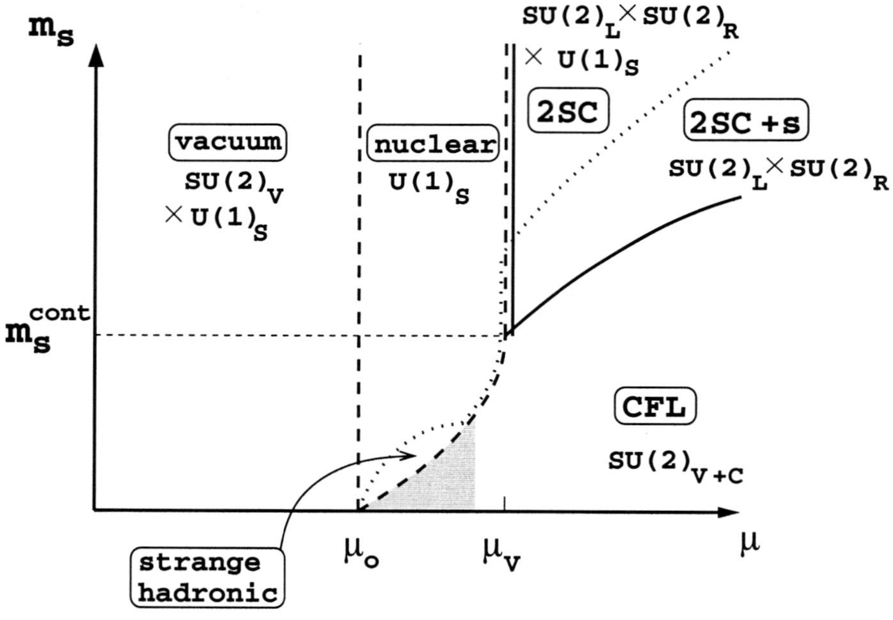

Fig. 4. Conjectured phase diagram for 2+1 flavor QCD at $T = 0$. The global symmetries of each phase are labelled. The solid line marks chiral symmetry breaking, the dashed line isospin breaking, and the dotted line strangeness breaking. The regions of the phase diagram labelled 2SC, 2SC+s and CFL denote color superconducting quark matter phases. The shading marks the region of quark-hadron continuity. A detailed explanation is given in the text

We characterize the phases using the $SU(2)_L \times SU(2)_R$ flavor rotations of the light quarks, and the $U(1)_S$ rotations of the strange quarks. The $U(1)_B$ symmetry associated with baryon number is a combination of $U(1)_S$, a $U(1)$

subgroup of isospin, and the gauged $U(1)_{\rm EM}$ of electromagnetism. Therefore, in our analysis of the global symmetries, once we have analyzed isospin and strangeness, considering baryon number adds nothing new.

4.1 Description of the Phase Diagram

To explain Fig. 4, we follow the phases that occur from low to high density, first for large m_s, then small m_s.

Heavy Strange Quark. For $\mu = 0$ the density is zero; isospin and strangeness are unbroken; Lorentz symmetry is unbroken; chiral symmetry is broken. Above a first-order transition [36] at an onset chemical potential $\mu_{\rm o} \sim$ 300 MeV, one finds nuclear matter. Lorentz symmetry is broken, leaving only rotational symmetry manifest. Chiral symmetry is broken, though perhaps with a reduced chiral condensate. We expect an instability of the nucleon Fermi surfaces to lead to Cooper pairing, and assume that, as is observed in nuclei, the pairing is *pp* and *nn*, breaking isospin (and perhaps also rotational invariance). Since there are no strange baryons present, $U(1)_S$ is unbroken. When μ is increased above $\mu_{\rm V}$, we find the "2SC" phase of color-superconducting matter consisting of up and down quarks only, paired in Lorentz singlet isosinglet channels. The full flavor symmetry $SU(2)_L \times SU(2)_R$ is restored. The phase transition at μ_V is first order according to NJL models with low cutoff [19,37,38,20] and random matrix models [39] as the chiral condensate competes with the superconducting condensate.

When μ exceeds the constituent strange quark mass $M_s(\mu)$, a strange quark Fermi surface forms, with a Fermi momentum far below that for the light quarks. The strange quarks pair with each other, in a color-spin locked phase [40] that we call "2SC+s". Strangeness is now broken, but the *ss* condensate is expected to be small [40].

Finally, when the chemical potential is high enough that the Fermi momenta for the strange and light quarks become comparable, we cross into the color-flavor locked (CFL) phase. There is an unbroken global symmetry constructed by locking the $SU(2)_V$ isospin rotations and an $SU(2)$ subgroup of color. Chiral symmetry is once again broken.

Light Strange Quark. Below $\mu_{\rm o}$, we have the vacuum, as before. At $\mu_{\rm o}$, one enters the nuclear matter phase, with the familiar *nn* and *pp* pairing at the neutron and proton Fermi surfaces breaking isospin.

At a somewhat larger chemical potential, strangeness is broken, first perhaps by kaon condensation [41–43] or by the appearance and Cooper pairing of strange baryons, Λ and Σ, and then Ξ, which pair with themselves in spin singlets. This phase is labelled "strange hadronic" in Fig. 4. The global symmetries $SU(2)_L \times SU(2)_R$ and $U(1)_S$ are all broken.

We can imagine two possibilities for what happens next as μ increases further, and we enter the shaded region of the figure. (1) Deconfinement: the baryonic Fermi surface is replaced by u, d, s quark Fermi surfaces, which are unstable against pairing, and we enter the CFL phase, described above. Isospin is locked to color and $SU(2)_{\text{color}+V}$ is restored, but chiral symmetry remains broken. (2) No deconfinement: the Fermi momenta of all of the octet baryons are now similar enough that pairing between baryons with differing strangeness becomes possible. At this point, isospin is restored: the baryons pair in rotationally invariant isosinglets ($p\Xi^-$, $n\Xi^0$, $\Sigma^+\Sigma^-$, $\Sigma^0\Sigma^0$, $\Lambda\Lambda$). The interesting point is that scenario (1) and scenario (2) are indistinguishable. Both look like the "CFL" phase of the figure: $U(1)_S$ and chirality are broken, and there is an unbroken vector $SU(2)$. This is the "continuity of quark and hadron matter" described by Schäfer and Wilczek [32]. We conclude that for low enough strange quark mass, $m_s < m_s^{\text{cont}}$, there may be a region where sufficiently dense baryonic matter has the same symmetries as quark matter, and there need not be any phase transition between them.

Color-flavor locking will always occur for sufficiently large chemical potential, for any nonzero, finite m_s. This follows from Son's model-independent analysis valid at very high densities [8]. As a consequence of color-flavor locking, chiral symmetry is spontaneously broken even at asymptotically high densities, in sharp contrast to the well-established restoration of chiral symmetry at high temperature.

Non-Zero Temperature. Finally, it is interesting to ask what we expect at non-zero temperature. There has been no comprehensive NJL study of this, but one can make the reasonable guess that quark pairing with a gap Δ at $T = 0$ will disappear in a phase transition at $T_c \approx 0.6\Delta$. This is the BCS result, which is also found to hold for quark pairing [12,11].

Assuming the zero-temperature phase structure given in Fig. 4, we can guess that the non-zero temperature μ-T phase diagram for strange quark masses varying from infinity to zero will be as shown in the diagrams of Fig. 5. These are assembled into a single three-dimensional diagram in Fig. 6, where for clarity only the chiral phase transition surface is shown: the thick line is tricritical, and the shaded region that it bounds is second-order.

The main features of the phase diagram are as follows.

- The second-order chiral phase transition (dashed line) that is present at low density and high temperature shrinks as the strange quark becomes lighter, until at $m_s = m_s^*$ the tricritical point arrives at $T = 0$. At lower masses, there is no second-order line.
- The strangeness-breaking line (dotted) and the high-density chiral symmetry breaking line (solid) do not exactly coincide because at low enough temperature there is a window of densities where strange quarks are present, but their Fermi momentum is too low to allow them to pair with the light quarks. This is the 2SC+s phase, where the strange quarks

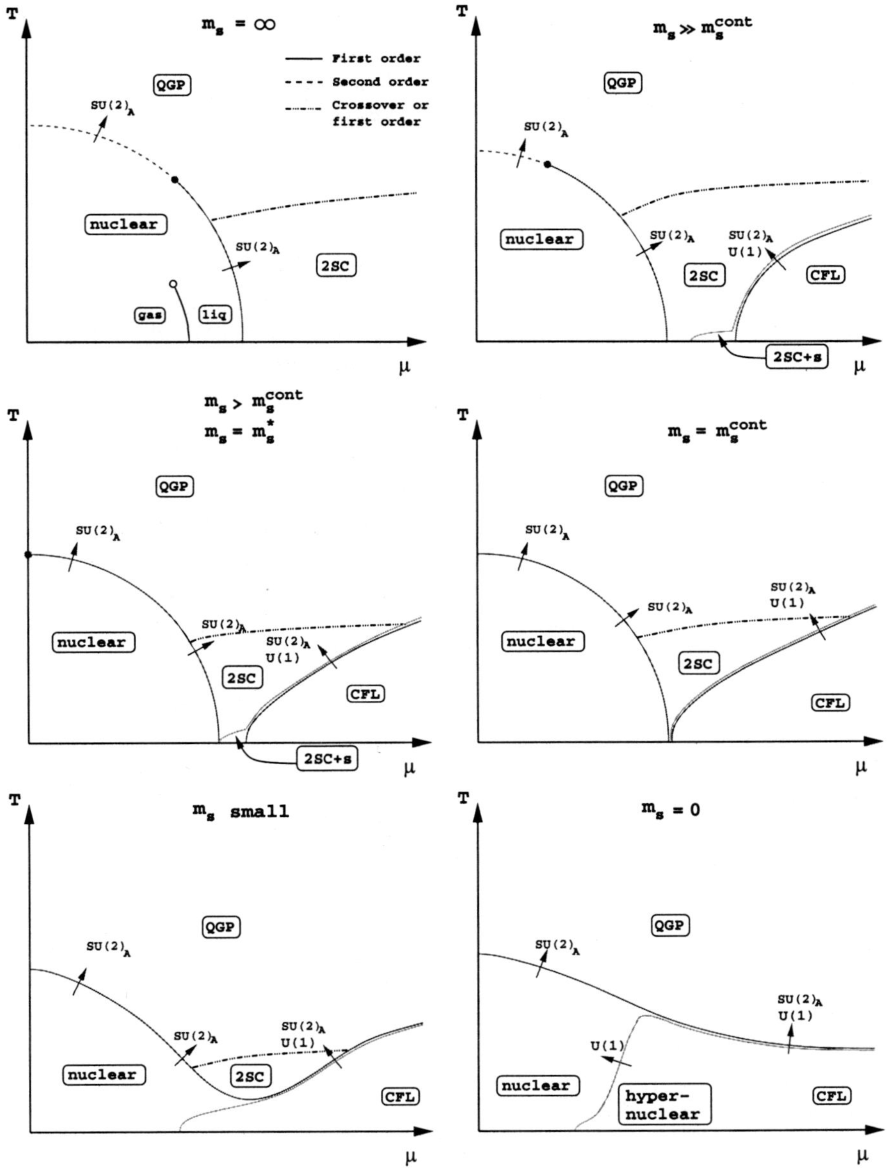

Fig. 5. The phases of QCD in the chemical potential and temperature plane, for various strange quark masses, interpolating from the two-flavor to the three-flavor theory. The up and down quark masses and electromagnetism are neglected. Lines of breaking of $SU(2)$ chiral symmetry and $U(1)$ strangeness are shown. The QGP and 2SC have the same global symmetries, but are probably separated by first-order transitions that end at critical points

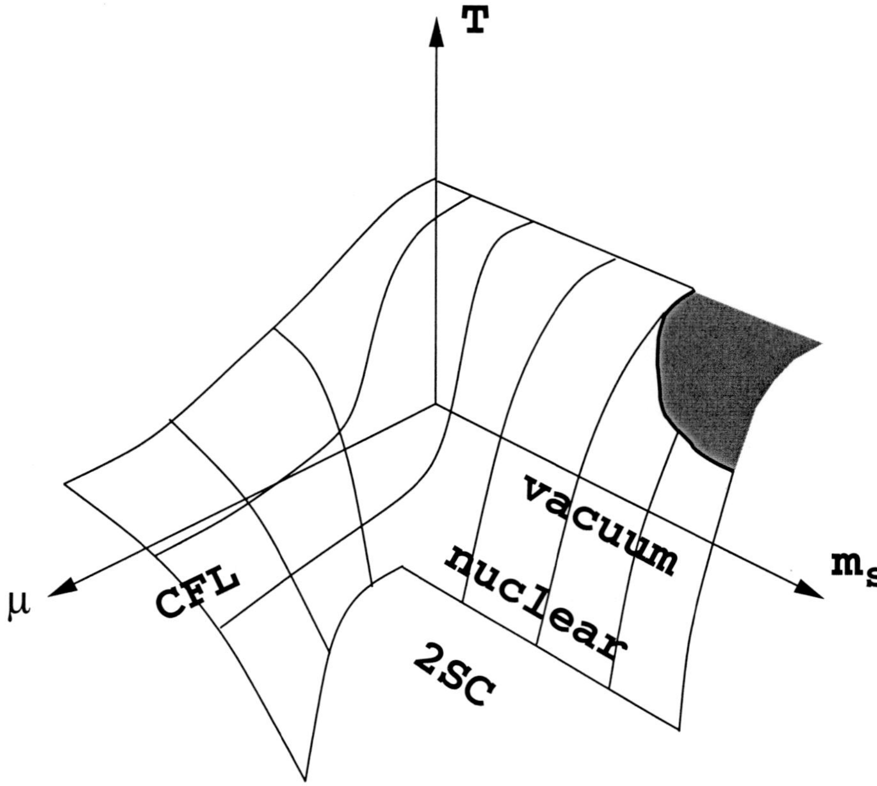

Fig. 6. The chiral phase transition surface as a function of chemical potential μ, strange quark mass m_s, and temperature T. The diagrams of Fig. 5 are a sequence of μ-T sections through this space. The shaded region at high m_s and T is the second-order part of the critical surface, which is bounded by a tricritical (thick) line. Everywhere else the phase transition is conjectured to be first order

pair with themselves, breaking strangeness/baryon number, in a color-spin locked phase whose gap and critical temperature are very small [40].

- At arbitrarily high densities, where the QCD gauge coupling is small, quark matter is always in the CFL phase with broken chiral symmetry. This is true independent of whether the "transition" to quark matter is continuous, or whether, as for larger m_s, there are two first-order transitions, from nuclear matter to the 2SC phase, and then to the CFL phase.
- Color-flavor locking survives for $M_s^2 \lesssim 2\sqrt{2}\mu\Delta$ (see below). Since the CFL state is $\tilde{Q}$-neutral, there are no electrons present in this phase [44], so introducing electromagnetism makes no difference to it.

Additional features, beyond those required by symmetry considerations alone, have been suggested by Pisarski [45], by analogy with scalar-gauge field theories.

4.2 Quark-Hadron Continuity

The shaded region in Fig. 4 is characterized by a definite global symmetry, $SU(2) \times [U(1)]$, but this can either be a hadronic (hyperonic) phase with unbroken isospin and electromagnetism, or a color-flavor locked quark-matter phase with an isospin+color symmetry and a rotated electromagnetism that allows a linear combination of the photon and a gluon to remain massless. In other words, in this regime there is no symmetry difference between hyperonic matter and quark matter. This raises the exciting possibility [32] that properties of sufficiently dense hadronic matter could be found by extrapolation from the quark matter regime where weak-coupling methods can be used.

Table 2. Quark hadron continuity: mapping between states in high-density hadronic and quark matter

Particle type	Hyperonic matter	$\Leftrightarrow$	CFL quark matter
Fermions	8 Baryons	$\Leftrightarrow$	9 Quarks
Chiral (pseudo)Goldstone	8 pion/kaons	$\Leftrightarrow$	8 pseudoscalars
Baryon number (pseudo)Goldstones	1	$\Leftrightarrow$	1
Vector Mesons	9	$\Leftrightarrow$	8 massive gluons

The most straightforward application of this idea is to relate the quark-gluon description of the spectrum to the hadronic description of the spectrum in the CFL phase [32]. The conjectured mapping is given in Table 2. Gluons in the CFL phase map to the octet of vector bosons; the Goldstone bosons associated with chiral symmetry breaking in the CFL phase map to the pions; and the quarks map onto baryons. Pairing occurs at the Fermi surfaces, and we therefore expect the gap parameters in the various quark channels to map to the gap parameters due to baryon pairing.

In Table 3 we show how this works for the fermionic states in 2+1 flavor QCD. There are nine states in the quark matter phase. We show how they transform under the unbroken "isospin" of $SU(2)_{\text{color}+V}$ and their charges under the unbroken "rotated electromagnetism" generated by $\tilde{Q}$, as described in Sect. 6.5. Table 3 also shows the baryon octet, and their transformation properties under the symmetries of isospin and electromagnetism that are unbroken in sufficiently dense hadronic matter. Clearly there is a correspondence between the two sets of particles (note that the final isosinglet has a gap Δ_+ twice as large as the others). As μ increases, the spectrum described in Table 3 may evolve continuously even as the language used to describe it changes from baryons, $SU(2)_V$ and Q to quarks, $SU(2)_{\text{color}+V}$ and $\tilde{Q}$.

If the spectrum changes continuously, then in particular so must the gaps. As discussed above, the quarks pair into rotationally invariant, $\tilde{Q}$-neutral,

Table 3. Mapping of fermionic states between high density quark and hadronic matter

Quark	$SU(2)_{\text{color}+V}$	$\tilde{Q}$	Hadron	$SU(2)_V$	Q
$\begin{pmatrix} bu \\ bd \end{pmatrix}$	**2**	$+1$ 0	$\begin{pmatrix} p \\ n \end{pmatrix}$	**2**	$+1$ 0
$\begin{pmatrix} gs \\ rs \end{pmatrix}$	**2**	0 -1	$\begin{pmatrix} \Xi^0 \\ \Xi^- \end{pmatrix}$	**2**	0 -1
$\begin{pmatrix} ru - gd \\ gu \\ rd \end{pmatrix}$	**3**	0 $+1$ -1	$\begin{pmatrix} \Sigma^0 \\ \Sigma^+ \\ \Sigma^- \end{pmatrix}$	**3**	0 $+1$ -1
$ru + gd + \xi_- bs$	**1**	0	Λ	**1**	0
$ru + gd - \xi_+ bs$	**1**	0	—		

$SU(2)_{\text{color}+V}$ singlets. The two doublets of Table 3 pair with each other, the triplet pairs with itself. Finally, the two singlets pair with themselves.

5 Color-Flavor Unlocking and the Crystalline Color Superconducting Phase

A prominent feature of the zero-temperature phase diagram in Fig. 4 is the "unlocking" phase transition between two-flavor pairing (2SC) and three-flavor pairing (CFL). At this phase transition, the Fermi momentum of free strange quarks is sufficiently different from that of the light quarks to disrupt pairing between them.

Such transitions are expected to be a generic feature of quark matter in nature. In the absence of interactions, the requirements of weak equilibrium and charge neutrality cause all three flavors of quark to have different Fermi momenta. In the extreme case where all three flavors had very different chemical potentials, each flavor would have to self-pair [46,40], but in the phenomenologically interesting density range we expect a rich and complex phase structure for cold dense matter as a function of quark masses and density. The CFL $\leftrightarrow$ 2SC transition of Fig. 4 is one example. Assuming that no other intermediate phases are involved, we now give a model independent argument that the unlocking phase transition between the CFL and 2SC phases in Fig. 4 must be first order. However, there is good reason to expect an intermediate state – the crystalline color-superconducting state, and we go on to discuss it in some detail. Note that another crystalline phase, the "chiral crsytal" has also been proposed [47], although it is not yet clear whether there is any window of density where it is favored.

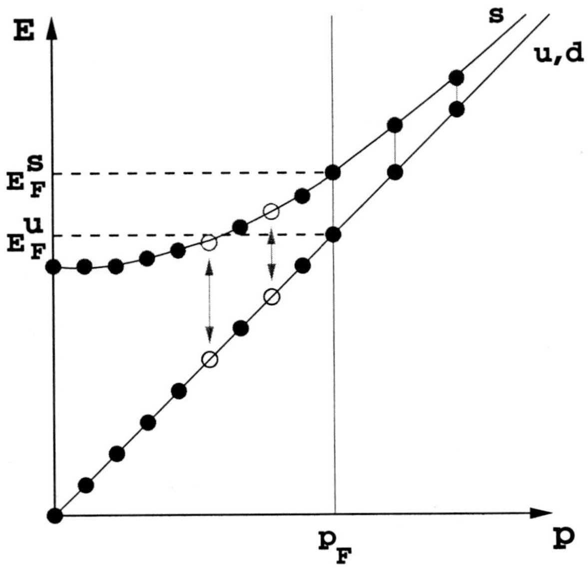

Fig. 7. How the strange quark mass interferes with a u-s condensate. The strange quark (upper curve) and light quark (straight line) dispersion relations are shown, with their Fermi seas filled up to a common Fermi momentum p_F. The horizontal axis is the magnitude of the spatial momentum; s-wave pairing occurs between particles (or holes) with the same p and opposite p. The energy gained by pairing stops the s quarks from decaying to u quarks (see text)

5.1 The (Un)locking Transition

Figure 7 shows part of the CFL pairing pattern: the quark states of the different flavors are filled up to a common Fermi momentum $p_F \approx \mu$, intermediate between the free-light-quark and free-strange-quark Fermi momenta. In the absence of interactions, this state would be unstable: weak interactions would turn strange quarks into light quarks, and there would be separate strange and light Fermi momenta, each filled up to the Fermi energy μ. However, pairing stabilizes it. For the paired state to be stable, it must be that the free energy gained from turning a strange quark into a light quark is less than the energy lost by breaking the Cooper pairs for the modes involved [44].

$$\begin{gathered}\sqrt{\mu^2 + M_s(\mu)^2} - \sqrt{\mu^2 + M_u(\mu)^2} \approx \frac{M_s(\mu)^2 - M_u(\mu)^2}{2\mu} \lesssim 2\Delta_{us} \ , \\ \text{i.e.,} \quad \frac{M_s(\mu)^2}{4\mu} \lesssim \Delta_{us} .\end{gathered} \tag{11}$$

Here $M_s(\mu)$ and $M_u(\mu)$ are the constituent quark masses in the CFL phase, and $M_u(\mu) \ll M_s(\mu)$. An additional factor of $1/\sqrt{2}$ on the RHS of (11) can be obtained requiring the paired state to have lower free energy [48,49,44].

Equation (11) implies that arbitrarily small values of Δ_{us} are impossible, which means that the phase transition must be first order: the gap cannot go continuously to zero. Such behavior has been found in calculations for unlocking phase transitions of this kind in electron superconductors [48] and nuclear superfluids [50] as well as QCD superconductors [35,51,52].

5.2 The Crystalline Color-Superconducting Phase

There is good reason to think that, in the region where the strange quark is just on the edge of decoupling from the light quarks, another form of pairing can occur. This is the "LOFF" state, first explored by Larkin and Ovchinnikov [53] and Fulde and Ferrell [54] in the context of electron superconductivity in the presence of magnetic impurities. They found that near the unpairing transition, it is favorable to form a state in which the Cooper pairs have nonzero momentum. This is favored because it gives rise to a region of phase space where each of the two quarks in a pair can be close to its Fermi surface, and such pairs can be created at low cost in free energy. Condensates of this sort spontaneously break translational and rotational invariance, leading to gaps which vary periodically in a crystalline pattern. The possible consequences for compact stars will be discussed in section 6.

In Ref. [49], the LOFF phase in QCD has been studied using a toy model in which the quarks interact via a four-fermion interaction with the quantum numbers of single-gluon exchange. The model only considers pairing between u and d quarks, with $\mu_d = \bar{\mu} + \delta\mu$ and $\mu_u = \bar{\mu} - \delta\mu$. For the rest of this section we will discuss properties of the model, but it is important to remember that in reality we expect a LOFF state wherever the difference between the Fermi momenta of any two quark flavors is near an unpairing transition, for example the unlocking phase transition between the 2SC and CFL phases.

The Nature of LOFF Pairing. Whereas the BCS state requires pairing between fermions with equal and opposite momenta, for some values of $\delta\mu$ it may be more favorable to form a condensate of Cooper pairs with *nonzero* total momentum. By pairing quarks with momenta which are not equal and opposite, some Cooper pairs are allowed to have both the up and down quarks on their respective Fermi surfaces even when $\delta\mu \neq 0$. LOFF found that within a range of $\delta\mu$ a condensate of Cooper pairs with momenta $\mathbf{k}_u = \mathbf{q} + \mathbf{p}$ and $\mathbf{k}_d = \mathbf{q} - \mathbf{p}$ (see Fig. 8) is favored over either the BCS condensate or the normal state. Here, our notation is such that $\mathbf{p}$ specifies a particular Cooper pair, while $\mathbf{q}$ is a fixed vector, the same for all pairs, which characterizes a given LOFF state. The magnitude $|\mathbf{q}|$ is determined by minimizing the free energy; the direction of $\mathbf{q}$ is chosen spontaneously. The resulting LOFF state breaks translational and rotational invariance. In position space, it describes a condensate which varies as a plane wave with wave vector $2\mathbf{q}$.

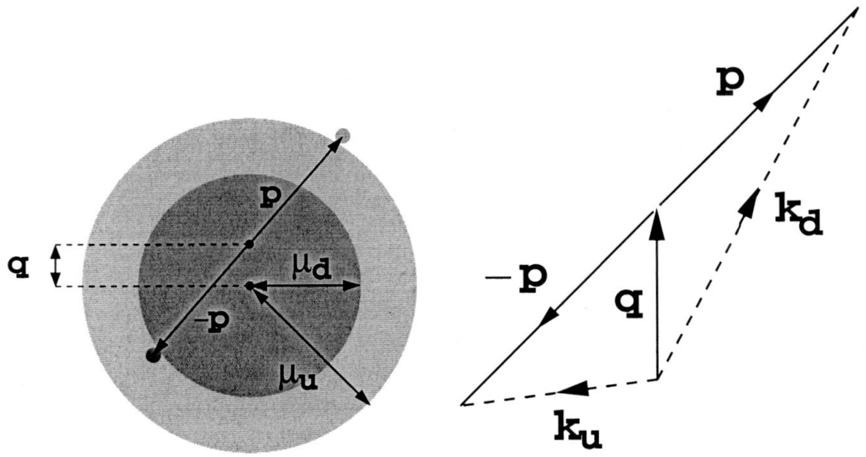

Fig. 8. The momenta $\mathbf{k}_u$ and $\mathbf{k}_d$ of the two members of a LOFF-state Cooper pair, with both quarks near their respective Fermi surfaces

Results from a Simplified Model. In the LOFF state, each Cooper pair carries momentum $2\mathbf{q}$, so the condensate and gap parameter vary in space with wavelength $\pi/|\mathbf{q}|$. In the range of $\delta\mu$, where the LOFF state is favored, $|\mathbf{q}| \approx 1.2\delta\mu$. In Ref. [49], we simplify the calculation of the gap parameter by assuming that the condensate varies in space like a plane wave, leaving the determination of the crystal structure of the QCD LOFF phase to future work. We make an ansatz for the LOFF wave function, and by variation obtain a gap equation which allows us to solve for the gap parameter Δ_A, the free energy and the values of the diquark condensates which characterize the LOFF state at a given $\delta\mu$ and $|\mathbf{q}|$. We then vary the momentum $|\mathbf{q}|$ of the ansatz, to find the preferred (lowest free energy) LOFF state at a given $\delta\mu$, and compare the free energy of the LOFF state to that of the BCS state with which it competes. We show results for one choice of parameters in Fig. 9(a).

In Fig. 9 the average quark chemical potential $\bar{\mu}$ has been set to 0.4 GeV, corresponding to a baryon density of about 4 to 5 times that in nuclear matter. A crude estimate [49] suggests that in quark matter at this density, $\delta\mu \sim 15 - 30$ MeV depending on the value of the density-dependent effective strange quark mass.

We find that the LOFF state is favored for values of $\delta\mu$ which satisfy $\delta\mu_1 < \delta\mu < \delta\mu_2$ as shown in Fig. 9(b), with $\delta\mu_1/\Delta_0 = 0.707$ and $\delta\mu_2/\Delta_0 = 0.754$ in the weak coupling limit $\Delta_0 \ll \mu$. (Δ_0 is the 2SC gap for $\delta\mu < \delta\mu_1$, and one can use it to parameterize the strength of the four fermion interaction G: small Δ_0 corresponds to a small G.) At weak coupling, the LOFF gap parameter decreases from $0.23\Delta_0$ at $\delta\mu = \delta\mu_1$ (where there is a first-order BCS-LOFF phase transition) to zero at $\delta\mu = \delta\mu_2$ (where there is a second-order LOFF-

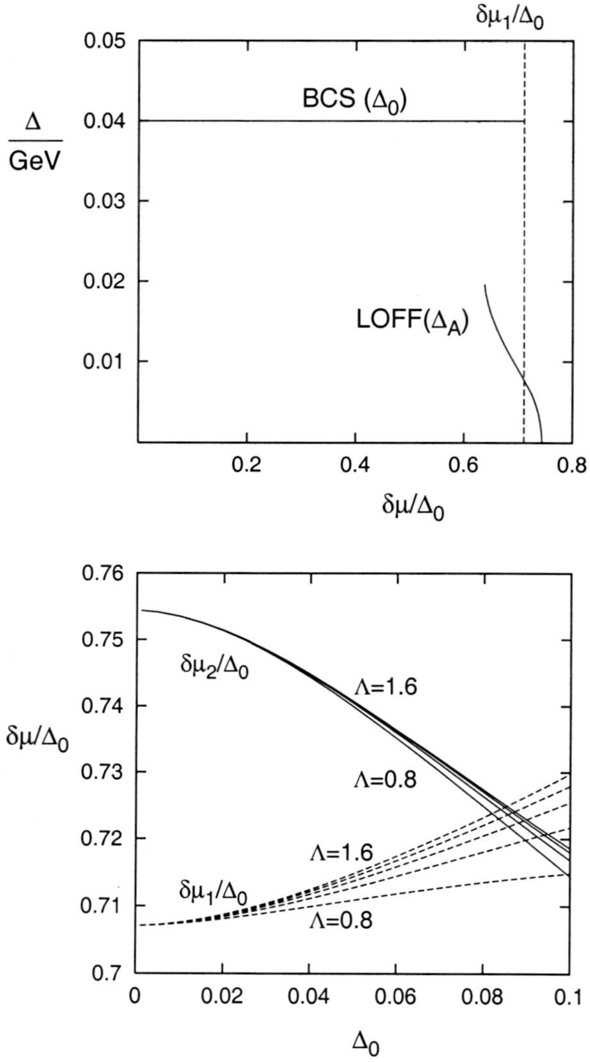

Fig. 9. (**a**) LOFF and BCS gap parameters as a function of $\delta\mu$, with coupling chosen so that $\Delta_0 = 40$ MeV, and $\Lambda = 1$ GeV. The vertical dashed line marks $\delta\mu = \delta\mu_1$, above which the LOFF state has lower free energy than BCS. (**b**) The interval of $\delta\mu$ within which the LOFF state occurs as a function of the coupling, parametrized by the BCS gap Δ_0 in GeV. Below the solid line, there is a LOFF state. Below the dashed line, the BCS state is favored. The different lines of each type correspond to different cutoffs on the four-fermion interaction: $\Lambda = 0.8$ GeV to 1.6 GeV. $\delta\mu_1/\Delta_0$ and $\delta\mu_2/\Delta_0$ show little cutoff-dependence, and the cutoff dependence disappears completely as $\Delta_0, \delta\mu \to 0$

normal transition). Except for very close to $\delta\mu_2$, the critical temperature above which the LOFF state melts will be much higher than typical neutron star temperatures. At stronger coupling the LOFF gap parameter decreases relative to Δ_0 and the window of $\delta\mu/\Delta_0$ within which the LOFF state is favored shrinks, as seen in Fig. 9(b).

6 Compact Stars and Color-Superconducting Quark Matter

Having described the interesting phenomena that we believe occur in cold quark matter, we now ask ourselves where in nature such phenomena might occur, and how we might see evidence of them.

The only place in the universe where we expect sufficiently high densities and low temperatures is compact stars, also known as "neutron stars", since it is often assumed that they are made primarily of neutrons (for a recent review, see [55]). A compact star is produced in a supernova. As the outer layers of the star are blown off into space, the core collapses into a very dense object. Typical compact stars have masses close to $1.4M_\odot$, and are believed to have radii of order 10 km. The density ranges from around nuclear density near the surface to higher values further in, although uncertainty about the equation of state leaves us unsure of the value in the core.

During the supernova, the core collapses, and its gravitational energy heats it to temperatures of order 10^{11} K (tens of MeV), but it cools rapidly by neutrino emission. Within a few minutes its internal temperature T drops to 10^9 K (100 keV), and reaches 10^7 K (1 keV) after a century. Neutrino cooling continues to dominate for the first million years of the life of the star. The effective temperature T_e of the X-ray emissions is lower than the internal temperature: $T_e/10^6\,\mathrm{K} \approx \sqrt{T/10^8\,\mathrm{K}}$ [56].

Color superconductivity gives mass to excitations around the ground state: it opens up a gap at the quark Fermi surface, and makes the gluons massive. One would therefore expect its main consequences to relate to transport properties, such as mean free paths, conductivities and viscosities. The influence of color superconductivity on the equation of state is an $\mathcal{O}((\Delta/\mu)^2)$ (few percent) effect, which is not phenomenologically interesting given the existing uncertainty in the equation of state at the relevant densities.

6.1 The Mixed Phase

Before surveying some of the suggested ways in which color superconductivity might manifest itself, we will briefly review the possibility of a mixed phase, in which regions of quark matter and nuclear matter are intermingled. To be concrete, consider the case where the strange quark is light enough so that quark pairing is always of the CFL type.

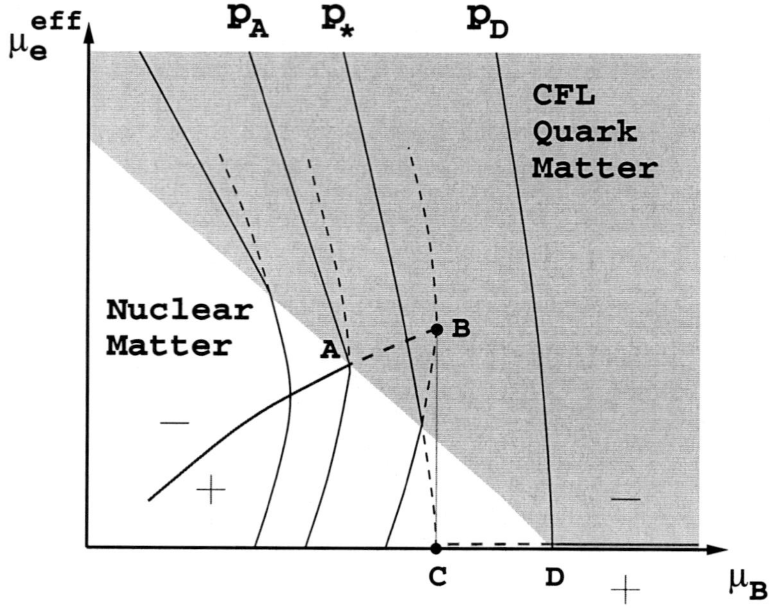

Fig. 10. A schematic form of the μ_B-μ_e phase diagram for nuclear matter and CFL quark matter, ignoring electromagnetism. For an explanation see the text

Figure 10 shows the μ_B-μ_e phase diagram, ignoring electromagnetism. The unshaded region is where nuclear matter (NM) has higher pressure. The shaded region is where quark matter (QM) has higher pressure. Where they meet is the coexistence line. The medium solid lines labelled by values of the pressure are isobars. Below the coexistence line they are given by the NM equation of state, above it by the QM equation of state.

The thick lines are the neutrality lines. Each phase is negatively charged above its neutrality line and positively charged below it. Dotted lines show extensions onto the unfavored sheet (NM above the coexistence line, QM below it).

The electric charge density is

$$Q = - \left. \frac{\partial p}{\partial \mu_e} \right|_{\mu_B} . \tag{12}$$

The neutrality line therefore goes through the right-most extremum of each isobar, since there the derivative of pressure with respect to μ_e is zero. For the CFL phase, the neutrality line is $\mu_e = 0$. Turning up μ_e introduces electrons, increasing the pressure.

Two possible paths from nuclear to CFL matter as a function of increasing μ are depicted. In the absence of electromagnetism and surface tension, the favored option is evidently a mixed phase made of negatively charged CFL matter and positively charged nuclear matter along the segment of the coexistence line from A to D. On this segment, positively charged nuclear matter coexists with negatively charged CFL matter, so for pressures in the range p_A to p_D an overall neutral mixed phase can be created by choosing an appropriate volume fraction of CFL relative to nuclear matter. If, on the other hand, Coulomb and surface energies are large, then the mixed phase is disfavored. The system remains on the nuclear neutrality line up to B, where there is a single interface between nuclear matter at B and CFL matter at C. This minimal interface, with its attendant charged boundary layers [57], occurs between phases with the same μ_e, $\mu = \mu_B = \mu_C$, and pressure P_*. The effective chemical potential μ_e^{eff} changes across the interface, though, as a result of the presence of the electric field. For more details on the single interface, the mixed phase, and the competition between them, see Ref. [57].

As yet, not much work has been done on signatures related to these features. The single interface creates a dramatic density discontinuity in the star: CFL quark matter at about four times nuclear density floats on nuclear matter at about twice nuclear density. This may affect the mass vs. radius relationship for neutron stars with quark matter cores. It may also have qualitative effects on the gravitational wave profile emitted during the inspiral and merger of two compact stars of this type. The mixed phase has distinctively short neutrino mean free paths, due to coherent scattering [58]. Also, the droplets form a crystal lattice that could pin vortices, leading to glitches.

6.2 Cooling by Neutrino Emission

As mentioned above, for its first million or so years, a neutron star cools by neutrino emission. The temperature is obtained from X-ray spectra of isolated compact stars, and is subject to many uncertainties, including emissions from plasma around the star, and distortion of the spectrum by a possible hydrogen atmosphere. The age, inferred from the spindown rate by assuming magnetic dipole radiation from a constant dipole moment, may also have large systematic errors. Even so, a consistent picture emerges [59,56] in which the youngest compact stars, about a thousand years old, have surface temperatures around 2×10^6 K (200 eV), falling to about 3×10^5 K (30 eV) after a million years.

The cooling rate is determined by the heat capacity and emissivity, both of which are dominated by quark modes whose energy is within T of the Fermi surface, and are therefore sensitive to the kind of gaps generated by color superconductivity [59,60,56].

In the CFL phase, all quarks and gluons have gaps $\Delta \gg T$, electrons are absent [44], and the transport properties are dominated by the only true Goldstone excitation, the superfluid mode arising from the breaking of the exact baryon number symmetry. The next lightest modes are the pseudo-Goldstone bosons associated with chiral symmetry breaking, which will only participate when the temperature is above their mass, which is of order tens of MeV [61]. This means that CFL quark matter has a much smaller neutrino emissivity and heat capacity than nuclear matter, and hence the cooling of a neutron star is likely to be dominated by the nuclear mantle rather than the CFL core [56]. A CFL core is therefore not detectable by cooling measurements.

We turn now to the 2SC quark matter phase, which occurs if the strange quarks are too heavy to pair with the light flavors. Up and down quarks of two of the colors (red and green, say) pair strongly with a gap much bigger than the temperature. This leaves the blue up and down, and the strange quarks (if present) with much more weakly attractive channels in which to pair. The strange quarks are believed to pair with each other in a "color-spin locked" condensate, with a gap of order hundreds of keV [40] or less [49]. The blue up and down quarks form $J = 1$ pairs, breaking rotational invariance [6], with a gap that was originally estimated to be in the keV range, but this estimate is not robust, and depends on details of the NJL model used [6].

This leads to potentially interesting phenomenology, since the blue and/or strange quarks have small gaps, so during the early life of the compact star they may participate in the cooling dynamics as long as the temperature is greater than their gap. Their effects would be dramatic, allowing high rates of neutrino emission via direct URCA processes such as $d \to u + e + \bar{\nu}$ and $u \to d + e^+ + \nu$, and leading to rapid cooling of the core [59,56]. The cooling would slow down suddenly when the temperature fell below the gap. Such a behavior would be observable, and if no sign of it is seen as our observations of neutron star temperatures improve then we will have to conclude that either 2SC matter does not occur, or the smallest gaps are larger than the observed temperatures.

6.3 The Neutrino Pulse at Birth

We have seen above that in the first seconds of a supernova, the inner regions ("protoneutron star") are heated to tens of MeV by the gain of a vast amount of energy from the gravitational collapse, and are consequently hot (tens of MeV). Over the next half-minute or so much of the energy is radiated off as neutrinos, whose detailed spectrum as a function of time is determined by the neutrino diffusion properties of the protoneutron star. Neutrinos from supernova 1987A were detected in terrestrial experiments, and the duration and mean energy of the pulse was measured. We can hope that neutrinos from future supernovae in our galaxy will be measured more precisely. It is therefore useful to study the effects of color superconductivity on neutrino diffusion, in order to see if it leads to any signature in the neutrino pulse.

Carter and Reddy [62] have performed a preliminary investigation of this question. They restricted themselves to two flavors, and studied the case where the core starts off as a hot quark-gluon plasma. Within seconds, thanks to neutrino emission, it cools into a superconducting phase, and they assumed that this occurred via a second-order phase transition. This leads to a striking two-stage signature. (1) Near the critical temperature T_c, the heat capacity rises, and the cooling of the star consequently slows. (2) Below the critical temperature, the quark modes are gapped, and the neutrino mean free path is enhanced by $\exp(\Delta/T)$, reflecting Boltzmann suppression of the population of quark quasiparticles. As a result, the core may suddenly empty itself of neutrinos, creating a final neutrino burst. There may be further processing of this burst on its way out of the supernova, but the suggestion is that it may survive to yield a noticeable signal in neutrino detectors on earth. The suggestion, then, is that the flux of supernova neutrinos detected on earth will not taper off, but show a final burst followed by no flux. Before that, there may be a plateau in the energy or flux of the neutrinos, as the cooling slows near the critical temperature.

There are many issues that require further investigation. It is not clear whether a second-order phase transition is to be expected, since the up - and down - quark Fermi surfaces will differ, and there may be a first-order unlocking transition [52]. Also, it is necessary to take into account the strange quark, and the processing of emitted neutrinos by the layers of neutrino-opaque hadronic matter that surround the core during the supernova explosion.

6.4 r-Mode Instability

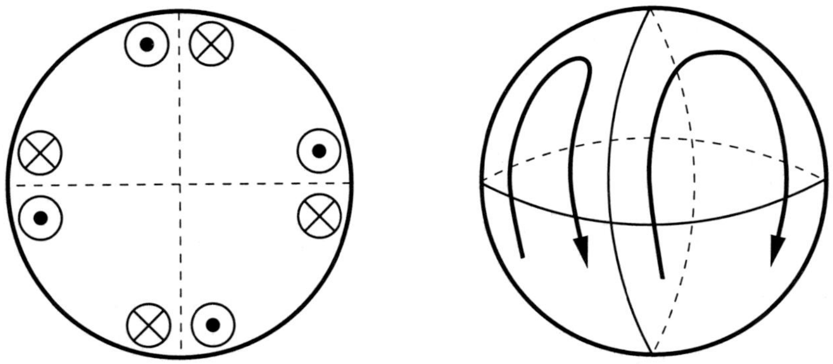

Fig. 11. The quadrupole pattern of r-mode bulk flows

The term "r-mode" (short for "rotational mode") refers to a bulk flow in a rotating star that radiates away energy and angular momentum in the form of gravitational waves (Fig. 11). If the rotation frequency f of the star is above a critical value f_*, the system becomes unstable to r-modes and will quickly spin down until its frequency drops to f_*, at which point the r-modes are damped out. The critical frequency depends on the sources of damping that could suppress the flows. These include shear and bulk viscosities, and also "surface rubbing" – the friction at the interface between the r-mode region and any rigid crust that cannot flow. Since the viscosities are sensitive functions of temperature, one calculates $f_*(T)$ as an upper limit on rotation frequencies, and thereby maps out an excluded high-f region in the T-f plane. Differently constituted compact stars (neutrons, quarks with various gaps due to pairing) have different excluded regions, and one can see whether any of them are ruled out by the observation of pulsars in nature with rotation rates and temperatures in their excluded region.

Madsen [63] has shown that gapless quark matter and neutron stars are not ruled out. However, color superconductivity creates gaps in the quark excitation spectrum, suppressing the viscosities by factors of order $\exp(-\Delta/T)$, and encouraging r-mode spindown. He found that for a compact star made *entirely* of quark matter in the CFL phase, even a quark gap as small as $\Delta = 1$ MeV reduces $f_*(T)$ dramatically to $\mathcal{O}(100$ Hz) for temperatures below 10^9 K (100 keV). This means that millisecond pulsars, with frequencies up to 640 Hz, cannot be CFL quark matter stars, making it questionable whether any compact stars are made entirely of CFL quark matter. Madsen found that 2SC quark matter stars were on the edge of being ruled out, so he was not able to say anything about them, either positive or negative.

Madsen included the additional damping from surface rubbing between the quark matter and a normal matter crust. Using the conventional picture, this is a very small effect, since the crust is separated from the quark matter by an electrostatic cushion of electrons, and so surface rubbing made no difference to the result for pure CFL stars. Actually, since CFL matter is neutral [44], it contains no electrons, so the cushioning mechanism may not be operative, and it is not clear that there is any such crust.

There are caveats to Madsen's conclusions. Firstly, the results are sensitive to the temperature of the inner regions of the star, which has to be inferred from the measured effective surface temperature using models of the heat flow, and is therefore not accurately known. However, this uncertainty is only important for unpaired or 2SC paired quark matter; pure CFL stars are ruled out for $T_e < 10^9$ K (100 keV), which is already at the upper end of conceivable temperatures. Secondly, as he points out, his calculations do not rule out the generic picture of how quark matter occurs in compact stars, namely as a quark matter core surrounded by a nuclear mantle. In this case substantial friction is expected at the core-mantle interface, and this may be enough [64,63] to stabilize the star irrespective of the viscosities of the quark matter. Furthermore, quark matter may contain a shell of LOFF crystal (see below), and the r-modes could be damped at the edges of that region rather than at the crust. We can hope that future work on hybrid stars will clarify the situation.

6.5 Magnetic Field Decay

The behavior of magnetic fields in quark matter is quite different from that in nuclear matter [65,66]. Nuclear matter is an electromagnetic superconductor (because of proton-proton pairing which breaks the $U(1)_Q$ gauge symmetry) and also a superfluid (because of neutron-neutron pairing). Magnetic fields are therefore restricted to Abrikosov flux tubes, and angular momentum is carried by rotational vortices. The magnetic flux tubes can be dragged about by the outward motion of the rotational vortices as the neutron star spins down [67–71], and can also be pushed outward if the gap at the proton Fermi surface increases with depth within the neutron star [72]. One therefore expects the magnetic field of an isolated pulsar to decay over billions of years as it spins down [68–71] or perhaps more quickly [72]. However, there is no observational evidence for the decay of the magnetic field of an isolated pulsar over periods of billions of years [69,73]

A color superconductor, on the other hand, leaves unbroken a rotated electromagnetism $U(1)_{\tilde{Q}}$, a mixture of photon and gluon, allowing long-range $\tilde{Q}$-magnetic fields. This is true of the CFL phase, and also of the 2SC phase as long as the temperature is high enough so that the blue quarks do not pair.

The new unbroken rotated electromagnetic field $A_{\tilde{Q}}$ is just a linear combination of the photon A_μ and one of the gluons G^8_μ,

$$A^{\tilde{Q}}_\mu = \cos\alpha_0\, A_\mu + \sin\alpha_0\, G^8_\mu, \tag{13}$$

the orthogonal combination A^X_μ is massive. The mixing angle α_0 is the analogue of the Weinberg angle in electroweak theory, in which the presence of the Higgs condensate causes the hypercharge and W_3 gauge bosons to mix to form the photon, A_μ, and the massive Z boson. $\sin(\alpha_0)$ is proportional to e/g and turns out to be about 1/20 in the 2SC phase and 1/40 in the CFL phase [66]. This means that the $\tilde{Q}$-photon which propagates in color-superconducting quark matter is mostly photon with only a small gluon admixture. If a color-superconducting neutron star core is subjected to an ordinary magnetic field, it will either expel the X component of the flux or restrict it to flux tubes, but it admits the great majority of the flux in the form of a $B_{\tilde{Q}}$ magnetic field satisfying Maxwell's equations. The decay in time of this "free field" (i.e. not in flux tubes) is limited by the $\tilde{Q}$-conductivity of the quark matter.

The CFL phase contains no electrons, and all its charged modes are gapped, making it an electromagnetic insulator. The 2SC phase has electrons as well as blue quasiquarks, and turns out to be a very good conductor. Thus the 2SC and CFL phases, while both allowing long-range $\tilde{Q}$-flux fields, react very differently to attempts to *change* the magnetic field. The CFL phase allows such changes, but the 2SC, as a near-perfect conductor, generates eddy currents that oppose the change, locking the magnetic field into the core with a decay time of order 10^{13} years [66]

This means that a 2SC quark matter core within a neutron star can act as an "anchor" for the magnetic field, preventing the flux-tube-dragging mechanism that can operate in ordinary nuclear matter. Even though this distinction is a qualitative one, it will be difficult to confront it with data since what is observed is the total dipole moment of the neutron star. A color-superconducting core can only anchor those magnetic flux lines which pass through the core, while in a neutron star with no quark matter core the entire internal magnetic field can decay over time. In both cases, however, the total dipole moment can change since the magnetic flux lines which do not pass through the core can move.

6.6 Glitches and the Crystalline Color Superconductor

The crystalline LOFF phase has been discussed above. It occurs when two different types of quark have different Fermi momenta (because their masses or chemical potentials are different) and are just barely able to pair.

Such situations are likely to be generic in nature, where, because of the strange quark mass, combined with requirements of weak equilibrium and charge neutrality, all three flavors of quark in general have different chemical potentials. To date the LOFF condensate has only been studied in simplified two-flavor models, so it is not clear whether it can be expected to occur in compact stars. However, in the model a LOFF phase occurred if the gap Δ_0 which characterizes the uniform color superconductor present at smaller values of $\delta\mu$ was about 40 MeV [49]. This is in the middle of the range of present estimates of superconducting gaps. It is therefore worthwhile to consider the consequences.

Glitches and Vortex Pinning. Glitches are sudden jumps in rotation frequency Ω of a pulsar, which may be as large as $\Delta\Omega/\Omega \sim 10^{-6}$, but may also be several orders of magnitude smaller. The frequency of observed glitches is statistically consistent with the hypothesis that all radio pulsars experience glitches [74]. Glitches are thought to originate in the rigid neutron star crust, typically somewhat more than a kilometer thick, where rotational vortices in a neutron superfluid are pinned to the crystal structure of the crust. As the pulsar's spin gradually slows, the vortices must gradually move outwards since the rotation frequency of a superfluid is proportional to the density of vortices. Models [75] differ in important respects as to how the stress associated with pinned vortices is released in a glitch: for example, the vortices may break and rearrange the crust, or a cluster of vortices may suddenly overcome the pinning force and move macroscopically outward, with the sudden decrease in the angular momentum of the superfluid within the crust resulting in a sudden increase in angular momentum of the rigid crust itself and hence a glitch. All the models agree that the fundamental requirements are the presence of rotational vortices in a superfluid and the presence of a rigid structure which impedes the motion of vortices and which encompasses enough of the volume of the pulsar to contribute significantly to the total moment of inertia.

It is reasonable to expect that a real-world LOFF phase is a superfluid, since it would involve pairing of the strange quarks with the light quarks, which is what makes the CFL phase a superfluid. This means that if it occurs within a pulsar it will be threaded by an array of rotational vortices. It is natural to expect that these vortices will be pinned in a LOFF crystal, in which the diquark condensate varies periodically in space. Indeed, one of the suggestions for how to look for a LOFF phase in terrestrial electron superconductors relies on the fact that the pinning of magnetic flux tubes (which, like the rotational vortices of interest to us, have normal cores) is expected to be much stronger in a LOFF phase than in a uniform BCS superconductor [76]. Note that the chiral crystal phase [47] is not a superfluid, so it will not contain rotational vortices.

Vortex Pinning in the LOFF Phase. A real calculation of the pinning force experienced by a vortex in a crystalline color superconductor must await the determination of the crystal structure of the LOFF phase. We can, however, attempt an order of magnitude estimate along the same lines as that done by Anderson and Itoh [77] for neutron vortices in the inner crust of a neutron star. In that context, this estimate has since been made quantitative [78,79,75]. For one specific choice of parameters [49], the resulting pinning force per unit length of vortex was estimated essentially by dimensional analysis at

$$\text{LOFF:} \quad f_p \sim (4\ \text{MeV})/(80\ \text{fm}^2). \tag{14}$$

It is premature to compare such a crude result to the results of serious calculations [78,79,75], but it is remarkable that they prove to be similar: the pinning force per unit length for neutron vortices in the inner crust is

$$\text{neutron star:} \quad f_p \approx (1 - 3\ \text{MeV})/(200 - 400\ \text{fm}^2). \tag{15}$$

This raises the possibility that pulsars might be strange stars after all [80,81]. Strange quark stars are made almost entirely of quark matter with either no hadronic matter content at all or perhaps a thin crust, of order one hundred meters thick, which contains no neutron superfluid [81,82]. No successful models of glitches in the crust of a strange quark star have been proposed, indicating that pulsars are not strange stars [83–85]. The possibility of a shell of crystalline LOFF quark matter inside a quark star revives the possibility that glitches could occur in quark stars, as a result of the pinning of quark-superfluid vortices to the LOFF crystal.

7 Conclusions

Quark pairing and color superconductivity are phenomena that have been discussed and speculated upon for some time. Only in recent years, however, has the resultant structure of the QCD phase diagram been considered in any detail. As outlined in Sects. 2 to 5 above, the phase diagram has been found to be full of unexpected richness. There are many directions that remain to be explored, from new pairing structures to detailed studies of the known 2SC, 2SC+s, CFL, and LOFF phases. It has even been suggested that zero-density QCD can be understood in terms of a quark-paired condensate in combination with an adjoint chiral condensate [86].

The search for signatures of color superconductivity is now proceeding in earnest. The most promising area is the phenomenology of neutron/quark stars, which are the only naturally occurring example of cold matter at densities where quark matter might occur. The first steps in this endeavor have been described in section 6. It would be a great stride forward if, at the same time as heavy-ion colliders map the high-temperature region of the QCD phase diagram, astrophysical observations and calculations could complement it by filling in details of the high-density region.

References

1. D. Hong: hep-ph/0101025; K. Rajagopal, F. Wilczek: hep-ph/0011333; T. Schäfer, E. Shuryak: nucl-th/0010049; K. Rajagopal: hep-ph/0009058; D. Rischke, R. Pisarski: Proceedings of the "Fifth Workshop on QCD", Villefranche, Jan. 3-7, 2000; S. Hsu: "Color Superconductivity in High-Density Quark Matter". In: *Dynamics of Gauge Fields, TMU-Yale Symposium, Tokyo, Japan, Dec. 13-15, 1999*, ed. by A. Chodos et. al. (Universal Academy Press, Tokyo 2000) pp. 247-261; M. Alford: "Color Superconductivity in Dense Quark Matter". In: *Dynamics of Gauge Fields, TMU-Yale Symposium, Tokyo, Japan, Dec. 13-15, 1999*, ed. by A. Chodos et. al. (Universal Academy Press, Tokyo 2000) pp. 227-246
2. J. Bardeen, L. Cooper, J. Schrieffer: Phys. Rev. **106**, 162 (1957); Phys. Rev. **108**, 1175 (1957)
3. B. Barrois: Nucl. Phys. **B129**, 390 (1977); S. Frautschi: "Asymptotic Freedom and Color Superconductivity in Dense Quark Matter". In: *Proceedings of 1978 Erice Workshop "Hadronic Matter at Extreme Density", Erice, Italy, Oct. 13-21, 1978*, ed. by N. Cabibbo and L. Sertorio, (Plenum, New York, 1980) pp. 19-27
4. B. Barrois: "Nonperturbative Effects in Dense Quark Matter", Cal. Tech. PhD thesis, UMI 79-04847-mc (1979)
5. D. Bailin, A. Love: Phys. Rep. **107**, 325 (1984)
6. M. Alford, K. Rajagopal, F. Wilczek: Phys. Lett. **B422**, 247 (1998)
7. R. Rapp, T. Schäfer, E. Shuryak, M. Velkovsky: Phys. Rev. Lett. **81**, 53 (1998)
8. D. Son: Phys. Rev. **D59**, 094019 (1999)
9. T. Schäfer, F. Wilczek: Phys. Rev. **D60**, 114033 (1999)

10. R. Pisarski, D. Rischke: Phys. Rev. **D61**, 074017 (2000)
11. D. Hong: Nucl. Phys. **B582**, 451 (2000); Phys. Lett. **B473**, 118 (2000)
12. R. Pisarski, D. Rischke: Phys. Rev. **D61**, 051501 (2000)
13. D. Hong, V. Miransky, I. Shovkovy, L. Wijewardhana: Phys. Rev. **D61**, 056001 (2000); Erratum Phys. Rev. **D62**, 059903 (2000)
14. W. Brown, J. Liu, H. Ren: Phys. Rev. **D61**, 114012 (2000); Phys. Rev. **D62**, 054016 (2000); Phys. Rev. **D62**, 054013 (2000)
15. S. Hsu, M. Schwetz: Nucl. Phys. **B572**, 211 (2000)
16. T. Schäfer: Nucl. Phys. **B575**, 269 (2000)
17. I. Shovkovy, L. Wijewardhana: Phys. Lett. **B470**, 189 (1999)
18. K. Rajagopal, E. Shuster: Phys. Rev. **D62**, 085007 (2000)
19. J. Berges, K. Rajagopal: Nucl. Phys. **B538**, 215 (1999)
20. G. Carter, D. Diakonov: Phys. Rev. **D60**, 016004 (1999)
21. M. Iwasaki, T. Iwado: Phys. Lett. **B350**, 163 (1995); M. Iwasaki: Prog. Theor. Phys. Suppl. **120**, 187 (1995)
22. M. Alford, K. Rajagopal, F. Wilczek: Nucl. Phys. **B537**, 443 (1999)
23. T. Schäfer, F. Wilczek: Phys. Lett. **B450**, 325 (1999)
24. N. Evans, S. Hsu, M. Schwetz: hep-ph/9810514, hep-ph/9808444
25. B. Vanderheyden, A. Jackson: Phys. Rev. **D62**, 094010 (2000); S. Pepin, A. Schäfer: hep-ph/0010225
26. D. Rischke, D. Son, M. Stephanov: hep-ph/0011379
27. M. Alford, J. Bowers, K. Rajagopal, in preparation
28. F. Sannino: Phys. Lett. **B480**, 280 (2000)
29. S. Hsu, F. Sannino, M. Schwetz: hep-ph/0006059
30. J. Berges: hep-ph/0012013
31. M. Srednicki, L. Susskind: Nucl. Phys. **B187**, 93 (1981)
32. T. Schäfer, F. Wilczek: Phys. Rev. Lett. **82**, 3956 (1999)
33. N. Evans, J. Hormuzdiar, S. Hsu, M. Schwetz: Nucl. Phys. **B581**, 391 (2000)
34. R. Pisarski, D. Rischke: "Why Color-Flavor Locking is Just Like Chiral Symmetry Breaking". To be published in: *Proceedings of the Judah Eisenberg Memorial Symposium, "Nuclear Matter, Hot and Cold", Tel Aviv, April 14 - 16, 1999*
35. M. Alford, J. Berges, K. Rajagopal: Nucl. Phys. **B558**, 219 (1999)
36. M. Halasz, A. Jackson, R. Shrock, M. Stephanov, J. Verbaarschot: Phys. Rev. **D58**, 096007 (1998)
37. J. Berges, U-D. Jungnickel, C. Wetterich: hep-ph/9811387
38. R. Pisarski, D. Rischke: Phys. Rev. Lett. **83**, 37 (1999)
39. M. Stephanov: Phys. Rev. Lett. **76**, 4472 (1996)
40. T. Schäfer: Phys. Rev. **D62**, 094007, (2000)
41. D. Kaplan, A. Nelson: Phys. Lett. **B175**, 57 (1986)
42. G. Brown, K. Kubodera, M. Rho: Phys. Lett. **B175**, 57 (1987)
43. T. Schäfer: Phys. Rev. Lett. **85**, 5531 (2000)
44. K. Rajagopal, F. Wilczek: hep-ph/0012039
45. R. Pisarski: Phys. Rev. **C62**, 035202 (2000)
46. M. Iwasaki, T. Iwado: Prog. Theor. Phys. **94**, 1073 (1995)
47. R. Rapp, E. Shuryak, I. Zahed: Phys. Rev. **D63**, 034008 (2001)
48. A. Clogston: Phys. Rev. Lett. **9**, 266 (1962); B. S. Chandrasekhar: Appl. Phys. Lett. **1**, 7 (1962)
49. M. Alford, J. Bowers, K. Rajagopal: Phys. Rev. **D63**, 074016 (2001)
50. A. Sedrakian, U. Lombardo: Phys. Rev. Lett. **84**, 602 (2000)

51. T. Schäfer, F. Wilczek: Phys. Rev. **D60**, 074014 (1999)
52. P. Bedaque: hep-ph/9910247, unpublished
53. A. Larkin, Y. Ovchinnikov: Zh. Eksp. Teor. Fiz. **47**, 1136 (1964); translation: Sov. Phys. JETP **20**, 762 (1965)
54. P. Fulde, R. Ferrell: Phys. Rev. **135**, A550 (1964)
55. H. Heiselberg, V. Pandharipande: Ann. Rev. Nucl. Part. Sci. **50**, 481 (2000)
56. D. Page, M. Prakash, J. Lattimer, A. Steiner: Phys. Rev. Lett. **85**, 2048 (2000)
57. M. Alford, K. Rajagopal, S. Reddy, and F. Wilczek: GUTPA/01/04/03, 2001.
58. S. Reddy, G. Bertsch and M. Prakash: Phys. Lett. **B475**, 1 (2000)
59. C. Schaab, et al: Astrophys. Lett. J. **480**, L111 (1997) and references therein
60. D. Blaschke, T. Klahn, D. Voskresensky: Astrophys. J. **533**, 406 (2000)
61. D. Son, M. Stephanov: Phys. Rev. **D61**, 074012 (2000); M. Rho, A. Wirzba, I. Zahed: Phys. Lett. **B473**, 126 (2000); R. Casalbuoni, R. Gatto: hep-ph/9911223, unpublished; D. Hong, T. Lee, D. Min: Phys. Lett. **B477**, 137 (2000); C. Manuel, M. Tytgat: Phys. Lett. **B479**, 190 (2000); S. Beane, P. Bedaque, M. Savage: Phys. Lett. **B483**, 131 (2000)
62. G. Carter, S. Reddy: Phys. Rev. **D62**, 103002 (2000)
63. J. Madsen: Phys. Rev. Lett. **85**, 10 (2000)
64. L. Bildsten, G. Ushomirsky: astro-ph/9911155, to appear in ApJ Letters
65. D. Blaschke, D. Sedrakian, K. Shahabasian: Astron. Astrophys. **350**, L47 (1999)
66. M. Alford, J. Berges, K. Rajagopal: Nucl. Phys. **B571**, 269 (2000)
67. J. Sauls: "Superfluidity in the Interiors of Neutron Stars". In: *Timing Neutron Stars*, ed. by J. Ogleman and E. van den Heuvel (Kluwer, Dordrecht 1989) pp. 457-489
68. G. Srinivasan, D. Bhattacharya, A. Muslimov, A. Tsyagan: Curr. Sci. **51**, 31 (1990)
69. D. Bhattacharya, G. Srinivasan, "The Magnetic Fields of Neutron Stars and their Evolution". In: *X-Ray Binaries*, ed. by W. Lewin, J. van Paradijs, E. van den Heuvel (Cambridge University Press, Cambridge, MA, 1995) pp. 495-522
70. M. Ruderman: Astrophys. J. **366**, 261 (1991); Astrophys. J. **382**, 576 (1991); Astrophys. J. **382**, 587 (1991)
71. M. Ruderman, T. Zhu, K. Chen: Astrophys. J. **492**, 267 (1998)
72. S. Hsu: Phys. Lett. **B469**, 161 (1999)
73. D. Lorimer, M. Bailes, P. Harrison: Mon. Not. R. Astron. Soc. **289**, 592 (1997)
74. M. Alpar, C. Ho: Mon. Not. R. Astron. Soc. **204**, 655 (1983) For a review, see A. Lyne: "Rotational Instabilities in Pulsars". In: *Pulsars: Problems and Progress*, ed. by S. Johnston, M. Walker, M. Bailes, (ASP, San Francisco, 1996) p. 73
75. For reviews, see D. Pines, A. Alpar: Nature **316**, 27 (1985); M. Alpar: "Models for Pulsar Glitches". In: *The Lives of Neutron Stars*, ed. by M. Alpar, U. Kiziloglu, J. van Paradijs (Kluwer, Dordrecht 1995) p. 185. For more recent developments and references to further work, see M. Ruderman: Astrophys. J. **382**, 587 (1991); R. Epstein, G. Baym: Astrophys. J. **387**, 276 (1992); M. Alpar, H. Chau, K. Cheng, D. Pines: Astrophys. J. **409**, 345 (1993); B. Link, R. Epstein: Astrophys. J. **457**, 844 (1996); M. Ruderman, T. Zhu, K. Chen: Astrophys. J. **492**, 267 (1998); A. Sedrakian, J. Cordes: Mon. Not. R. Astron. Soc. **307**, 365 (1999)
76. R. Modler, et al: Phys. Rev. Lett. **76**, 1292 (1996)
77. P. Anderson, N. Itoh: Nature **256**, 25 (1975)

78. M. Alpar: Astrophys. J. **213**, 527 (1977)
79. M. Alpar, P. Anderson, D. Pines, J. Shaham: Astrophys. J. **278**, 791 (1984)
80. P. Haensel, J. Zdunik, R. Schaeffer: Astron. Astrophys. **160**, 121 (1986)
81. C. Alcock, E. Farhi, A. Olinto: Phys. Rev. Lett. **57**, 2088 (1986); Astrophys. J. **310**, 261 (1986)
82. N. Glendenning, F. Weber: Astrophys. J. **400**, 647 (1992)
83. A. Alpar: Phys. Rev. Lett. **58**, 2152 (1987)
84. J. Madsen: Phys. Rev. Lett. **61**, 2909 (1988)
85. R. Caldwell, J. Friedman: Phys. Lett. **B264**, 143 (1991)
86. J. Berges, C. Wetterich: hep-ph/0012311
87. N. Agasian, B. Kerbikov, V. Shevchenko: Phys. Rep. **320**, 131 (1999)
88. D. Ebert, K. Klimenko, H. Toki: hep-ph/0011273

Theory of the Quark-Gluon Plasma

Jean-Paul Blaizot

Service de Physique Théorique, CEA Saclay
F-91191 Gif-sur-Yvette Cedex, France

1 Introduction

In spite of what the title might suggest, I shall not try to cover in this article all interesting aspects of the theory of the quark-gluon plasma. Rather, I shall focus on progress made in recent years in understanding the high temperature phase of QCD by using weak coupling techniques. Such techniques go far beyond strict perturbation theory viewed as an expansion in powers of the gauge coupling. In fact such an expansion becomes meaningless as soon as the coupling is not vanishingly small. However, we shall see that a rather simple structure emerges from weak coupling studies, with a characteristic hierarchy of scales and degrees of freedom. The interactions renormalize the properties of these elementary degrees of freedom, but do not destroy the simple picture of the high temperature quark-gluon plasma as a system of weakly interacting quasiparticles. As we shall see at the end of this article, this picture is supported by a first principle calculation of the entropy which reproduces accurately lattice data above 2 or 3 times the critical temperature.

Some of the material presented here is borrowed from the recent review [1], and complements can also be found in [2–6]. Another perspective on some of the topics discussed here can be found in the article by A. Rebhan in this volume.

The outline of the article is the following. In order to get a first rough picture of the phase diagram of hadronic matter I use the bag model to describe the quark-hadron phase transition: this exercise will give us some familiarity with the thermodynamics of massless, non-interacting, particles. Then I briefly recall some techniques of quantum field theory at finite temperature needed to treat the interactions [7–12], and introduce the concept of effective theory in a simple case of a scalar field. Then I proceed to an analysis of the various important scales and degrees of freedom of the quark-gluon plasma and focus on the effective theory for the collective modes which develop at the particular momentum scale gT, where g is the gauge coupling and T the temperature. A powerful technique to construct the effective theory is based on kinetic equations which govern the dynamics of the hard degrees of freedom. Some of the collective phenomena that are described by this effective theory are briefly mentioned. Then I turn to the calculation of the entropy and show how the information coded in the effective theory can be exploited in (approximately) self-consistent calculations [13–15].

2 The Quark-Hadron Transition in the Bag Model

The phase diagram of dense hadronic matter has the expected shape indicated in Fig. 1. There is a low density, low temperature region, corresponding to the world of ordinary hadrons, and a high density, high temperature region, where the dominant degrees of freedom are quarks and gluons. The precise determination of the transition line requires elaborate non-perturbative techniques, such as those of lattice gauge theories (see the article by F. Karsch in this volume). However, but one can get rough orders of magnitude for the transition temperature and density using a simple model dealing mostly with non-interacting particles [3,5].

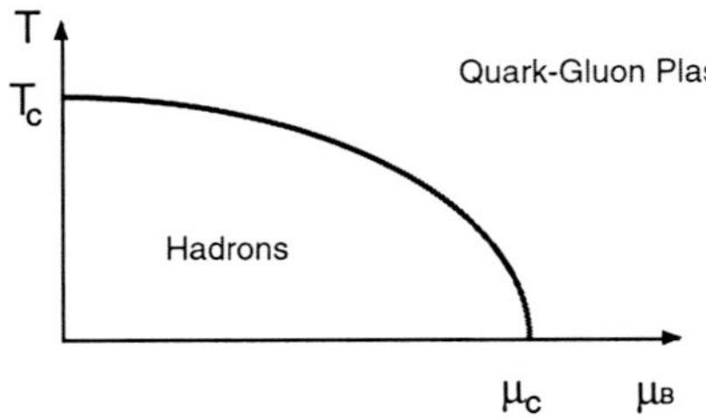

Fig. 1. The expected phase diagram of hot and dense hadronic matter in the plane (μ_B, T), where T is the temperature and μ_B the baryon chemical potential

Let us first consider the transition in the case where $\mu_B = 0$. At low temperature this baryon free matter is composed of the lightest mesons, i.e. mostly the pions. At sufficiently high temperature one should also take into account heavier mesons, but in the present discussion this is an inessential complication. We shall even make a further approximation by treating the pion as a massless particle. At very high temperature, we shall consider that hadronic matter is composed only of quarks and antiquarks (in equal numbers), and gluons, forming a quark-gluon plasma. In both the high temperature and the low temperature phases, interactions are neglected (except for the bag constant to be introduced below). The description of the transition will therefore be dominated by entropy considerations, i.e. by counting the degrees of freedom.

The energy density ε and the pressure P of a gas of massless pions are given by:

$$\varepsilon = 3 \cdot \frac{\pi^2}{30} T^4 , \qquad P = 3 \cdot \frac{\pi^2}{90} T^4 , \tag{1}$$

where the factors 3 account for the 3 types of pions (π^+, π^-, π^0).

The energy density and pressure of the quark-gluon plasma are given by similar formulae:

$$\varepsilon = 37 \cdot \frac{\pi^2}{30} T^4 + B,$$

$$P = 37 \cdot \frac{\pi^2}{90} T^4 - B, \tag{2}$$

where $37 = 2\times8+\frac{7}{8}\times2\times2\times2\times3$ is the effective number of degrees of freedom of gluons (8 colors, 2 spin states) and quarks (3 colors, 2 spins, 2 flavors, q and $\bar{q}$). The quantity B, which is added to the energy density, and subtracted from the pressure, summarizes interaction effects which are responsible for a change in the vacuum structure between the low temperature and the high temperature phases. It was introduced first in the "bag model" of hadron structure as a restoring force needed to equilibrate the pressure generated by the kinetic energy of the quarks inside the bag [16]. Roughly, the energy of the bag is

$$E(R) = \frac{4\pi}{3} R^3 B + \frac{C}{R}, \tag{3}$$

where C/R is the kinetic energy of massless quarks. Minimizing with respect to R, one finds that the energy at equilibrium is $E\,(R_0) = 4BV_0$, where $V_0 = 4\pi R_0^3/3$ is the equilibrium volume. For a proton with $E_0 \approx 1$ GeV and $R_0 \approx 0.7$ fm, one finds $E_0/V_0 \simeq 0.7$ GeV/fm^3, which corresponds to a "bag constant" $B \approx 175$ MeV/fm^3, or $B^{1/4} \approx 192$ MeV.

We can now compare the two phases as a function of the temperature. Fig. 2 shows how P varies as a function of T^4. One sees that there exists a transition temperature

$$T_c = \left(\frac{45}{17\pi^2}\right)^{1/4} B^{1/4} \approx 0.72\, B^{1/4}, \tag{4}$$

beyond which the quark-gluon plasma is thermodynamically favored (has largest pressure) compared to the pion gas. For $B^{1/4} \approx 200$ MeV, $T_c \approx 150$ MeV.

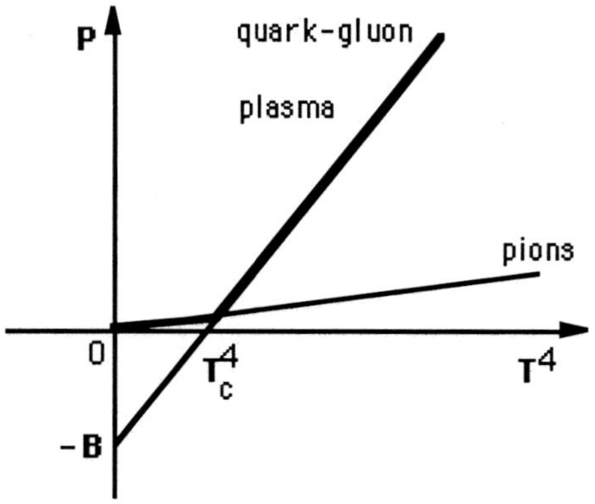

Fig. 2. The pressure of the massless pion gas compared to that of a quark-gluon plasma, showing the transition temperature T_c

The variation of the entropy density $s = \partial P/\partial T$ as a function of the temperature is displayed in Fig. 3. Note that the bag constant B does not enter explicitly the expression of the entropy. However, B is involved in Fig. 3 indirectly, via the temperature T_c where the discontinuity Δs occurs. One verifies easily that the jump in entropy density $\Delta s = \Delta\varepsilon/T_c$ is directly proportional to the change in the number of active degrees of freedom when T crosses T_c.

In order to extend these considerations to the case where $\mu_B \neq 0$, we note that the transition is taking place when the total pressure approximately vanishes, that is when the kinetic pressure of quarks and gluons approximately equilibrates the bag pressure. Taking as a criterion for the phase transition the condition $P = 0$, one replaces the value (4) for T_c by the value $(90/37\pi^2)^{1/4}B^{1/4} \approx 0.70B^{1/4}$, which is nearly identical to (4). We shall then assume that for any value of μ_B and T, the phase transition occurs when $P(\mu_B, T) = B$, where B is the bag constant and $P(\mu_B, T)$ is the kinetic pressure of quarks and gluons:

$$P(\mu_B, T) = \frac{37}{90}\pi^2 T^4 + \frac{\mu_B^2}{9}(T^2 + \frac{\mu_B^2}{9\pi^2}). \tag{5}$$

The transition line is then given by $P(\mu_c, T_c) = B$, and it has indeed the shape illustrated in Fig. 1.

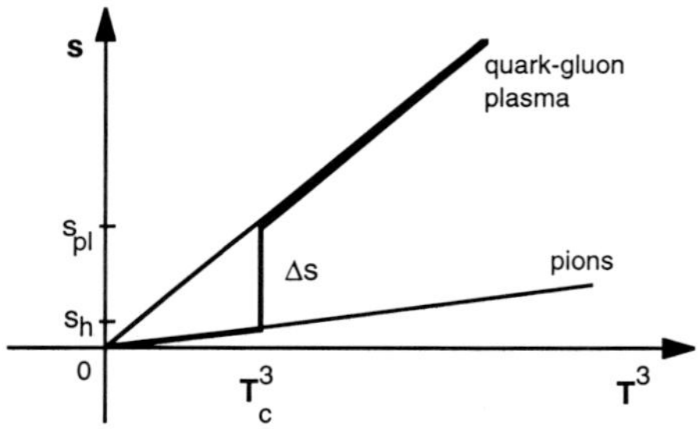

Fig. 3. The entropy density. The jump Δs at the transition is proportional to the increase in the number of active degrees of freedom

The model that we have just described reproduces some of the bulk features of the equation of state obtained through lattice gauge calculations (see the article by F. Karsch in this volume). In particular, it exhibits the characteristic increase of the entropy density at the transition which corresponds to the emergence of a large number of new degrees of freedom associated with quarks and gluons. Its simplicity has made it popular, for instance, among the practitioners of hydrodynamic calculations with which one tries to simulate the behavior of matter produced in high energy nuclear collisions. As such it has been very useful. One should be cautious, however, when attempting to draw too detailed conclusions about the nature of the phase transitions from such simple models. In particular this model predicts (by construction!)

a discontinuous transition; but this prediction should not be trusted. Further discussion of this model can be found in [3].

3 Quantum Fields at Finite Temperature

The effects of interactions among quarks and gluons at finite temperature can be calculated by using the tools of quantum field theory at finite temperature. We shall briefly recall some essential formalism, and emphasize in particular the periodicity properties of the propagators. At the end of this section we discuss, with a simple example of a scalar field, the method of effective field theory which proves useful in problems where various scales can be separated. In the example that we shall consider, the separation of scale is provided by the Matsubara frequencies. As we shall see, in some cases, one is lead to single out the mode with vanishing Matsubara frequency. The corresponding effective theory is a classical field in three dimensions, and the procedure commonly called 'dimensional reduction'.

3.1 Finite Temperature Calculations

All thermodynamic observables can be deduced from the partition function:

$$\mathcal{Z} = \mathrm{tr}\, e^{-\beta H}. \tag{6}$$

Thus the energy density and the pressure are given by:

$$\epsilon = -\frac{1}{V}\frac{\partial}{\partial\beta}\ln Z, \qquad\qquad P = \frac{1}{\beta}\frac{\partial}{\partial V}\ln Z. \tag{7}$$

In order to calculate the partition function, one may observe that $e^{-\beta H}$ is like an evolution operator in imaginary time:

$$t \to -i\beta, \qquad\qquad e^{-iHt} \to e^{-\beta H}. \tag{8}$$

One may then take advantage of all the techniques developed to evaluate matrix elements of the evolution operator in quantum mechanics or field theory.

For instance, one may use a perturbative expansion. We assume that one can split the Hamiltonian into $H = H_0 + H_1$ with $H_1 \ll H_0$, and define the following "interaction representation" of the perturbation H_1:

$$H_1(\tau) = e^{\tau H_0} H_1 e^{-\tau H_0}, \tag{9}$$

and similarly for other operators. Using standard techniques, one can then obtain the following expression for the partition function $\mathcal{Z}$:

$$\mathcal{Z} = \mathcal{Z}_0 \,\langle \mathrm{T}\exp\left\{ -\int_0^\beta d\tau H_1(\tau) \right\}\rangle_0. \tag{10}$$

In this equation, the symbol T implies an ordering of the operators on its right, from left to right in decreasing order of their time arguments; $\mathcal{Z}_0 = \mathrm{tr}\, e^{-\beta H_0}$ and, for any operator $\mathcal{O}$,

$$\langle \mathcal{O} \rangle_0 \equiv \mathrm{Tr}\left(\frac{e^{-\beta H_0}}{\mathcal{Z}_0}\mathcal{O}\right). \tag{11}$$

One commonly refers to τ as the "imaginary time" (τ is real). This τ has no direct physical interpretation: its role here is to properly keep track of ordering of operators in the perturbative expansion.

In field theory, it is often more convenient to use the formalism of path integrals. Let us recall, for instance, that for one particle in one dimension the matrix element of the evolution operator can be written as

$$\langle q_2 | e^{-iHt} | q_1 \rangle = \int_{q(0)=q_1}^{q(t)=q_2} \mathcal{D}\left(q\left(t\right)\right) \quad e^{i\int_{t_1}^{t_2}\left(\frac{1}{2}m\dot{q}^2 - V(q)\right)dt}, \tag{12}$$

where q_1 and q_2 denote the positions of the particle at times 0 and t respectively. Changing $t \to -i\tau$, and taking the trace, one obtains the following formula for the partition function:

$$\mathcal{Z} = \mathrm{tr}\, e^{-\beta H} = \int_{q(\beta)=q(0)} \mathcal{D}(q) \exp\left\{-\int_0^\beta \left(\frac{1}{2}m\dot{q}^2 + V(q)\right)\right\}. \tag{13}$$

This expression immediately generalizes to the case of a scalar field, for which the Lagrangian is of the form:

$$\begin{aligned}\mathcal{L} &= \frac{1}{2}\partial_\mu\phi\partial^\mu\phi - \frac{m^2}{2}\phi^2 - V(\phi)\\ &= \frac{1}{2}(\partial_0\phi)^2 - \frac{1}{2}(\nabla\phi)^2 - \frac{m^2}{2}\phi^2 - V(\phi).\end{aligned} \tag{14}$$

Again, we replace t by $-i\tau$, $\partial_0 = \partial_t$ by $i\partial_\tau$, so that $(\partial_0\phi)^2 \to -(\partial_\tau\phi)^2$. The partition function becomes then (integrations over spatial coordinates are implicit):

$$\mathcal{Z} = \int \mathcal{D}(\phi) \exp\left\{-\int_0^\beta d\tau \left(\frac{1}{2}(\partial_\tau\phi)^2 + \frac{1}{2}(\nabla\phi)^2 + \frac{m^2}{2}\phi^2 + V(\phi)\right)\right\}, \tag{15}$$

where the integral is over periodic fields: $\phi(0) = \phi(\beta)$.

Remarks. i) The partition function (15) may be viewed formally as a sum over classical field configurations in four dimensions, with particular boundary conditions in the (imaginary) time direction.

ii) At high temperature, $\beta \to 0$, the time dependence of the fields play no role. The partition function becomes that of a classical field theory in three dimensions:

$$\mathcal{Z} = \int \mathcal{D}(\phi) \exp\left\{-\beta \int \mathrm{d}^3 r \left(\frac{1}{2}(\nabla\phi)^2 + \frac{m^2}{2}\phi^2 + V(\phi)\right)\right\}. \qquad (16)$$

Ignoring the time dependence of the fields amounts to take into account only the Matsubara frequency $i\omega_\nu = 0$. We shall discuss later explicit examples of this "dimensional reduction".
iii) Note the Euclidean metric in (15). Since the integrand is the exponential of a negative definite quantity, it is well suited to numerical evaluations, using, for instance, the lattice technique.

3.2 Free Propagators

An important feature of the path integral representation of the partition function is the boundary conditions to be imposed on the fields over which one integrates. For the scalar case considered here, the field has to be periodic in imaginary time, with a period β. Similar conditions hold for the fermion fields, which are antiperiodic in imaginary time, with the same period β. It is instructive to see how these periodicity conditions emerge in the operator formalism, and for this reason we consider now the free propagators, first in the simple case of the non-relativistic many body problem. The generalization to relativistic fields is straightforward.

Let us consider a system with unperturbed Hamiltonian:

$$H_0 = \sum_k \epsilon_k \, a_k^\dagger a_k, \qquad (17)$$

where k denotes the set of quantum numbers necessary to specify a single particle state, for instance the three components of the momentum. We define time-dependent creation and annihilation operators in the interaction picture:

$$\begin{aligned} a_k^\dagger(\tau) &\equiv e^{\tau H_0} a_k^\dagger e^{-\tau H_0} = e^{\epsilon_k \tau} a_k^\dagger \\ a_k(\tau) &\equiv e^{\tau H_0} a_k e^{-\tau H_0} = e^{-\epsilon_k \tau} a_k . \end{aligned} \qquad (18)$$

The last equalities follow (for example) from the commutation relations:

$$[H_0, a_k^\dagger] = \epsilon_k a_k^\dagger, \qquad [H_0, a_k] = -\epsilon_k a_k \qquad (19)$$

which hold for bosons and fermions. The single-particle propagator can then be obtained by a direct calculation:

$$\begin{aligned} G_k(\tau_1 - \tau_2) &= \langle \mathrm{T} a_k(\tau_1) a_k^\dagger(\tau_2)\rangle_0 \\ &= e^{-\epsilon_k(\tau_1-\tau_2)} \left[\theta(\tau_1 - \tau_2)(1 \pm n_k) \pm n_k \theta(\tau_2 - \tau_1)\right], \end{aligned} \qquad (20)$$

where

$$n_k \equiv \langle a_k^\dagger a_k \rangle_0 = \frac{1}{e^{\beta\epsilon_k} \mp 1}, \qquad (21)$$

and the upper (lower) sign is for bosons (fermions). One can verify on the expression (20) that, in the interval $-\beta < \tau = \tau_1 - \tau_2 < \beta$, $G_k(\tau)$ is a periodic (boson) or antiperiodic (fermion) function of τ:

$$G_k(\tau - \beta) = \pm G_k(\tau) \qquad (0 \leq \tau \leq \beta). \tag{22}$$

(To show this relation note that $\mathrm{e}^{\beta\epsilon_k} n_k = 1 \pm n_k$.) It can therefore be represented by a Fourier series

$$G_k(\tau) = \frac{1}{\beta}\sum_\nu e^{-i\omega_\nu\tau} G_k(i\omega_\nu), \tag{23}$$

where the ω_ν's are called the Matsubara frequencies:

$$\begin{array}{lll} \omega_\nu = 2\nu\pi/\beta & \text{bosons,} \\ \omega_\nu = (2\nu+1)\pi/\beta & \text{fermions.} \end{array} \tag{24}$$

The inverse transform is given by

$$G(i\omega_\nu) = \int_0^\beta \mathrm{d}\tau \, e^{i\omega_\nu\tau} G(\tau) = \frac{1}{H_0 - i\omega_\nu}. \tag{25}$$

Using the property

$$\delta(\tau) = \frac{1}{\beta}\sum_\nu e^{-i\omega_\nu\tau}, \qquad -\beta < \tau < \beta \tag{26}$$

and (23), it is easily seen that $G(\tau)$ satisfies the differential equation

$$(\partial_\tau + H_0)\, G(\tau) = \delta(\tau), \tag{27}$$

which may be also verified directly from (20). Alternatively, the single propagator at finite temperature may be obtained as the solution of this equation with periodic (bosons) or antiperiodic (fermions) boundary conditions.

Remark. The periodicity or antiperiodicity that we have uncovered on the explicit form of the unperturbed propagator is, in fact, a general property of the propagators of a many-body system in thermal equilibrium. It is a consequence of the commutation relations of the creation and annihilation operators and the cyclic invariance of the trace.

The propagator of the free scalar field $\Delta(\tau) = \langle \mathrm{T}\phi(\tau_1)\phi(\tau_2)\rangle$, where $\tau \equiv \tau_1 - \tau_2$, satisfies the differential equation

$$\left[-\partial_{\tau_1}^2 - \nabla_1^2 + m^2\right] \Delta(\tau_1\mathbf{r_1};\tau_2\mathbf{r_2}) = \delta(\tau_1 - \tau_2)\delta(\mathbf{r_1} - \mathbf{r_2}), \tag{28}$$

and obeys periodic boundary conditions. It admits the Fourier representation

$$\Delta(\tau) = \frac{1}{\beta}\sum_n e^{-i\omega_n\tau}\Delta(i\omega_n), \tag{29}$$

where $\omega_n = 2\pi n/\beta$ and

$$\Delta(i\omega_n) = \frac{1}{\epsilon_k^2 - \omega_n^2}. \tag{30}$$

By inverting the Fourier transform (30), one gets

$$\Delta(\tau) = \frac{1}{2\epsilon_k}\left\{(1+N_k)e^{-\epsilon_k|\tau|} + N_k e^{\epsilon_k|\tau|}\right\}, \tag{31}$$

with $N_k = 1/(\mathrm{e}^{\beta\epsilon_k} - 1)$.

3.3 Classical Field Approximation and Dimensional Reduction

In the high temperature limit, $\beta \to 0$, the imaginary-time dependence of the fields frequently becomes unimportant and can be ignored in a first approximation. The integration over imaginary time becomes then trivial and the partition function (15) reduces to:

$$Z \approx \mathcal{N} \int \mathcal{D}(\phi) \exp\left\{-\beta \int \mathrm{d}^3x\, \mathcal{H}(\mathbf{x})\right\}, \tag{32}$$

where $\phi \equiv \phi(\mathbf{x})$ is now a three-dimensional field, and

$$\mathcal{H} = \frac{1}{2}(\boldsymbol{\nabla}\phi)^2 + \frac{m^2}{2}\phi^2 + V(\phi). \tag{33}$$

The functional integral in (32) is recognized as the partition function for static three-dimensional field configurations with energy $\int \mathrm{d}^3x\, \mathcal{H}(x)$. We shall refer to this limit as the *classical field approximation.*

Ignoring the time dependence of the fields is equivalent to retaining only the zero Matsubara frequency in their Fourier decomposition. Then the Fourier transform of the free propagator is simply:

$$G_0(\mathbf{k}) = \frac{T}{\varepsilon_k^2}. \tag{34}$$

This may be obtained directly from (29) keeping only the term with $\omega_\nu = 0$, or from Eq. (31) by ignoring the time dependence and using the approximation

$$N(\varepsilon_k) = \frac{1}{\mathrm{e}^{\beta\varepsilon_k} - 1} \approx \frac{T}{\varepsilon_k}. \tag{35}$$

Both approximations make sense only for $\varepsilon_k \ll T$, implying $N(\varepsilon_k) \gg 1$. In this limit, the energy per mode is $\propto \varepsilon_k N(\varepsilon_k) \approx T$, as expected from the classical equipartition theorem.

The classical field approximation may be viewed as the leading term in a systematic expansion. To see that, let us expand the field variables in the path integral (15) in terms of their Fourier components:

$$\phi(\tau) = \frac{1}{\beta}\sum_\nu \mathrm{e}^{-i\omega_\nu\tau}\phi(i\omega_\nu), \tag{36}$$

where the ω_ν's are the Matsubara frequencies. The path integral (15) can then be written as:

$$Z = \mathcal{N}_1 \int \mathcal{D}(\phi_0) \exp\{-S[\phi_0]\} , \tag{37}$$

where $\phi_0 \equiv \phi(\omega_\nu = 0)$ depends only on spatial coordinates, and

$$\exp\{-S[\phi_0]\} = \mathcal{N}_2 \int \mathcal{D}(\phi_{\nu\neq 0}) \exp\left\{ -\int_0^\beta \mathrm{d}\tau \int \mathrm{d}^3x \, \mathcal{L}_E(x) \right\} . \tag{38}$$

The quantity $S[\phi_0]$ may be called the effective action for the "zero mode" ϕ_0. Aside from the direct classical field contribution that we have already considered, this effective action receives also contributions from integrating out the non-vanishing Matsubara frequencies. Diagrammatically, $S[\phi_0]$ is the sum of all the connected diagrams with external lines associated to ϕ_0, and in which the internal lines are the propagators of the non-static modes $\phi_{\nu\neq 0}$. Thus, a priori, $S[\phi_0]$ contains operators of arbitrarily high order in ϕ_0, which are also non-local. In practice, however, one wishes to expand $S[\phi_0]$ in terms of *local* operators, i.e., operators with the schematic structure $a_{m,n}\nabla^m\phi_0^n$ with coefficients $a_{m,n}$ to be computed in perturbation theory.

To implement this strategy, it is useful to introduce an intermediate scale Λ ($\Lambda \ll T$) which separates *hard* ($k \gtrsim \Lambda$) and *soft* ($k \lesssim \Lambda$) momenta. All the non-static modes, as well as the static ones with $k \gtrsim \Lambda$ are *hard* (since $K^2 \equiv \omega_\nu^2 + k^2 \gtrsim \Lambda^2$ for these modes), while the static ($\omega_\nu = 0$) modes with $k \lesssim \Lambda$ are *soft*. Thus, strictly speaking, in the construction of the effective theory along the lines indicated above, one has to integrate out also the static modes with $k \gtrsim \Lambda$. The benefits of this separation of scales are that (a) the resulting effective action for the soft fields can be made *local* (since the initially non-local amplitudes can be expanded out in powers of p/K, where $p \ll \Lambda$ is a typical external momentum, and $K \gtrsim \Lambda$ is a hard momentum on an internal line), and (b) the effective theory is now used exclusively at soft momenta, where classical approximations such as (35) are expected to be valid. This strategy, which consists in integrating out the non-static modes in perturbation theory in order to obtain an effective three-dimensional theory for the soft static modes (with $\omega_\nu = 0$ and $k \equiv |\mathbf{k}| \lesssim \Lambda$), is generally referred to as "dimensional reduction" [17–22].

As an illustration let us consider a massless scalar theory with quartic interactions; that is, $m = 0$ and $V(\phi) = (g^2/4!)\phi^4$ in (14). The ensuing effective action for the soft fields (which we shall still denote as ϕ_0) reads

$$S[\phi_0] = \beta\mathcal{F}(\Lambda) + \int \mathrm{d}^3x \left\{ \frac{1}{2}(\boldsymbol{\nabla}\phi_0)^2 + \frac{1}{2} M^2(\Lambda)\phi_0^2 + \frac{g_3^2(\Lambda)}{4!}\,\phi_0^4 + \frac{h(\Lambda)}{6!}\,\phi_0^6 + \Delta\mathcal{L} \right\} , \tag{39}$$

where $\mathcal{F}(\Lambda)$ is the contribution of the hard modes to the free-energy, and $\Delta\mathcal{L}$ contains all the other local operators which are invariant under rotations and under the symmetry $\phi \to -\phi$, i.e., all the local operators which are consistent with the symmetries of the original Lagrangian. We have changed the normalization of the field ($\phi_0 \to \sqrt{T}\phi_0$) with respect to (32)–(33), so as to absorb the factor β in front of the effective action. The effective "coupling constants" in (39), i.e. $M^2(\Lambda)$, $g_3^2(\Lambda)$, $h(\Lambda)$ and the infinitely many parameters in $\Delta\mathcal{L}$, are computed in perturbation theory, and depend upon the separation scale Λ, the temperature T and the original coupling g^2. To lowest order in g, $g_3^2 \approx g^2 T$, $h \approx 0$ (the first contribution to h arises at order g^6, via one-loop diagrams), and $M \sim gT$, as we shall see shortly. Note that eq. (39) involves in general non-renormalizable operators, via $\Delta\mathcal{L}$. This is not a difficulty, however, since this is only an effective theory, in which the scale Λ acts as an explicit ultraviolet (UV) cutoff for the loop integrals. Since, however, the scale Λ is arbitrary, the dependence on Λ coming from such soft loops must cancel against the dependence on Λ of the parameters in the effective action.

Fig. 4. One-loop tadpole diagram for the self-energy of the scalar field

Let us verify this cancellation explicitly in the case of the thermal mass M of the scalar field, and to lowest order in perturbation theory. To this order, the scalar self-energy is given by the tadpole diagram in Fig. 4. The mass parameter $M^2(\Lambda)$ in the effective action is obtained by integrating over hard momenta within the loop in Fig. 4:

$$\begin{aligned} M^2(\Lambda) &= \frac{g^2}{2} T \sum_\nu \int \frac{d^3k}{(2\pi)^3} \frac{(1-\delta_{\nu 0}) + \theta(k-\Lambda)\delta_{\nu 0}}{\omega_\nu^2 + k^2} \\ &= \frac{g^2}{2} \int \frac{d^3k}{(2\pi)^3} \left\{ \frac{N(k)}{k} + \frac{1}{2k} - \theta(\Lambda - k)\frac{T}{k^2} \right\}, \end{aligned} \tag{40}$$

where the θ-function in the second line has been generated by writing $\theta(k - \Lambda) = 1 - \theta(\Lambda - k)$. The first term, involving the thermal distribution, gives the contribution

$$\hat{M}^2 \equiv \frac{g^2}{2} \int \frac{d^3k}{(2\pi)^3} \frac{N(k)}{k} = \frac{g^2}{24} T^2 . \tag{41}$$

As it will turn out, this is the leading-order (LO) scalar thermal mass, and also the simplest example of what will be called "hard thermal loops" (HTL).

The second term, involving $1/2k$, in (40) is quadratically UV divergent, but independent of the temperature; the standard renormalization procedure at $T=0$ amounts to simply removing this term. The third term, involving the θ-function, is easily evaluated. One finally gets:

$$M^2(\Lambda) = \hat{M}^2 - \frac{g^2}{4\pi^2}\Lambda T \equiv \frac{g^2T^2}{24}\left(1 - \frac{6}{\pi^2}\frac{\Lambda}{T}\right). \tag{42}$$

The Λ-dependent term above is subleading, by a factor $\Lambda/T \ll 1$.

The one-loop correction to the thermal mass within the effective theory is given by the same diagram as in Fig. 4, but where the internal field is static and soft, with the massive propagator $1/(k^2 + M^2(\Lambda))$, and coupling constant $g_3^2 \approx g^2T$. Since the typical momenta in the integral will be $k \gtrsim M$, and $M \sim \hat{M} \sim gT$, we choose $\Lambda \gg gT$. We then obtain

$$\begin{aligned}\delta M^2(\Lambda) &= \frac{g^2}{2}\int \frac{\mathrm{d}^3k}{(2\pi)^3}\,\Theta(\Lambda - k)\,\frac{T}{k^2 + M^2(\Lambda)} \\ &= \frac{g^2T\Lambda}{4\pi^2}\left(1 - \frac{\pi M}{2\Lambda}\arctan\frac{\Lambda}{M}\right) \simeq \frac{g^2T\Lambda}{4\pi^2} - \frac{g^2}{8\pi}\hat{M}T,\end{aligned} \tag{43}$$

where the terms neglected in the last step are of higher order in $\hat{M}/\Lambda$ or Λ/T.

As anticipated, the Λ-dependent terms cancel in the sum $M^2 \equiv M^2(\Lambda) + \delta M^2(\Lambda)$, which then provides the physical thermal mass within the present accuracy:

$$M^2 = M^2(\Lambda) + \delta M^2(\Lambda) = \frac{g^2T^2}{24} - \frac{g^2}{8\pi}\hat{M}T. \tag{44}$$

The LO term, of order g^2T^2, is the HTL $\hat{M}$. The next-to-leading order (NLO) term, which involves the resummation of the thermal mass $M(\Lambda)$ in the soft propagator, is of order $g^2\hat{M}T \sim g^3T^2$, and therefore non-analytic in g^2. This non-analyticity is related to the fact that the integrand in (43) cannot be expanded in powers of M^2/k^2 without generating infrared divergences.

4 Effective Theories for the Quark-Gluon Plasma

We return now to the quark-gluon plasma and analyze the various scales and degrees of freedom which are relevant in the weak coupling regime. We show that there is a hierarchy of scales controlled by powers of the gauge coupling g. We focus in this article on two particular momentum scales, the 'hard' one which is that of the plasma particles with momenta $k \sim T$, and the 'soft' one with $k \sim gT$ at which collective phenomena develop. We shall be in particular interested in the effective theory obtained when the hard degrees of freedom are 'integrated out'. The resulting effective theory describes long wave length, low frequency collective phenomena; that is, it accounts for time dependent fields, in contrast to the example discussed in the previous section which concerned only static fields. As we shall see later, getting a complete description of the dynamics of the collective excitations turns out to be important also for the calculation of the equilibrium properties of the quark-gluon plasma.

4.1 Scales and Degrees of Freedom in Ultrarelativistic Plasmas

A property of QCD which is essential in the present discussion is that of asymptotic freedom, according to which the coupling constant depends on the scale $\bar{\mu}$ as

$$\alpha_s(\bar{\mu}) \equiv \frac{g^2}{4\pi} \propto \frac{1}{\ln(\bar{\mu}/\Lambda_{QCD})}. \tag{45}$$

At high temperature, the natural scale is $\bar{\mu} = 2\pi T$, so that the coupling becomes weak when $2\pi T \gg \Lambda_{QCD}$. At extremely high temperature the interactions become negligible and hadronic matter turns into an ideal gas of quarks and gluons: this is the quark-gluon plasma. As we shall see an important effect of the interactions is to turn free quarks and gluons into weakly interacting quasiparticles.

In the absence of interactions, the plasma particles are distributed in momentum space according to the Bose-Einstein or Fermi-Dirac distributions:

$$N_k = \frac{1}{\mathrm{e}^{\beta\varepsilon_k} - 1}, \qquad n_k = \frac{1}{\mathrm{e}^{\beta\varepsilon_k} + 1}, \tag{46}$$

where $\varepsilon_k = k \equiv |\mathbf{k}|$ (massless particles), $\beta \equiv 1/T$, and chemical potentials are assumed to vanish. In such a system, the particle density n is determined by the temperature: $n \propto T^3$. Accordingly, the mean interparticle distance $n^{-1/3} \sim 1/T$ is of the same order as the thermal wave length $\lambda_T = 1/k$ of a typical particle in the thermal bath for which $k \sim T$. Thus the particles of an ultrarelativistic plasma are quantum degrees of freedom for which in particular the Pauli principle can never be ignored.

In the weak coupling regime ($g \ll 1$), the interactions do not alter significantly the picture. The hard degrees of freedom, i.e. the plasma particles, remain the dominant degrees of freedom and since the coupling to gauge fields occurs typically through covariant derivatives, $D_x = \partial_x + igA(x)$, the effect of interactions on particle motion is a small perturbation unless the fields are very large, i.e., unless $A \sim T/g$, where g is the gauge coupling: only then do we have $\partial_X \sim T \sim gA$, where ∂_X is a space-time gradient. We should note here that we rely on considerations, based on the magnitude of the gauge fields, which depend on the choice of a gauge. What is meant is that there exists a large class of gauge choices for which they are valid. And we shall verify a posteriori that within such a class, the final results are gauge invariant.

Considering now more generally the effects of the interactions, we note that these depend both on the strength of the gauge fields and on the wave length of the modes under study. A measure of the strength of the gauge fields in typical situations is obtained from the magnitude of their thermal fluctuations, that is $\bar{A} \equiv \sqrt{\langle A^2(t,\mathbf{x})\rangle}$. In equilibrium $\langle A^2(t,\mathbf{x})\rangle$ is independent of t and $\mathbf{x}$ and given by $\langle A^2\rangle = G(t=0,\mathbf{x}=\mathbf{0})$ where $G(t,\mathbf{x})$ is the gauge field propagator. In the non-interacting case we have (with $\varepsilon_k = k$):

$$\langle A^2\rangle = \int \frac{\mathrm{d}^3k}{(2\pi)^3}\frac{1}{2\varepsilon_k}(1+2N_k). \tag{47}$$

Here we shall use this formula also in the interacting case, assuming that the effects of the interactions can be accounted for simply by a change of ε_k. We shall also ignore the (divergent) contribution of the vacuum fluctuations (the term independent of the temperature in (47)).

For the plasma particles $\varepsilon_k = k \sim T$ and $\langle A^2\rangle_T \sim T^2$. The associated electric (or magnetic) field fluctuations are $\langle E^2\rangle_T \sim \langle(\partial A)^2\rangle_T \sim k^2\langle A^2\rangle_T \sim T^4$ and are a dominant contribution to the plasma energy density. As already mentioned, these short wave length, or *hard*, gauge field fluctuations produce a small perturbation on the motion of a plasma particle. However, this is not so for an excitation at the momentum scale $k \sim gT$, since then the two terms in the covariant derivative ∂_X and $g\bar{A}_T$ become comparable. That is, the properties of an excitation with momentum gT are expected to be non-perturbatively renormalized by the hard thermal fluctuations. And indeed, the scale gT is that at which collective phenomena develop. The emergence of the Debye screening mass $m_D \sim gT$ is one of the simplest examples of such phenomena.

Let us now consider the fluctuations at this scale $gT \ll T$, to be referred to as the *soft* scale. These fluctuations can be accurately described by classical fields. In fact the associated occupation numbers N_k are large, and accordingly one can replace N_k by T/ε_k in (47). Introducing an upper cut-off gT in the momentum integral, one then gets:

$$\langle A^2\rangle_{gT} \sim \int^{gT} \mathrm{d}^3k\,\frac{T}{k^2} \sim gT^2. \tag{48}$$

Thus $\bar{A}_{gT} \sim \sqrt{g}T$ so that $g\bar{A}_{gT} \sim g^{3/2}T$ is still of higher order than the kinetic term $\partial_X \sim gT$. In that sense the soft modes with $k \sim gT$ are still perturbative, i.e. their self-interactions can be ignored in a first approximation. Note, however, that they generate contributions to physical observables which are not analytic in g^2, as shown by the example of the order g^3 contribution to the energy density of the plasma:

$$\epsilon^{(3)} \sim \int_0^{\omega_{pl}} \mathrm{d}^3k\,\omega_{pl}\,\frac{1}{\mathrm{e}^{\omega_{pl}/T}-1} \sim \omega_{pl}^3\,\omega_{pl}\,\frac{T}{\omega_{pl}} \sim g^3T^4, \tag{49}$$

where $\omega_{pl} \sim gT$ is the typical frequency of a collective mode.

Moving down to a lower momentum scale, one meets the contribution of the unscreened magnetic fluctuations which play a dominant role for $k \sim g^2T$. At that scale, to be referred to as the *ultrasoft* scale, it becomes necessary to distinguish the electric and the magnetic sectors (which provide comparable contributions at the scale gT). The electric fluctuations are damped by the Debye screening mass ($\varepsilon_k^2 = k^2 + m_D^2 \approx m_D^2$ when $k \sim g^2T$) and their contribution is negligible, of order g^4T^2. However, because of the absence of static screening in the magnetic sector, we have here $\varepsilon_k \sim k$ and

$$\langle A^2 \rangle_{g^2T} \sim T \int_0^{g^2T} \mathrm{d}^3k \frac{1}{k^2} \sim g^2T^2, \tag{50}$$

so that $g\bar{A}_{g^2T} \sim g^2T$ is now of the same order as the ultrasoft derivative $\partial_X \sim g^2T$: the fluctuations are no longer perturbative. This is the origin of the breakdown of perturbation theory in high temperature QCD.

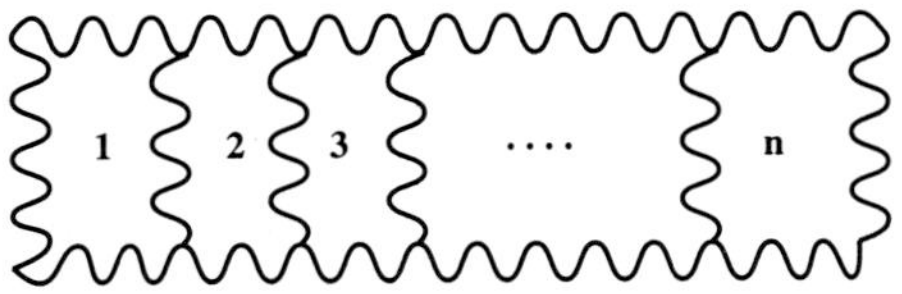

Fig. 5. Example of a multiloop diagram which is infrared divergent

To appreciate the difficulty from another perspective, let us first observe that the dominant contribution to the fluctuations at scale g^2T comes from the zero Matsubara frequency:

$$\langle A^2 \rangle_{g^2T} = T \sum_n \int_0^{g^2T} \mathrm{d}^3k \, \frac{1}{\omega_n^2 + k^2} \sim T \int_0^{g^2T} \mathrm{d}^3k \, \frac{1}{k^2}. \tag{51}$$

Thus the fluctuations that we are discussing are those of a three dimensional theory of static fields. Following Linde [23,24] consider then the higher order corrections to the pressure in hot Yang-Mills theory. Because of the strong static fluctuations most of the diagrams of perturbation theory are infrared (IR) divergent. By power counting, the strongest IR divergences arise from ladder diagrams, like the one depicted in Fig. 5, in which all the propagators are static, and the loop integrations are three-dimensional. Such n-loop diagrams can be estimated as (μ is an IR cutoff):

$$g^{2(n-1)} \left(T \int \mathrm{d}^3k \right)^n \frac{k^{2(n-1)}}{(k^2 + \mu^2)^{3(n-1)}}, \tag{52}$$

which is of the order $g^6T^4 \ln(T/\mu)$ if $n = 4$ and of the order $g^6T^4 \left(g^2T/\mu \right)^{n-4}$ if $n > 4$. (The various factors in (52) arise, respectively, from the $2(n-1)$

three-gluon vertices, the n loop integrations, and the $3(n-1)$ propagators.) According to this equation, if $\mu \sim g^2T$, all the diagrams with $n \geq 4$ loops contribute to the same order, namely to $\mathcal{O}(g^6)$. In other words, the correction of $\mathcal{O}(g^6)$ to the pressure cannot be computed in perturbation theory.

4.2 Effective Theory at Scale gT

Having identified the main scales and degrees of freedom, our task will be to construct appropriate effective theories at the various scales, obtained by eliminating the degrees of freedom at higher scales. We shall consider here the effective theory at the scale gT obtained by eliminating the hard degrees of freedom with momenta $k \sim T$.

The soft excitations at the scale gT can be described in terms of *average fields* [25,26]. Such average fields develop, for example, when the system is exposed to an external perturbation, such as an external electromagnetic current. In QED, we can summarize the effective theory for the soft modes by the equations of motion:

$$\partial_\mu F^{\mu\nu} = j^\nu_{ind} + j^\nu_{ext} \tag{53}$$

that is, Maxwell equations with a source term composed of the external perturbation j^ν_{ext}, and an extra contribution j^ν_{ind} which we shall refer to as the *induced current*. The induced current is generated by the collective motion of the charged particles, i.e. the hard degrees of freedom. It may be regarded itself as a functional of the average gauge fields and, once this functional is known, the equations above constitute a closed system of equations for the soft fields.

The main problem is to calculate j_{ind}. This is done by considering the dynamics of the hard particles in the background of the soft fields. For QED, the induced current can be obtained using linear response theory. To be more specific, consider as an example a system of charged particles on which is acting a perturbation of the form $\int \mathrm{d}x\, j_\mu(x) A^\mu(x)$, where $j_\mu(x)$ is the current operator and $A^\mu(x)$ some applied gauge potential. Linear response theory leads to the following relation for the induced current:

$$j^{ind}_\mu = \int \mathrm{d}^4y\, \Pi^R_{\mu\nu}(x-y) A^\nu(y),$$
$$\Pi^R_{\mu\nu}(x-y) = -i\theta(x_0-y_0)\langle [j_\mu(x), j_\nu(y)]\rangle_{eq.}, \tag{54}$$

where the (retarded) response function $\Pi^R_{\mu\nu}(x-y)$ is also referred to as the polarization operator. Note that in (54), the expectation value is taken in the equilibrium state. Thus, within linear response, the task of calculating the basic ingredients of the effective theory for soft modes reduces to that of calculating appropriate equilibrium correlation functions.

In fact we shall need the response function only in the weak coupling regime, and for particular kinematic conditions which allow for important simplifications. In leading order in weak coupling, the polarization tensor is given by the one-loop approximation. In the kinematic regime of interest, where the incoming momentum is soft while the loop momentum is hard, we can write $\Pi(\omega, p) = g^2T^2 f(\omega/p, p/T)$ with f a dimensionless function, and in leading order in $p/T \sim g$, Π is of the form $g^2T^2 f(\omega/p)$. This particular contribution of the one-loop polarization tensor is an example of what has been called a "hard thermal loop" [27–32,25,26]; for photons in QED, this is the only one. It turns out that this hard thermal loop can be obtained from simple *kinetic theory*, and the corresponding calculation is done in the next subsection.

In non-Abelian theory, linear response is not sufficient: constraints due to gauge symmetry force us to take into account specific non-linear effects and a more complicated formalism needs to be worked out. Still, simple kinetic equations can be obtained in this case also, but in contrast to QED, the resulting induced current is a non linear functional of the gauge fields. As a result, it generates an infinite number of "hard thermal loops".

5 Kinetic Equations for the Plasma Particles

The hard degrees of freedom enter the equations of motion (53) for the soft collective excitations only through their average density or current, that is, through the induced current. This induced current can be calculated by studying the dynamics of the plasma particles in the background of soft external gauge fields. This is what we now turn to. In order to keep the discussion at an elementary level, we shall merely analyze the main steps involved in the derivation of the corresponding QCD equations in the simpler context of non-relativistic electromagnetic plasmas. The QCD equations are presented at the end of this section.

5.1 One-Loop Polarization Tensor from Kinetic Theory

As indicated above, in the kinematic regime considered, the dominant contribution to the one-loop polarization tensor can be obtained using elementary kinetic theory, and we present now this calculation. We consider an electromagnetic plasma and momentarily assume that we can describe its charged particles in terms of classical distribution functions $f_q(\mathbf{p}, x)$ giving the density of particles of charge q ($q = \pm e$) and momentum $\mathbf{p}$ at the space-time point $x = (t, \mathbf{r})$ [33]. We consider then the case where collisions among the charged particles can be neglected and where the only relevant interactions are those of particles with average electric ($\mathbf{E}$) and magnetic ($\mathbf{B}$) fields. Then the distribution functions obey the following simple kinetic equation, known as the Vlasov equation [33]:

$$\frac{\partial f_q}{\partial t} + \mathbf{v}\frac{\partial f_q}{\partial \mathbf{r}} + \mathbf{F}\frac{\partial f_q}{\partial \mathbf{p}} = 0, \tag{55}$$

where $\mathbf{v} = \mathrm{d}\varepsilon_p/\mathrm{d}\mathbf{p}$ is the velocity of a particle with momentum $\mathbf{p}$ and energy ε_p (for massless particles $\mathbf{v} = \hat{\mathbf{p}}$), and $\mathbf{F} = q(\mathbf{E} + \mathbf{v} \wedge \mathbf{B})$ is the Lorentz force. The average fields $\mathbf{E}$ and $\mathbf{B}$ depend themselves on the distribution functions f_q. Indeed, the induced current

$$j^{\mu}_{ind}(x) = e \int \frac{\mathrm{d}^3p}{(2\pi)^3}\, v^{\mu}\, \left(f_+(\mathbf{p}, x) - f_-(\mathbf{p}, x)\right), \tag{56}$$

where $v^{\mu} \equiv (1, \mathbf{v})$, is the source term in the Maxwell equations (53) for the mean fields.

When the plasma is in equilibrium, the distribution functions, denoted as $f^0_q(p) \equiv f^0(\varepsilon_p)$, are isotropic in momentum space and independent of space-time coordinates; the induced current vanishes, and so do the average fields $\mathbf{E}$ and $\mathbf{B}$. When the plasma is weakly perturbed, the distribution functions deviate slightly from their equilibrium values, and we can write: $f_q(\mathbf{p}, x) = f^0(\varepsilon_p) + \delta f_q(\mathbf{p}, x)$. In the linear approximation, δf obeys

$$(v \cdot \partial_x)\delta f_q(\mathbf{p}, x) = -q\mathbf{v} \cdot \mathbf{E}\frac{\mathrm{d}f^0}{\mathrm{d}\varepsilon_p}, \tag{57}$$

where $v \cdot \partial_x \equiv \partial_t + \mathbf{v} \cdot \nabla$. The magnetic field does not contribute because of the isotropy of the equilibrium distribution function.

It is convenient here to set

$$\delta f_q(\mathbf{p}, x) \equiv -qW(x, \mathbf{v})\, \frac{\mathrm{d}f^0}{\mathrm{d}\varepsilon_p}, \tag{58}$$

thereby introducing a notation which will be useful later for the QCD case. Since

$$f_q(\mathbf{p}, x) = f^0(\varepsilon_p) - qW(x, \mathbf{v})\,\frac{\mathrm{d}f^0}{\mathrm{d}\varepsilon_p} \simeq f^0(\varepsilon_p - qW(x, \mathbf{v})), \tag{59}$$

$W(x, \mathbf{v})$ may be viewed as a local distortion of the momentum distribution of the plasma particles. The equation for W is simply:

$$(v \cdot \partial_x)W(x, \mathbf{v}) = \mathbf{v} \cdot \mathbf{E}(x). \tag{60}$$

Contrary to (55), the linearized equations (57) or (60) do not involve the derivative of f with respect to $\mathbf{p}$, and they can be solved by the method of characteristics: $v \cdot \partial_x$ is the time derivative of $\delta f(\mathbf{p}, x)$ along the characteristic defined by $\mathrm{d}\mathbf{x}/\mathrm{d}t = \mathbf{v}$. Assuming then that the perturbation is introduced adiabatically so that the fields and the fluctuations vanish as $\mathrm{e}^{\eta t_0}$ $(\eta \to 0^+)$ when $t_0 \to -\infty$, we obtain the retarded solution:

$$W(x, \mathbf{v}) = \int_{-\infty}^{t} \mathrm{d}t'\, \mathbf{v} \cdot \mathbf{E}(\mathbf{x} - \mathbf{v}(t - t'), t'), \tag{61}$$

and the corresponding induced current:

$$j^{\mu}_{ind}(x) = -2e^2 \int \frac{\mathrm{d}^3p}{(2\pi)^3}\, v^{\mu} \frac{\mathrm{d}f^0}{\mathrm{d}\varepsilon_p} \int_0^{\infty} \mathrm{d}\tau\, \mathbf{v} \cdot \mathrm{E}(x - v\tau). \tag{62}$$

Since $\mathbf{E} = -\nabla A^0 - \partial \mathbf{A}/\partial t$, the induced current is a linear functional of A^{μ}. At this point we assume explicitly that the particles are massless. In this case, $\mathbf{v}$ is a unit vector, and the angular integral over the direction of $\mathbf{v}$ factorizes in (62). Then, using (54) as definition for the polarization tensor $\Pi^{\mu\nu}(x-y)$, and the fact that the Fourier transform of $\int_0^{\infty} \mathrm{d}\tau\, \mathrm{e}^{-\eta\tau} f(x - v\tau)$ is $i\, f(Q)/(v \cdot Q + i\eta)$, with $Q^{\mu} = (\omega, \mathbf{q})$ and $f(Q)$ the Fourier transform of $f(x)$, one gets, after a simple calculation [34] :

$$\Pi_{\mu\nu}(\omega, \mathbf{q}) = m_D^2 \left\{ -\delta_{\mu 0}\delta_{\nu 0} + \omega \int \frac{\mathrm{d}\Omega}{4\pi} \frac{v_{\mu} v_{\nu}}{\omega - \mathbf{v}\cdot\mathbf{q} + i\eta} \right\}, \tag{63}$$

where the angular integral $\int \mathrm{d}\Omega$ runs over all the orientations of $\mathbf{v}$, and m_D is the Debye screening mass

$$m_D^2 = -\frac{2e^2}{\pi^2} \int_0^{\infty} \mathrm{d}p\, p^2 \frac{\mathrm{d}f^0}{\mathrm{d}\varepsilon_p} \,. \tag{64}$$

It turns out that (63) is the dominant contribution at high temperature to the one-loop polarization tensor in QED, provided one substitutes for f^0 the actual quantum equilibrium distribution function, that is, $f^0(\varepsilon_p) = n_p$, with n_p given in (46). After insertion in (64), this yields $m_D^2 = e^2T^2/3$.

In the next subsection, we shall address the question of how simple kinetic equations emerge in the description of systems of quantum particles, and under which conditions such systems can be described by seemingly classical distribution functions where both positions and momenta are simultaneously specified.

We shall later find that the expression obtained for the polarization tensor using simple kinetic theory generalizes to the non-Abelian case. This is so in particular because the kinematic regime remains that of the linear Vlasov equation, with straight line characteristics.

5.2 Kinetic Equations for Quantum Particles

In order to discuss in a simple setting how kinetic equations emerge in the description of collective motions of quantum particles, we consider in this subsection a system of non-relativistic fermions coupled to classical gauge fields. Since we are dealing with a system of independent particles in an external field, all the information on the quantum many-body state is encoded in the one-body density matrix [9,10] :

$$\rho(\mathbf{r}, \mathbf{r}', t) = \langle \Psi^{\dagger}(\mathbf{r}', t) \Psi(\mathbf{r}, t) \rangle \,, \tag{65}$$

where Ψ and $\Psi^\dagger$ are the annihilation and creation operators, and the average is over the initial equilibrium state. It is on this object that we shall later implement the relevant kinematic approximations. To this aim, we introduce the *Wigner transform* of $\rho(\mathbf{r}, \mathbf{r}', t)$ [35,36]:

$$f(\mathbf{p}, \mathbf{R}, t) = \int \mathrm{d}^3 s\, \mathrm{e}^{-i\mathbf{p}\cdot\mathbf{s}}\, \rho\left(\mathbf{R} + \frac{\mathbf{s}}{2}, \mathbf{R} - \frac{\mathbf{s}}{2}, t\right). \tag{66}$$

The Wigner function has many properties that one expects of a classical phase space distribution function as may be seen by calculating the expectation values of simple one-body observables. For instance the average density of particles $n(\mathbf{R})$ is given by:

$$n(\mathbf{R}, t) = \rho(\mathbf{R}, \mathbf{R}, t) = \int \frac{\mathrm{d}^3 p}{(2\pi)^3}\, f(\mathbf{p}, \mathbf{R}, t). \tag{67}$$

Similarly, the current operator: $(1/2mi)\left(\psi^\dagger \boldsymbol{\nabla}\psi - (\boldsymbol{\nabla}\psi^\dagger)\psi\right)$ has for expectation value:

$$\mathbf{j}(\mathbf{R}, t) = \frac{1}{2mi}\left(\boldsymbol{\nabla}_y - \boldsymbol{\nabla}_x\right)\rho(\mathbf{y}, \mathbf{x}, t)|_{|\mathbf{y}-\mathbf{x}|\to 0} = \int \frac{\mathrm{d}^3 p}{(2\pi)^3}\, \frac{\mathbf{p}}{m}\, f(\mathbf{p}, \mathbf{R}, t). \tag{68}$$

These results are indeed those one would obtain in a classical description with $f(\mathbf{p}, \mathbf{R}, t)$ the probability density to find a particle with momentum $\mathbf{p}$ at point $\mathbf{R}$ and time t. Note, however, that while f is real, due to the hermiticity of ρ, it is not always positive as a truly classical distribution function would be. Of course, f contains the same quantum information as ρ, and it does not make sense to specify quantum mechanically both the position and the momentum. However, f behaves as a classical distribution function in the calculation of one-body observables for which the typical momenta p that are involved in the integration are large in comparison with the scale $1/\lambda$ characterizing the range of spatial variations of f, i.e. $p\lambda \gg 1$.

By using the equations of motion for the field operators, $i\dot{\Psi}(\mathbf{r}, t) = [H, \Psi]$, where H is the single-particle Hamiltonian, one obtains easily the following equation of motion for the density matrix

$$i\partial_t \rho = [H, \rho]. \tag{69}$$

In fact we shall need the Wigner transform of this equation in cases where the gradients with respect to R are small compared to the typical values of p. Under such conditions, the equation of motion reduces to

$$\frac{\partial}{\partial t} f + \boldsymbol{\nabla}_p H \cdot \boldsymbol{\nabla}_R f - \boldsymbol{\nabla}_R H \cdot \boldsymbol{\nabla}_p f = 0, \tag{70}$$

where we have kept only the leading terms in an expansion in $\boldsymbol{\nabla}_R$. For particles interacting with gauge potentials $A^\mu(X)$, the Wigner transform of the single-particle Hamiltonian in (70) takes the form:

$$H(\mathbf{R}, \mathbf{p}, t) = \frac{\mathbf{p}^2}{2m} - \frac{e}{m}\mathbf{A}\cdot\mathbf{p} + \frac{e^2}{m}\mathbf{A}^2(\mathbf{R}, t) + eA_0(\mathbf{R}, t). \tag{71}$$

Assuming that the field is weak and neglecting the term in A^2, one can write (70) in the form:

$$\partial_t f + \mathbf{v}\cdot\boldsymbol{\nabla}_R f + e(\mathbf{E}+\mathbf{v}\wedge\mathbf{B})\cdot\boldsymbol{\nabla}_p f + \frac{e}{m}(p_j\partial_j A^i)\nabla_p^i f = 0, \tag{72}$$

where we have set $\mathbf{v} = (\mathbf{p}-\mathbf{eA})/m$. This equation is almost the Vlasov equation (55): it differs from it by the last term which is not gauge invariant. The presence of such a term, and the related gauge dependence of the Wigner function, obscure the physical interpretation. It is then convenient to define a gauge invariant density matrix:

$$\acute{\rho}(\mathbf{r},\mathbf{r}',t) = \langle\psi^\dagger(\mathbf{r}',t)\psi(\mathbf{r},t)\rangle U(\mathbf{r},\mathbf{r}',t), \tag{73}$$

where ($\mathbf{s}=\mathbf{r}-\mathbf{r}'$)

$$U(\mathbf{r},\mathbf{r}') = \exp\left(-ie\int_{\mathbf{r}'}^{\mathbf{r}} \mathrm{d}\mathbf{z}\cdot\mathbf{A}(\mathbf{z},t))\right) \approx \exp\left(-ie\mathbf{s}\cdot\mathbf{A}(\mathbf{R})\right), \tag{74}$$

and the integral is along an arbitrary path going from $\mathbf{r}'$ to $\mathbf{r}$. Actually, in the last step we have used an approximation which amounts to choose for this path the straight line between $\mathbf{r}'$ to $\mathbf{r}$; furthermore, we have assumed that the gauge potential does not vary significantly between $\mathbf{r}'$ to $\mathbf{r}$. (Typically, $\rho(\mathbf{r},\mathbf{r}')$ is peaked at $s=0$ and drops to zero when $s\gtrsim\lambda_T$ where λ_T is the thermal wave length of the particles. What we assume is that over a distance of order λ_T the gauge potential remains approximately constant.) Note that in the calculation of the current, only the limit $s\to 0$ is required, and that is given correctly by (74) (see also (75) below). With the approximate expression (74) the Wigner transform of (73) is simply $\acute{f}(\mathbf{R},\mathbf{k}) = f(\mathbf{R},\mathbf{k}+e\mathbf{A})$. By making the substitution $f(\mathbf{R},\mathbf{p}) = \acute{f}(\mathbf{R},\mathbf{p}-e\mathbf{A})$ in (72), one verifies that the non-covariant term cancels out and that the covariant Wigner function $\acute{f}$ obeys indeed Vlasov's equation.

In the presence of a gauge field, the previous definition (68) of the current suffers from the lack of gauge covariance. It is, however, easy to construct a gauge invariant expression for the current operator,

$$\mathbf{j} = \frac{1}{2m}\left(\psi^\dagger(\frac{1}{i}\boldsymbol{\nabla} - \mathbf{eA})\psi - \left((\frac{1}{i}\boldsymbol{\nabla} + \mathbf{eA})\psi^\dagger\right)\psi\right), \tag{75}$$

whose expectation value in terms of the Wigner transforms reads:

$$\mathbf{j}(\mathbf{R},t) = \int\frac{\mathrm{d}^3p}{(2\pi)^3}\left(\frac{\mathbf{p}-e\mathbf{A}}{m}\right) f(\mathbf{R},\mathbf{p},t) = \int\frac{\mathrm{d}^3k}{(2\pi)^3}\left(\frac{\mathbf{k}}{m}\right)\acute{f}(\mathbf{R},\mathbf{k},t). \tag{76}$$

The last expression involving the covariant Wigner function makes it clear that $\mathbf{j}(\mathbf{R},t)$ is gauge invariant, as it should. The momentum variable of the gauge covariant Wigner transform is often referred to as the *kinetic* momentum. It is directly related to the velocity of the particles: $\mathbf{k}=m\mathbf{v}=\mathbf{p}-e\mathbf{A}$.

As for $\mathbf{p}$, the argument of the non-covariant Wigner function, it is related to the gradient operator and is often referred to as the *canonical* momentum.

In order to understand the structure of the equations that we shall obtain for the QCD plasma, it is finally instructive to consider the case where the particles possess internal degrees of freedom (such spin, isospin, or colour). The density matrix is then a matrix in internal space. As a specific example, consider a system of spin 1/2 fermions. The Wigner distribution reads [37]:

$$f(\mathbf{p},\mathbf{R}) = f_0(\mathbf{p},\mathbf{R}) + f_a(\mathbf{p},\mathbf{R})\,\sigma_a, \tag{77}$$

where the σ_a are the Pauli matrices, and the f_a are three independent distributions which describe the excitations of the system in the various spin channels; together they form a vector that we can interpret as a local spin density, $\mathbf{f} = (1/2)\mathrm{Tr}(f\boldsymbol{\sigma})$. When the system is in a magnetic field with Hamiltonian $H = -\mu_0\,\boldsymbol{\sigma}\cdot\mathbf{B}$ the equation of motion for $\mathbf{f}$ acquires a new component, $\partial_t\mathbf{f} = 2\mu_0\mathbf{B}\wedge\mathbf{f}$, which accounts for the spin precession in the magnetic field. In the linear approximation this precession may be viewed as an extra time dependence of the distribution function along the characteristics:

$$\frac{\mathrm{d}}{\mathrm{d}t} = \frac{\partial}{\partial t} + \mathbf{v}\cdot\boldsymbol{\nabla}_R + 2\mu_0\mathbf{B}\wedge\ . \tag{78}$$

It is important to realize that all the differential operators above and in the Vlasov equation apply to the arguments of distribution functions, and not to the coordinates of the actual particles. Note, however, that equations similar to the ones presented here can be obtained for classical spinning particles. When the angular momentum of such particles is large, it can indeed be treated as a classical degree of freedom, and the corresponding equations of motion have been written by Wong [38]. After replacing spin by colour, these equations have been used by Heinz [39,40] in order to write down transport equations for classical coloured particles. By implementing the relevant kinematic approximations one then recovers [41] the non-Abelian Vlasov equations to be derived below, i.e., (79) and (80). (See also [42,43] for related work.)

5.3 QCD Kinetic Equations and Hard Thermal Loops

We are now ready to present the equations that are obtained for the QCD plasma. These are equations for generalized one-body density matrices describing the long wave length collective motions of colour particles (quarks and gluons), and possible excitations involving oscillations of fermionic degrees of freedom. They look formally as the Vlasov equation, the main ones being [26,25]:

$$[v\cdot D_x,\, \delta n_\pm(\mathbf{k},x)] = \mp g\,\mathbf{v}\cdot\mathbf{E}(x)\,\frac{\mathrm{d}n_k}{\mathrm{d}k}, \tag{79}$$

$$[v\cdot D_x,\, \delta N(\mathbf{k},x)] = -\,g\,\mathbf{v}\cdot\mathbf{E}(x)\frac{\mathrm{d}N_k}{\mathrm{d}k}, \tag{80}$$

$$(v \cdot D_x)\not{\Lambda}(\mathbf{k}, x) = -igC_f\,(N_k + n_k)\,\not{v}\,\Psi(x). \tag{81}$$

In these equations, $v^\mu = (1, \mathbf{v})$, $\mathbf{v} = \mathbf{k}/k$, $\Psi(x)$ is an average (relativistic) fermionic field, and $\delta n_\pm$, δN and $\not{\Lambda}$ are gauge-covariant Wigner functions for the hard particles. The first two Wigner functions are those of the density matrices of the quarks and the gluons, respectively; the last one is that of a more exotic density matrix which mixes boson and fermion degrees of freedom, $\Lambda \sim \langle \psi A \rangle$. The right hand sides of the equations specify the quantum numbers of the excitations that they are describing: gluon for the first two, quark for the last one. One of the major difference between the QCD equations above and the linear Vlasov equation for QED is the presence of covariant derivatives in the left hand sides of the equations. These play a role similar to that of the magnetic field in (78) for the distribution functions of particles with spin. (Note that the equation for $\not{\Lambda}$ holds for QED, with a covariant derivative there as well.)

The equations (79)–(81) have a number of interesting properties which are reviewed in [1]. In particular, they are covariant under local gauge transformations of the classical fields, and independent of the gauge-fixing in the underlying quantum theory.

By solving these equations, one can express the induced sources as functionals of the background fields. To be specific, consider the colour current:

$$j_a^\mu(x) \equiv 2g \int \frac{\mathrm{d}^3k}{(2\pi)^3}\, v^\mu\, \mathrm{Tr}\left(T^a \delta N(\mathbf{k}, x)\right), \tag{82}$$

where δN is the gluon density matrix. Quite generally, the induced colour current may be expanded in powers of A_μ, thus generating the one-particle irreducible amplitudes of the gauge fields [26]:

$$j_\mu^a \;=\; \Pi_{\mu\nu}^{ab} A_b^\nu + \frac{1}{2}\,\Gamma_{\mu\nu\rho}^{abc} A_b^\nu A_c^\rho + \ldots \tag{83}$$

Here, $\Pi_{\mu\nu}^{ab} = \delta^{ab}\Pi_{\mu\nu}$ is the polarization tensor, and the other terms represent vertex corrections. These amplitudes are "hard thermal loops" (HTL) [30–32,25,26] which define the effective theory for the soft fields at the scale gT. It is worth noticing that the kinetic equations isolate directly these hard thermal loops, in a gauge invariant manner, without further approximations.

The gluon density matrix can be parametrized as in (58)

$$\delta N_{ab}(\mathbf{k}, x) \;=\; -gW_{ab}(x, \mathbf{v})\,(\mathrm{d}N_k/\mathrm{d}k), \tag{84}$$

where $N_k \;\equiv\; 1/(\mathrm{e}^{\beta k} - 1)$ is the Bose-Einstein thermal distribution, and $W(x, \mathbf{v}) \equiv W_a(x, \mathbf{v})T^a$ is a colour matrix in the adjoint representation which depends upon the velocity $\mathbf{v} = \mathbf{k}/k$ (a unit vector), but not upon the magnitude $k = |\mathbf{k}|$ of the momentum. Then the colour current takes the form:

$$j_{ind}^\mu(x) = m_D^2 \int \frac{\mathrm{d}\Omega}{4\pi} v^\mu W(x, \mathbf{v}) \tag{85}$$

with $m_D^2 \sim g^2T^2$. A similar representation holds for the quark density matrices $\delta n_\pm(\mathbf{k}, x)$. The kinetic equations for δN_{ab} and $\delta n_\pm$ can then be written as an equation for $W_a(x, \mathbf{v})$:

$$(v \cdot D_x)^{ab} W_b(x, \mathbf{v}) = \mathbf{v} \cdot \mathbf{E}^a(x). \tag{86}$$

They differ from the corresponding Abelian equation (60) merely by the replacement of the ordinary derivative $\partial_x \sim gT$ by the covariant one $D_x = \partial_x + igA$. Accordingly, the soft gluon polarization tensor derived from (85)–(86), i.e., the "hard thermal loop" $\Pi_{\mu\nu}$, is formally identical to the photon polarization tensor obtained from (60) and given by (63) [27,28]. The reason for the existence of an infinite number of hard thermal loops in QCD is the presence of the covariant derivative in the left hand side of (86). A similar observation can be made by writing the induced electromagnetic current in the form:

$$\begin{aligned} j^\mu_{ind}(x) &= m_D^2 \int \frac{\mathrm{d}\Omega}{4\pi} v^\mu \int \mathrm{d}^4y \, \langle x | \frac{1}{v \cdot \partial} | y \rangle \, \mathbf{v} \cdot \mathbf{E}(y) \\ &= \int \mathrm{d}^4y \, \sigma^{\mu j}(x, y) E^j(y). \end{aligned} \tag{87}$$

This expression, which is easily obtained from the expression (57) of δf, defines the conductivity tensor $\sigma^{\mu\nu}$. The generalization of this expression to QCD amounts essentially to replacing the ordinary derivative by a covariant one.

6 Collective Phenomena in the Quark-Gluon Plasma

At the classical level, the effective theory at the scale gT is summarized by the generalized Yang-Mills equations

$$D_\nu F^{\nu\mu} = \hat{m}_D^2 \int \frac{\mathrm{d}\Omega}{4\pi} \frac{v^\mu v^i}{v \cdot D} E^i \equiv \hat{\Pi}^{ab}_{\mu\nu} A^\nu_b + \frac{1}{2} \hat{\Gamma}^{abc}_{\mu\nu\rho} A^\nu_b A^\rho_c + ..., \tag{88}$$

where the induced current in the right hand side describes the polarization of the hard particles by the soft colour fields A^μ_a. In this equation, $\hat{m}_D \sim gT$ is the Debye mass, E^i_a is the soft electric field, $v^\mu \equiv (1, \mathbf{v})$, and the angular integral $\int \mathrm{d}\Omega$ runs over the orientations of the unit vector $\mathbf{v}$. The induced current is non-local and gauge symmetry, which forces the presence of the covariant derivative $D^\mu = \partial^\mu + igA^\mu$ in the denominator of (88), makes it also non-linear.

Similarly, the soft fermionic fields obey the following generalized Dirac equation [26] (with $\hat{M} \sim gT$ and $\not{v} = \gamma_\mu v^\mu$) :

$$i\not{D}\psi = \hat{M}^2 \int \frac{\mathrm{d}\Omega}{4\pi} \frac{\not{v}}{i(v \cdot D)} \psi \equiv \hat{\Sigma}\psi + \hat{\Gamma}^a_\mu A^\mu_a \psi + ... \tag{89}$$

These equations allow the description of a variety of collective phenomena. We discuss briefly here some of them (collective modes, Debye screening and Landau damping). More details can be found in the article by A. Rebhan in this volume. See also [12,4].

6.1 Collective Modes

The collective plasma waves are propagating solutions to (88) or (89). We restrict ourselves in this subsection to the weak field limit where these equations become linear and essentially Abelian.

The solutions can then be analyzed with the help of the propagator. We consider here the gluon propagator ${}^*G_{\mu\nu}$, in Coulomb's gauge, where it has the following non-trivial components, corresponding to longitudinal (or electric) and transverse (or magnetic) degrees of freedom:

$$ {}^*G_{00}(\omega,\mathbf{p}) \equiv {}^*\Delta_L(\omega,p), \qquad {}^*G_{ij}(\omega,\mathbf{p}) \equiv (\delta_{ij}-\hat{p}_i\hat{p}_j)\,{}^*\Delta_T(\omega,p), \tag{90} $$

where:

$$ {}^*\Delta_L(\omega,p) = \frac{-1}{p^2+\Pi_L(\omega,p)}, \qquad {}^*\Delta_T(\omega,p) = \frac{-1}{\omega^2-p^2-\Pi_T(\omega,p)}, \tag{91} $$

and the electric (Π_L) and magnetic (Π_T) polarization functions are defined as:

$$ \Pi_L(\omega,p) \equiv -\Pi_{00}(\omega,p)\,, \qquad \Pi_T(\omega,p) \equiv \frac{1}{2}\,(\delta^{ij}-\hat{p}^i\hat{p}^j)\Pi_{ij}(\omega,\mathbf{p})\,. \tag{92} $$

Explicit expressions for these functions can be found in [1].

The dispersion relations for the modes are obtained from the poles of the propagators, that is,

$$ p^2+\Pi_L(\omega_L,p)=0, \qquad \omega_T^2=p^2+\Pi_T(\omega_T,p), \tag{93} $$

for longitudinal and transverse excitations, respectively. The solutions to these equations, $\omega_L(p)$ and $\omega_T(p)$, are displayed in Fig. 6.b. The longitudinal mode is the analogue of the familiar plasma oscillation. It corresponds to a collective oscillation of the hard particles, and disappears when $p \gg gT$. Both dispersion relations are time-like ($\omega_{L,T}(p) > p$), and show a gap at zero momentum (the same for transverse and longitudinal modes since, when $p \to 0$, we recover isotropy). With increasing momentum, the transverse branch becomes that of a relativistic particle with an effective mass $m_\infty \equiv m_D/\sqrt{2}$ (commonly referred to as the "asymptotic mass"). Although, strictly speaking, the HTL approximation does not apply at hard momenta, the above dispersion relation $\omega_T(p)$ remains nevertheless correct for $p \sim T$ where it coincides with the light-cone limit of the full one-loop result [44] :

$$ m_\infty^2 \equiv \Pi_T^{1-loop}(\omega^2=p^2) = \frac{m_D^2}{2}. \tag{94} $$

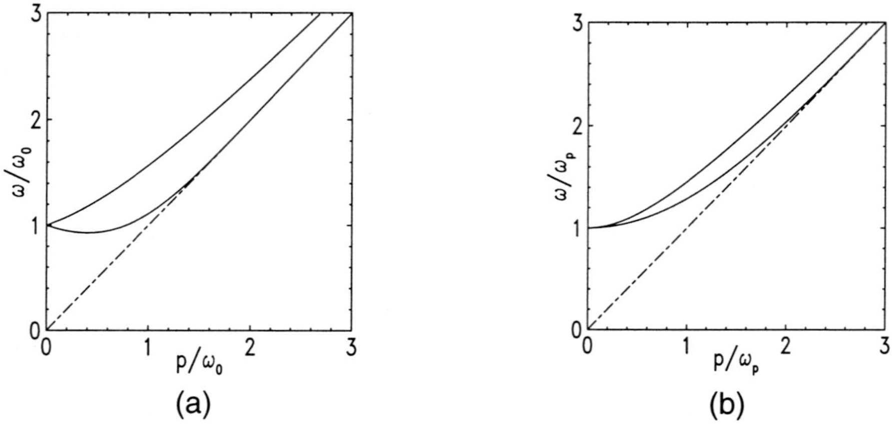

Fig. 6. Dispersion relation for soft excitations in the linear regime: (a) soft fermions; (b) soft gluons (or linear plasma waves), with the upper (lower) branch corresponding to transverse (longitudinal) polarization

The dispersion relations of soft fermionic excitations exhibit also collective feature with a characteristic splitting at low momenta (see Fig. 6.a). We shall not discuss here this interesting phenomenon (see [4] and references therein).

We note finally that particular solutions of the *non-linear* equations (88) have also been found, in [45,46,4]. These solutions describe non-linear plane waves propagating through the plasma, and represent truly non-Abelian collective excitations.

6.2 Debye Screening

The screening of a static chromoelectric field by the plasma constituents is the natural non-Abelian generalization of the Debye screening, a familiar phenomenon in classical plasma physics [33]. In coordinate space, screening reduces the range of the gauge interactions. In momentum space, it contributes to regulate the infrared behaviour of the various n-point functions.

Screening properties can be inferred from an analysis of the effective photon (or gluon) propagators (91) in the static limit $\omega \to 0$. We have:

$$\Pi_L(0,p) = m_D^2\,, \qquad \Pi_T(0,p) = 0, \tag{95}$$

and therefore:

$${}^*\!\Delta_L(0,p) = \frac{-1}{p^2 + m_D^2}\,, \qquad {}^*\!\Delta_T(0,p) = \frac{1}{p^2}\,, \tag{96}$$

which clearly shows that the Debye mass m_D acts as an infrared cut-off $\sim gT$ in the electric sector, while there is no such cut-off in the magnetic sector.

6.3 Landau Damping

For time-dependent fields, there exists a different screening mechanism associated to the energy transfer to the plasma constituents. In Abelian plasmas, this mechanism is known as *Landau damping* [33]. The mechanical work done by a long-wave-length electromagnetic field acting on the charged particles leads to an energy transfer [33]:

$$\frac{\mathrm{d}\,\mathrm{E}_W(t)}{\mathrm{d}\,t} = \int \mathrm{d}^3\mathbf{x}\,\mathbf{E}(t,\,\mathbf{x})\cdot\mathbf{j}(t,\,\mathbf{x}), \tag{97}$$

where $j^i(p) = \Pi_R^{i\nu}(p)A_\nu(p)$ is the induced current. One can then show that the average energy loss is related to the imaginary part of the retarded polarization tensor. From (63) we get:

$$\mathrm{Im}\,\Pi_R^{\mu\nu}(\omega,\mathbf{p}) = -\,\pi m_D^2\,\omega \int \frac{\mathrm{d}\Omega}{4\pi}\, v^\mu v^\nu\,\delta(\omega - \mathbf{v}\cdot\mathbf{p})\,. \tag{98}$$

The δ-function in (98) shows that the particles which absorb energy are those moving in phase with the field (i.e., the particles whose velocity component along $\mathbf{p}$ is equal to the field phase velocity: $\mathbf{v}\cdot\hat{\mathbf{p}} = \omega/p$). Since in ultrarelativistic plasmas $\mathbf{v}$ is a unit vector, only *space-like* ($|\omega| < p$) fields are damped in this way.

To see how this mechanism leads to screening, consider the effective photon (or gluon) propagator (91), and focus on the magnetic propagator. For small but non-vanishing frequencies the corresponding polarization function $\Pi_T(\omega,p)$ is dominated by its imaginary part:

$$\Pi_T(\omega \ll p) = -i\,\frac{\pi}{4}\,m_D^2\,\frac{\omega}{p} + \mathrm{O}(\omega^2/p^2)\,, \tag{99}$$

and therefore

$$^*\!\Delta_T(\omega \ll p) \simeq \frac{1}{p^2 - i\,(\pi\omega/4p)\,m_D^2}. \tag{100}$$

Thus $\mathrm{Im}\,\Pi_T(p)$ acts as a frequency-dependent IR cutoff at momenta $p \sim (\omega m_D^2)^{1/3}$. That is, as long as the frequency ω is different from zero, the soft momenta are dynamically screened by Landau damping [47].

7 The Entropy of the Quark-Gluon Plasma

We come now to the last part of this article which will be mainly devoted to an introduction to the recent progress made in the calculation of the entropy of the quark-gluon plasma. We first comment on various aspects of perturbation theory and show that it is not appropriate for calculating the thermodynamics of the quark-gluon plasma, even at high temperature where the coupling is weak. The main source of difficulties is that the contributions of the collective modes, for which we have constructed an effective theory in the previous sections, are non-perturbative and cannot be expanded in powers of the coupling constant. We then show that these contributions can be included by using self-consistent approximations familiar in many-body physics. These are best formulated for the entropy of the plasma, for which we obtain a simple approximation which provides an accurate description of lattice gauge calculations.

7.1 Results from Perturbation Theory

The free energy has been calculated up to order g^5, including the contribution of fermions [48]. However, since our purpose here is mostly pedagogical, we shall limit our discussion to the gluon contribution at order g^4, in an SU(N) gauge theory. The pressure $P = -F/V$ can then be written:

$$P = P_0 \left[1 + a_2 g^2 + a_3 g^3 + (a_4(\mu/T) + a_4' \ln g)\, g^4 + O(g^5)\right], \tag{101}$$

with

$$a_2 = -5\left(\frac{\sqrt{N}}{4\pi}\right)^2, \quad a_3 = \frac{80}{\sqrt{3}}\left(\frac{\sqrt{N}}{4\pi}\right)^3, \quad a_4' = 240\left(\frac{\sqrt{N}}{4\pi}\right)^4 \ln\frac{\sqrt{N}}{2\pi\sqrt{3}}$$

$$a_4 = -5\left(\frac{\sqrt{N}}{4\pi}\right)^4 \left[\frac{22}{3}\ln\frac{\mu}{4\pi T} + \frac{38}{3}\frac{\zeta'(-3)}{\zeta(-3)} - \frac{148}{3}\frac{\zeta'(-1)}{\zeta(-1)} - 4\gamma_E + \frac{64}{5}\right], \tag{102}$$

where ζ is Riemann's zeta function, and μ the renormalization scale.

The first term in the expansion is P_0, the pressure of an ideal gas of gluons:

$$P_0 = (N^2 - 1)\, T^4 \frac{\pi^2}{45}. \tag{103}$$

The next term, of order g^2, is a genuine perturbative correction, and so is the term of order g^4. The contributions of order g^3 can be interpreted as a contribution of the collective modes to the pressure, and the odd power reflects the fact that the calculation of this contribution requires resummations. Similar resummations are responsible for the term in $g^4 \ln g$.

We note that some of the coefficients in (102) depend on the renormalization scale μ. However, the pressure itself should not depend on μ. It obeys a renormalization group equation:

$$\left[\mu^2 \frac{\partial}{\partial \mu^2} + \left(\mu^2 \frac{\mathrm{d}\alpha}{\mathrm{d}\mu^2}\right)\frac{\partial}{\partial \alpha}\right] P = 0. \tag{104}$$

In this equation, $\alpha(\mu) \equiv g^2(\mu)/4\pi$ is the running coupling constant which satisfies the equation:

$$\mu^2 \frac{\mathrm{d}\alpha}{\mathrm{d}\mu^2} = \beta(\alpha) = -\beta_2 \alpha^2 - \beta_3 \alpha^3, \tag{105}$$

with

$$\beta_2 = \frac{11N}{12\pi}, \qquad \beta_3 = \frac{17N^2}{24\pi^2}. \tag{106}$$

One can then show that, indeed, P is independent of μ: the explicit μ dependence of the coefficients cancels with that of the running coupling. Look indeed at the following combination of terms coming from the contributions of $a_2 g^2$ and the μ dependent part of $a_4 g^4$:

$$\frac{N}{4\pi}\left\{\alpha + \frac{N}{4\pi}\alpha^2 \frac{22}{3} \ln \frac{\mu}{4\pi T}\right\}. \tag{107}$$

By taking the derivative of this expression with respect to μ^2 one gets:

$$\mu^2 \frac{\mathrm{d}}{\mathrm{d}\mu^2}\{\} = \mu^2 \frac{\mathrm{d}\alpha}{\mathrm{d}\mu^2} + \frac{N}{4\pi}\alpha^2 \frac{11}{3} + \text{higher order terms.} \tag{108}$$

By using the leading order expression of the β-function given in (105), one then obtains, as announced:

$$-\frac{11}{12\pi} N\,\alpha^2 + \frac{N}{4\pi}\alpha^2 \frac{11}{3} = 0. \tag{109}$$

Note, however, that the pressure is only *formally* independent of μ at order g^4, in the sense that its derivative with respect to μ involves terms of order g^5 at least. But the approximate expression (101) for P does depend on μ. As in all perturbative calculations, one is then led to look for the best value of μ, i.e. the one which minimizes the higher order corrections. In the present context, a "natural choice" is to fix $\mu = 2\pi T$, where $2\pi T$ is the scale provided by the basic Matsubara frequency. This choice makes the running coupling decrease with increasing temperature, and leads in particular to the expectation that the quark-gluon plasma becomes perturbative at very large temperature.

By calculating explicitly the various coefficients in (102) for $N = 3$, one can write (101) thus:

$$P = P_0 \left[1 - 0.095 g^2 + 0.12 g^3 + \left(0.09 \ln g - 0.007 - 0.013 \ln\left(\frac{\mu}{2\pi T}\right)\right) g^4 + O(g^5)\right]. \tag{110}$$

Then, if for example one fixes $\mu = 2\pi T$ and chooses a large temperature such that $\alpha(2\pi T) = 0.1$, one gets $g = 1.12$, and

$$P = P_0 \left[1 - 0.12 + 0.17 + 0.004\right], \tag{111}$$

which shows no sign of convergence, with the term of order g^3 larger than the term of order g^2. Furthermore, if one analyzes the dependence of P on the renormalization scale, on finds large variations as μ runs within the interval $\pi T < \mu < 4\pi T$.

Attempts have been made to extract information from the first terms of this series using Padé approximants [53,54] or Borel summation techniques [55,56]. The resulting expression of the pressure becomes indeed a smooth function of the coupling, better behaved than the polynomial approximation (101). These techniques, however, which are in some situations very powerful, provide little physical insight, and we shall not discuss them further here.

The behavior of perturbation theory does not improve as one takes into account the higher order terms that one can calculate (namely orders g^4 and g^5). Furthermore, at order g^6, as we have already mentioned, perturbation theory becomes inapplicable because of infrared divergences. It has been shown in [49–51] how, in principle, an effective theory could be constructed to overcome this particular problem by marrying analytical techniques (to determine the coefficients of the effective theory) and numerical ones (to solve the non-perturbative 3-dimensional effective theory). The resulting effective theory is a 3-dimensional theory of static fields, with Lagrangian:

$$\mathcal{L}_{eff} = \frac{1}{4}(F_{ij}^a)^2 + \frac{1}{2}(D_i A_0^a)^2 + \frac{1}{2}m_D^2(A_0^a)^2 + \lambda(A_0^a)^4 + \delta\mathcal{L}, \tag{112}$$

with $D_i = \partial_i - ig\sqrt{T}A_i$. This strategy has been applied recently to the calculation of the free energy of the quark-gluon plasma a high temperature [52]. The slow convergence of the pressure towards the ideal gas value that is seen in lattice calculations above T_c, is well reproduced. It is worth-emphasizing that this technique of dimensional reduction puts a special weight on the static sector (it singles out the contributions of the zero Matsubara frequency). However, as we shall see, it may be advantageous to keep, even in the calculation of equilibrium thermodynamic properties, the full spectral information that one has about the plasma excitations.

There are indeed indications that lattice data are well accounted for by simple phenomenological models of weakly interacting quasiparticles [57,58]. In the case of the scalar field, the dominant effect of the interactions is to give a mass to the excitations. And indeed a perturbative expansion in terms of screened propagators (that is keeping the screening mass $\sim gT$ as a parameter, not considered as a term of order g entering the expansion) has been shown to be quite stable with good convergence properties [59]. In the case of gauge theory, the effect of the interactions is more complicated than just generating a mass. But we know how to determine the dominant corrections to the self-energies. When the momenta are soft, these are given by the hard thermal loops discussed above. By adding these corrections to the tree level Lagrangian, and subtracting them from the interaction part, one generated the so-called hard thermal loop perturbation theory [60]. The resulting perturbative expansion is made complicated, however, by the non-local nature of the hard thermal loop action, and by the necessity of introducing temperature dependent counter terms. At the expense of some extra formalism, some of these difficulties can be avoided. This is discussed now.

7.2 Skeleton Expansion for Thermodynamic Potential and Entropy

In this section we recall the formalism of propagator renormalization that allows systematic rearrangements of the perturbative expansion while avoiding double-counting. We shall see in particular how self-consistent approximations can be used to obtain a simple expression for the entropy which isolates the contribution of the elementary excitations as a leading contribution. For pedagogical purposes, we shall mainly consider in these lectures the example of the scalar field.

The thermodynamic potential $\Omega = -PV$ of the scalar field can be written as the following functional of the full propagator D [61,62]:

$$\beta\Omega[D] = -\log Z = \frac{1}{2}\,\mathrm{T}r\,\log D^{-1} - \frac{1}{2}\,\mathrm{T}r\,\Pi D + \Phi[D]\,, \tag{113}$$

where $\mathrm{T}r$ denotes the trace in configuration space, $\beta = 1/T$, Π is the self-energy related to D by Dyson's equation (D_0 denotes the bare propagator):

$$D^{-1} = D_0^{-1} + \Pi, \tag{114}$$

and $\Phi[D]$ is the sum of the 2-particle-irreducible "skeleton" diagrams

$-\Phi[D] =$ 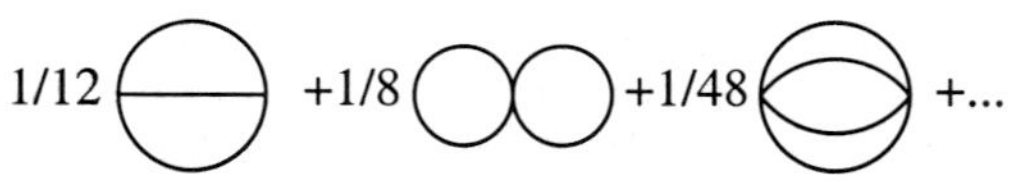 (115)

The essential property of the functional $\Omega[D]$ is to be stationary under variations of D (at fixed D_0) around the physical propagator. The physical pressure is then obtained as the value of $\Omega[D]$ at its extremum. The stationarity condition,

$$\delta\Omega[D]/\delta D = 0, \tag{116}$$

implies the following relation

$$\delta\Phi[D]/\delta D = \frac{1}{2}\Pi, \tag{117}$$

which, together with (114), defines the physical propagator and self-energy in a self-consistent way. The equation (117) expresses the fact that the skeleton diagrams contributing to Π are obtained by opening up one line of a two-particle-irreducible skeleton. Note that while the diagrams of the bare perturbation theory, i.e., those involving bare propagators, are counted once and only once in the expression of Π given above, the diagrams of bare perturbation theory contributing to the thermodynamic potential are counted several times in Φ. The extra terms in (113) precisely correct for this double-counting.

Self-consistent (or variational) approximations, i.e., approximations which preserve the stationarity property (116), are obtained by selecting a class of skeletons in $\Phi[D]$ and calculating Π from (117). Such approximations are commonly called "Φ-derivable" [62].

The traces over configuration space in (113) involve integration over imaginary time and over spatial coordinates. Alternatively, these can be turned into summations over Matsubara frequencies and integrations over spatial momenta:

$$\int_0^\beta \mathrm{d}\tau \int \mathrm{d}^3x \to \beta V \int [\mathrm{d}k], \tag{118}$$

where V is the spatial volume, $k^\mu = (i\omega_n, \mathbf{k})$ and $\omega_n = n\pi T$, with n even (odd) for bosonic (fermionic) fields (the fermions will be discussed later). We have introduced a condensed notation for the measure of the loop integrals (i.e., the sum over the Matsubara frequencies ω_n and the integral over the spatial momentum $\mathbf{k}$):

$$\int [\mathrm{d}k] \equiv T \sum_{n,even} \int \frac{\mathrm{d}^3k}{(2\pi)^3}. \tag{119}$$

Strictly speaking, the sum-integrals in equations like (113) contain ultraviolet divergences, which requires regularization (e.g., by dimensional continuation). Since, however, most of the forthcoming calculations will be free of ultraviolet problems, we do not need to specify here the UV regulator (see, however, Sect. 7.3 for explicit calculations).

For the purpose of developing approximations for the entropy it is convenient to perform the summations over the Matsubara frequencies. One obtains then integrals over real frequencies involving discontinuities of propagators or self-energies which have a direct physical significance. Using standard contour integration techniques, one gets:

$$\Omega/V = \int \frac{d^4k}{(2\pi)^4}\, n(\omega) \left(\mathrm{Im}\log(-\omega^2 + \mathrm{k}^2 + \Pi) - \mathrm{Im}\Pi \mathrm{D}\right) + T\Phi[D]/V, \tag{120}$$

where $n(\omega) = 1/(\mathrm{e}^{\beta\omega} - 1)$.

The analytic propagator $D(\omega, k)$ can be expressed in terms of the spectral function

$$D(\omega, k) = \int_{-\infty}^{\infty} \frac{dk_0}{2\pi} \frac{\rho(k_0, k)}{k_0 - \omega}, \tag{121}$$

and we define, for ω real,

$$\mathrm{ImD}(\omega, \mathrm{k}) \equiv \mathrm{ImD}(\omega + \mathrm{i}\epsilon, \mathrm{k}) = \frac{\rho(\omega, \mathrm{k})}{2}. \tag{122}$$

The imaginary parts of other quantities are defined similarly.

We are now in the position to calculate the entropy density:

$$\mathcal{S} = -\partial(\Omega/V)/\partial T\,. \tag{123}$$

The thermodynamic potential, as given by (120) depends on the temperature through the statistical factors $n(\omega)$ and the spectral function ρ, which is determined entirely by the self-energy. Because of (116) the temperature derivative of the spectral density in the dressed propagator cancels out in the entropy density and one obtains [63,64]:

$$\mathcal{S} = -\int \frac{d^4k}{(2\pi)^4} \frac{\partial n(\omega)}{\partial T} \mathrm{Im} \log \mathrm{D}^{-1}(\omega, \mathrm{k}) + \int \frac{d^4k}{(2\pi)^4} \frac{\partial n(\omega)}{\partial T} \mathrm{Im}\Pi(\omega, \mathrm{k}) \mathrm{Re}\mathrm{D}(\omega, \mathrm{k}) + \mathcal{S}' \tag{124}$$

with

$$\mathcal{S}' \equiv -\frac{\partial(T\Phi)}{\partial T}\bigg|_D + \int \frac{d^4k}{(2\pi)^4} \frac{\partial n(\omega)}{\partial T} \mathrm{Re}\Pi\, \mathrm{Im}\mathrm{D}. \tag{125}$$

For the two-loop skeletons, we have:

$$\mathcal{S}' = 0. \tag{126}$$

Loosely speaking, the first two terms in (124) represent essentially the entropy of "independent quasiparticles", while $\mathcal{S}'$ accounts for a residual interaction among these quasiparticles [64].

The vanishing of $\mathcal{S}'$ holds whether the propagator are the self-consistent propagators or not. That is, only the relation (117) is used in the proof which does not require D to satisfy the self-consistent Dyson equation (114). A general analysis of the contributions to $\mathcal{S}'$ and their physical interpretation can be found in [65].

We emphasize now a few attractive features of the formula (124) with $\mathcal{S}' = 0$, which makes the entropy a privileged quantity to study the thermodynamics of ultrarelativistic plasmas. We note first that the formula for $\mathcal{S}$ at 2-loop order involves the self-energy only at 1-loop order. Besides this important simplification, this formula for $\mathcal{S}$, in contrast to the pressure, has the advantage of manifest ultra-violet finiteness, since $\partial n/\partial T$ vanishes exponentially for both $\omega \to \pm\infty$. Also, any multiplicative renormalization $D \to ZD$, $\Pi \to Z^{-1}\Pi$ with real Z drops out from (124). Finally, the entropy has a more direct quasiparticle interpretation than the pressure. This will be illustrated explicitly in the simple model of the next subsection.

7.3 A Simple Model

In this section we shall present the self-consistent solution for the $(\lambda/4!)\phi^4$ theory, keeping in Φ only the two-loop skeleton. Anticipating the fact that the fully dressed propagator will be that of a massive particle, we write the spectral function as $\rho(k_0, \mathbf{k}) = 2\pi\, \epsilon(k_0)\, \delta(k_0^2 - \mathbf{k}^2 - m^2)$, and consider m as a variational parameter. The thermodynamic potential (113), or equivalently the pressure, becomes then a simple function of m. By Dyson's equation, the self-energy is simply $\Pi = m^2$. We set:

$$I(m) \equiv \frac{1}{2}\int [\mathrm{d}k]\, D(k) = \frac{1}{2}\int [\mathrm{d}k]\, \frac{1}{\omega_n^2 + \mathbf{k}^2 + m^2}. \tag{127}$$

Then the pressure $P = -\Omega/V$ can be written as:

$$-P = \frac{1}{2}\int \frac{\mathrm{d}^3k}{(2\pi)^3}\varepsilon_k + \frac{1}{\beta}\int \frac{\mathrm{d}^3k}{(2\pi)^3}\log(1-\mathrm{e}^{-\beta\varepsilon_k}) - m^2 I(m) + \frac{\lambda_0}{2}I^2(m), \tag{128}$$

where $\varepsilon_k^2 \equiv k^2 + m^2$. By demanding that P be stationary with respect to m one obtains the self-consistency condition which takes here the form of a "gap equation":

$$m^2 = \lambda_0\, I(m). \tag{129}$$

The pressure in the two-loop Φ-derivable approximation, as given by (127)–(129), is formally the same as the pressure per scalar degree of freedom in the (massless) N-component model with the interaction term written as $\frac{3}{N+2}(\lambda/4!)(\phi_i\phi_i)^2$ in the limit $N \to \infty$ [66]. From the experience with this latter model, we know that (127)–(129) admit an exact, renormalizable solution which we recall now.

At this stage, we need to specify some properties of the loop integral $I(m)$ which we can write as the sum of a vacuum piece $I_0(m)$ and a finite temperature piece $I_T(m)$ such that, at fixed m, $I_T(m) \to 0$ as $T \to 0$. We use dimensional regularization to control the ultraviolet divergences present in I_0, which implies $I_0(0) = 0$. Explicitly one has:

$$\mu^\epsilon I(m) = -\frac{m^2}{32\pi^2}\left(\frac{2}{\epsilon} + \log\frac{\bar{\mu}^2}{m^2} + 1\right) + I_T(m) + \mathrm{O}(\epsilon), \tag{130}$$

with

$$I_T(m) = \int \frac{\mathrm{d}^3k}{(2\pi)^3}\,\frac{n(\varepsilon_k)}{2\varepsilon_k}, \tag{131}$$

and $\varepsilon_k \equiv (k^2+m^2)^{1/2}$. In (130), μ is the scale of dimensional regularization, introduced, as usual, by rewriting the bare coupling λ_0 as $\mu^\epsilon\hat{\lambda}_0$, with dimensionless $\hat{\lambda}_0$; furthermore, $\epsilon = 4 - n$, with n the number of space-time dimensions, and $\bar{\mu}^2 = 4\pi\mathrm{e}^{-\gamma}\mu^2$.

We use the modified minimal subtraction scheme ($\overline{\mathrm{MS}}$) and define a dimensionless renormalized coupling λ by:

$$\frac{1}{\lambda} = \frac{1}{\lambda_0\mu^{-\epsilon}} + \frac{1}{16\pi^2\epsilon}. \tag{132}$$

When expressed in terms of the renormalized coupling, the gap equation becomes free of ultraviolet divergences. It reads:

$$m^2 = \frac{\lambda}{2}\int \frac{\mathrm{d}^3k}{(2\pi)^3}\,\frac{n(\varepsilon_k)}{\varepsilon_k} + \frac{\lambda m^2}{32\pi^2}\left(\log\frac{m^2}{\bar{\mu}^2} - 1\right). \tag{133}$$

The renormalized coupling constant satisfies

$$\frac{\mathrm{d}\lambda}{\mathrm{d}\log\bar{\mu}} = \frac{\lambda^2}{16\pi^2}, \tag{134}$$

which ensures that the solution m^2 of (133) is independent of $\bar{\mu}$. The expression (134) coincides with the exact β-function in the large-N limit, but gives only one third of the lowest-order perturbative β-function for $N = 1$. This is no actual fault since the running of the coupling affects the thermodynamic potential only at order λ^2 which is beyond the perturbative accuracy of the 2-loop Φ-derivable approximation. In order to see the correct one-loop β-function at finite N, the approximation for Φ would have to be pushed to 3-loop order.

Note also that, in the present approximation, the renormalization (132) of the coupling constant is sufficient to make the pressure (128) finite. Indeed, in dimensional regularization the sum of the zero point energies $\varepsilon_k/2$ in (128) reads:

$$\mu^\epsilon \int \frac{\mathrm{d}^{n-1}k}{(2\pi)^{n-1}} \frac{\varepsilon_k}{2} = -\frac{m^4}{64\pi^2}\left(\frac{2}{\epsilon} + \log\frac{\bar{\mu}^2}{m^2} + \frac{3}{2}\right) + \mathrm{O}(\epsilon), \tag{135}$$

so that

$$\mu^\epsilon \int \frac{\mathrm{d}^{n-1}k}{(2\pi)^{n-1}} \frac{\varepsilon_k}{2} - \frac{\Pi^2}{2\hat{\lambda}_0} = -\frac{m^4}{2\lambda} - \frac{m^4}{64\pi^2}\left(\log\frac{\bar{\mu}^2}{m^2} + \frac{3}{2}\right) + \mathrm{O}(\epsilon) \tag{136}$$

is indeed UV finite as $n \to 4$. After also using the gap equation (133), one obtains the $\bar{\mu}$-independent result

$$P = -T\int \frac{\mathrm{d}^3k}{(2\pi)^3} \log(1 - \mathrm{e}^{-\beta\varepsilon_k}) + \frac{m^2}{2} I_T(m) + \frac{m^4}{128\pi^2}. \tag{137}$$

We now compute the entropy according to (124). Since $\mathrm{Im}\Pi = 0$ and $\mathrm{Re}\Pi = \mathrm{m}^2$, we have simply:

$$\mathcal{S} = -\int \frac{d^4k}{(2\pi)^4} \frac{\partial n(\omega)}{\partial T} \mathrm{Im}\log(\mathrm{k}^2 - \omega^2 + \mathrm{m}^2). \tag{138}$$

Using

$$\mathrm{Im}\log(\mathrm{k}^2 - \omega^2 + \mathrm{m}^2) = -\pi\epsilon(\omega)\theta(\omega^2 - \varepsilon_\mathrm{k}^2), \tag{139}$$

and the identity,

$$\frac{\partial n(\omega)}{\partial T} = -\frac{\partial\sigma(\omega)}{\partial\omega}, \qquad \sigma(\omega) \equiv -n\log n + (1+n)\log(1+n), \tag{140}$$

one can rewrite (138) in the form (with $n_k \equiv n(\varepsilon_k)$):

$$\mathcal{S} = \int \frac{d^3k}{(2\pi)^3} \Big\{(1+n_k)\log(1+n_k) - n_k\log n_k\Big\}. \tag{141}$$

This formula shows that, in the present approximation, the entropy of the interacting scalar gas is formally identical to the entropy of an ideal gas of massive bosons, with mass m.

It is instructive to observe that such a simple interpretation does not hold for the pressure. The pressure of an ideal gas of massive bosons is given by:

$$P^{(0)}(m) = \int \frac{d^3k}{(2\pi)^3} \int_{\epsilon_k}^{\infty} \mathrm{d}\omega \left(n(\omega) + \frac{1}{2} \right)$$
$$= - \int \frac{d^3k}{(2\pi)^3} \left\{ T \log(1 - \mathrm{e}^{-\epsilon_k/T}) + \frac{\epsilon_k}{2} \right\}, \qquad (142)$$

which differs indeed from (128) by the term m^4/λ which corrects for the double-counting of the interactions included in the thermal mass.

7.4 Comparison with Thermal Perturbation Theory

In view of the subsequent application to QCD, where a fully self-consistent determination of the gluonic self-energy seems prohibitively difficult, we shall be led to consider approximations to the gap equation. These will be constructed such that they reproduce (but eventually transcend) the perturbative results up to and including order $\lambda^{3/2}$ or g^3, which is the maximum perturbative accuracy allowed by the approximation $\mathcal{S}' = 0$.

In view of this it is important to understand the perturbative content of the self-consistent approximations for m^2, P and $\mathcal{S}$. In this section we shall demonstrate that, when expanded in powers of the coupling constant, these approximations reproduce the correct perturbative results up to order $\lambda^{3/2}$ [11]. This will also elucidate how perturbation theory gets reorganized by the use of the skeleton representation together with the stationarity principle.

For the scalar theory with only $(\lambda/4!)\,\phi^4$ self-interactions, we write[1] $\lambda \equiv 24g^2$, and compute the corresponding self-energy $\Pi = m^2$ by solving the gap equation (133) in an expansion in powers of g, up to order g^3. Since we anticipate m to be of order gT, we can ignore the second term $\propto \lambda m^2 \sim g^4$ in the r.h.s. of (133), and perform a high-temperature expansion of the integral $I_T(m)$ in the first term (cf. (131)) up to terms linear in m. This gives the following, approximate, gap equation:

$$m^2 \simeq g^2T^2 - \frac{3}{\pi}\, g^2 T m\,. \qquad (143)$$

The first term in the r.h.s. arises as

$$24g^2 I_T(0) = 12g^2 \int \frac{d^3k}{(2\pi)^3} \frac{n(k)}{k} = g^2T^2 \equiv \hat{m}^2. \qquad (144)$$

[1] This normalization for g is chosen in view of the subsequent extension to QCD since it makes the scalar thermal mass in (144) equal to the leading-order Debye mass in pure-glue QCD.

This is also the leading-order result for m^2, commonly dubbed the "hard thermal loop".

The second term, linear in m, in (143) comes from

$$12g^2 \int \frac{d^3k}{(2\pi)^3} \left(\frac{n(\varepsilon_k)}{\varepsilon_k} - \frac{n(k)}{k} \right) \simeq \tag{145}$$

$$12g^2 T \int \frac{d^3k}{(2\pi)^3} \left(\frac{1}{k^2+m^2} - \frac{1}{k^2} \right) = -\frac{3g^2}{\pi} mT \,, \tag{146}$$

where we have used the fact that the momentum integral is saturated by soft momenta $k \sim gT$, so that to the order of interest $n(\varepsilon_k) \simeq T/\varepsilon_k$ (and similarly $n(k) \simeq T/k$). This provides the next-to-leading order (NLO) correction to the thermal mass

$$\delta m^2 \equiv -\frac{3g^2}{\pi} \hat{m} T = -\frac{3}{\pi} g^3 T^2 \,. \tag{147}$$

Thus, to order g^3, one has $m^2 = \hat{m}^2 + \delta m^2$. In standard perturbation theory [11,12], the first term arises as the one-loop tadpole diagram evaluated with a bare massless propagator, while the second term comes from the same diagram where the internal line is soft and dressed by the HTL, that is $\hat{D}(\omega, k) \equiv -1/(\omega^2 - k^2 - \hat{m}^2)$.

Consider similarly the perturbative estimates for the pressure and entropy, as obtained by evaluating (128) and (141) with the perturbative self-energy $\Pi = m^2 \simeq \hat{m}^2 + \delta m^2$, and further expanding in powers of g, to order g^3. The renormalized version of (128) yields, to this order (recall that $m \sim gT$ and $\lambda \sim g^2$),

$$P \simeq \frac{\pi^2 T^4}{90} - \frac{m^2 T^2}{24} + \frac{m^3 T}{12\pi} + \cdots + \frac{m^4}{2\lambda} . \tag{148}$$

The first terms before the dots represent the pressure of massive bosons, i.e. (142) expanded up to third order in powers of m/T. From (148), it can be easily verified that the above perturbative solution for m^2 ensures the stationarity of P up to order g^3, as it should. Indeed, if we denote

$$P_2(m) \equiv -\frac{m^2 T^2}{24} + \frac{m^4}{2\lambda} \,, \qquad P_3(m) \equiv \frac{m^3 T}{12\pi} \,, \tag{149}$$

then the following identities hold:

$$\left.\frac{\partial P_2}{\partial m}\right|_{\hat{m}} = 0, \qquad \left.\frac{\partial P_2}{\partial m}\right|_{\hat{m}+\delta m} + \left.\frac{\partial P_3}{\partial m}\right|_{\hat{m}} = 0. \tag{150}$$

This shows that the NLO mass correction $\delta m^2 \sim g^3 T^2$ can be also obtained as

$$\delta m^2 = -\left.\frac{(\partial P_3/\partial m)}{(\partial^2 P_2/\partial m^2)}\right|_{\hat{m}} = -\frac{3g}{\pi} \hat{m}^2 \,, \tag{151}$$

in agreement with (147). Moreover, $P_2 \equiv P_2(\hat{m}) = -g^2T^2/48$ and $P_3 \equiv P_3(\hat{m}) = \hat{m}^3T/12\pi$ are indeed the correct perturbative corrections to the pressure, to orders g^2 and g^3, respectively [11]. In fact, the pressure to this order can be written as:

$$\begin{aligned} P &= \frac{\pi^2T^4}{90} - \frac{\hat{m}^2T^2}{24}(1-\frac{3}{\pi}g) + \frac{\hat{m}^3T}{12\pi} + \cdots + \frac{\hat{m}^4}{2\lambda}(1-\frac{3}{\pi}g)^2 + \mathcal{O}(g^4) \\ &= \frac{\pi^2T^4}{90} - \frac{\hat{m}^2}{48}T^2 + \frac{\hat{m}^3T}{12\pi}\,. \end{aligned} \tag{152}$$

Note that the term of order g^2 is only *half* of that one would obtain from (142) by replacing m by $\hat{m}$. This is due to the mismatch between (142) and the correct expression (128) for the pressure. In fact the net order g^2 contribution to the pressure comes from Φ evaluated with bare propagators: the order g^2 contributions in the other two terms mutually cancel indeed. This is to be expected: there is a single diagram of order g^2; this is a skeleton diagram, counted therefore once and only once in Φ. Observe also that the terms of order g^3 originating from the terms $\hat{m}^2$ and $\hat{m}^4$ mutually cancel; that is, the NLO mass correction δm drops out from the pressure up to order g^3. This is no accident: the cancellation results from the stationarity of P at order g^2, the first equation (150).

Consider now the entropy density. The correct perturbative result up to order g^3 may be obtained directly by taking the total derivative of the pressure, (152) with respect to T. One then obtains:

$$\mathcal{S} = \frac{4}{T}\left(\frac{\pi^2T^4}{90} - \frac{\hat{m}^2T^2}{48} + \frac{\hat{m}^3T}{12\pi}\right) + \mathcal{O}(g^4). \tag{153}$$

We wish, however, to proceed differently, using (141), or equivalently, since $\partial P/\partial m = 0$ when m is a solution of the gap equation, by writing:

$$\mathcal{S} = \left.\frac{\partial P}{\partial T}\right|_m. \tag{154}$$

This yields:

$$\mathcal{S} = \frac{4}{T}\left(\frac{\pi^2T^4}{90} - \frac{m^2T^2}{48} + \frac{m^3T}{48\pi}\right) + \mathcal{O}(m^4/T), \tag{155}$$

which coincides as expected with the expression obtained by expanding the entropy (141) of massive bosons, up to order $(m/T)^3$. If we now replace m by its leading order value $\hat{m}$, the resulting approximation for $\mathcal{S}$ reproduces the perturbative effect of order $\sim g^2$, but it underestimates the correction of order g^3 by a factor of 4. This is corrected by changing m to $\hat{m} + \delta m$ with $\delta m = -3g\hat{m}/2\pi$ in the second order term of (155). Note that although it makes no difference to enforce the gap equation to order g^3 in the pressure (because of the cancellation discussed above), there is no such cancellation in the entropy.

7.5 Approximately Self-consistent Solutions

As we have seen, the 2-loop Φ-derivable approximation provides an expression for the entropy $\mathcal{S}$ as a functional of the self-energy Π which has a simple quasiparticle interpretation and is manifestly ultraviolet finite for any (finite) Π. These attractive features of the formula (124) are independent of the specific form of the self-energy, and will be shown to hold in QCD as well. Of course, within this approximation, the self-energy is uniquely specified: by the stationarity principle, this is given by the self-consistent solution to the one-loop gap equation. In the scalar ϕ^4-model, it was easy to give the exact solution to this equation. In QCD, however, it will turn out that a fully self-consistent solution is both prohibitively difficult (because of the non-locality of the gap equation), and not really desirable (because gauge symmetry implies relations between the renormalization of the propagators and that of the vertices, and the present approximation deals only with propagator renormalization). This leads us to consider *approximately self-consistent* resummations, which are obtained in two steps: (a) An approximation is constructed for the solution Π to the gap equation, and (b) the entropy (124) is evaluated *exactly* (i.e., numerically) with this approximate self-energy. While step (b) above is unambiguous and inherently non-perturbative, step (a), on the other hand, will be constrained primarily by the requirement of preserving the maximum possible perturbative accuracy, of order g^3. In addition to that, we shall add the qualitative requirement that the approximation for Π, and the ensuing one for $\mathcal{S}$, are well defined and physically meaningful for all the values of g of interest, and not only for small g—that is, for all the values of g where the fully self-consistent calculation makes sense a priori. Finally, in the case of QCD, relaxing the requirement of complete self-consistency allows us to construct gauge invariant approximations.

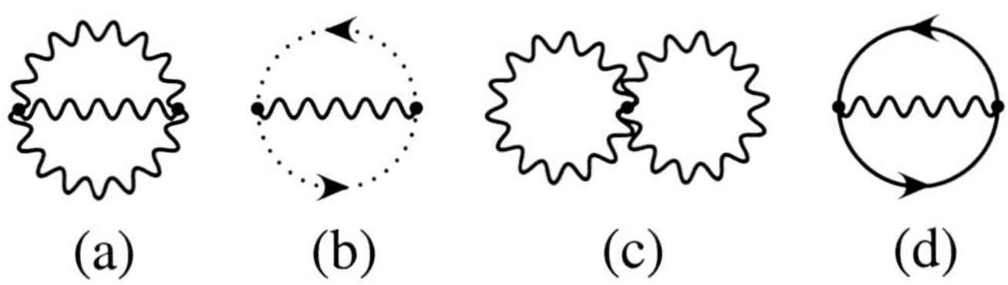

Fig. 7. QCD skeletons contributing to Φ at 2-loop order. Wiggly, plain and dotted lines refer, respectively, to gluons, quarks, and ghosts

We shall now, in the rest of this article, outline the main steps that are involved in the implementation of these approximations in the case of QCD. Details can be found in the original publications [13–15].

At 2-loop order, the relevant skeletons are displayed in Fig. 7. By itself, the corresponding self-consistent truncation is not a gauge invariant approximation. Our strategy then will be to use gauge-invariant approximations to self-energies, in place of the self-consistent ones. These self-energies are then used to compute the entropy without further approximations. In complete analogy with the example of the scalar case that we have discussed in the previous section, these approximations are such that, when expanded in powers of the coupling the entropy is identical to that given by perturbation theory up to and including order g^3.

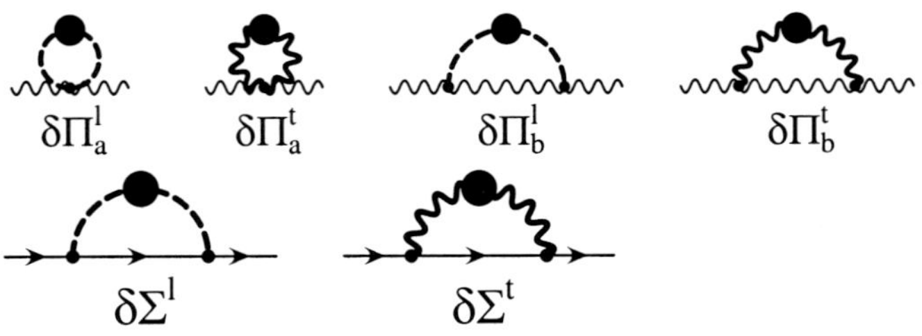

Fig. 8. Next to leading order contribution to $\delta\Pi_T$ (top) and to $\delta\Sigma$ (bottom) at hard momentum. Thick dashed and wiggly lines with a blob represent HTL-resummed longitudinal and transverse propagators, respectively

The approximate self-energies that we use are the hard thermal loops discussed above. Namely, for soft momenta ω, $p \sim gT$, we take $\Pi_{soft} \approx \Pi_{HTL}$ and $\Sigma_{soft} \approx \Sigma_{HTL}$, for gluons and quarks, respectively. We shall also need an approximation valid for ω, $p \sim T$: $\Pi_{hard}(\omega^2 \sim p^2)$ and similarly for Σ. It turns out that this is accurately given by the hard thermal loop, even though the momenta are not soft [44]. All these approximations are gauge invariant. The corresponding diagrams are displayed in Figs. 8.

We can then proceed exactly as in the scalar case. As a first approximation one may simply use the hard thermal loops $\Pi = \Pi_{HTL}$ and $\Sigma \sim \Sigma_{HTL}$ at all momenta; we refer to the corresponding entropy as $\mathcal{S} = \mathcal{S}_{HTL}$. The perturbative content of this approximation is schematically $\mathcal{O}(g^2) + \frac{1}{4}\,\mathcal{O}(g^3)$; that is, the approximation fully accounts of the order g^2, but reproduces only a quarter of the g^3 order, exactly as in the scalar case. In the next-to-leading approximation, we correct the hard degrees of freedom by their interaction with the soft modes. That is, we continue to use the hard thermal loops at small momenta, but use at hard momenta the corrections corresponding to the diagrams displayed in Fig. 8. The resulting approximation to the entropy, $\mathcal{S} = \mathcal{S}_{NLA}$ accounts then fully for the orders g^2 and g^3. But, of course, these expressions are not limited to values of the coupling as small as required for the validity of perturbation theory.

7.6 Some Results for QCD

As an illustration of the quality of the results that are obtained within that scheme, we show in Fig. 9 the entropy of pure SU(3) gauge theory. The bands delimiting the various lines in this figure correspond to varying the $\overline{\text{MS}}$ renormalization scale $\bar{\mu}$, which defines the renormalized coupling constant $g(\bar{\mu})$, from $\bar{\mu} = \pi T$ to $4\pi T$. One sees that in contrast to ordinary perturbation theory, going from one level of approximation to the next one is indeed a small correction. In particular, the effects of the soft modes are here a small contribution. This is to be contrasted with perturbation theory where the order g^3 contribution is large for moderate values of the coupling. The comparison with the lattice data [67] is quite good down to $T \gtrsim 2.5T_c$.

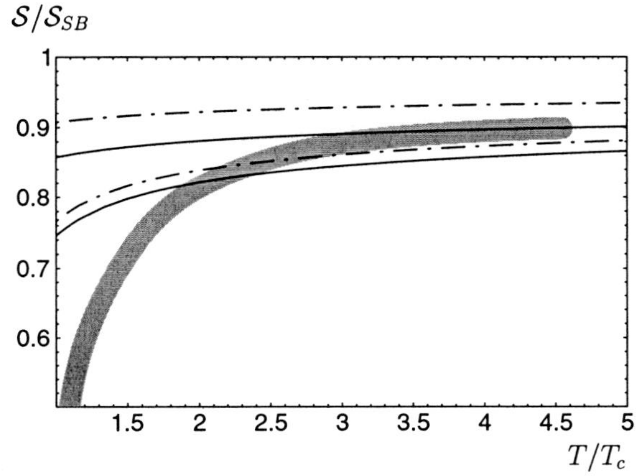

Fig. 9. The entropy of pure SU(3) gauge theory normalized to the ideal gas entropy $\mathcal{S}_0$. Full lines: $\mathcal{S}_{HTL}$. Dashed-dotted lines: $\mathcal{S}_{NLA}$. 2-loop β-function $\rightarrow$ the running coupling constant $\alpha_s(\bar{\mu})$. The $\overline{\text{MS}}$ renormalization scale: $\bar{\mu} = \pi T \cdots 4\pi T$. The dark grey band: lattice result by Boyd et al. [67]

The quality of the agreement between the self-consistent approximation and the lattice data supports the quasiparticle picture of the quark-gluon plasma: the dominant effect of the interactions at high temperature seems to be to change the bare quarks and gluons into massive quasiparticles, with small residual interactions between the quasiparticles. It should be emphasized that, in contrast to the approximations based on dimensional reduction, the method makes full use of the spectral information on the quasiparticles contained in particular in the hard thermal loops.

The approach is easily extended to finite chemical potential, and the calculation of the baryonic density can be done using approximations similar to those we used for the entropy. Furthermore, from the knowledge of $N(\mu, T)$ and $S(\mu, T)$ one can reconstruct $P(\mu, T)$, using lattice data to fix the integration constant (e.g. $P(\mu = 0, T)$). Such investigations are under way.

Acknowledgements

I would like to thank the organizers of the school "Dense Matter" for what has been a very enjoyable meeting. I also express my gratitude to Edmond Iancu and Anton Rebhan: much of the material presented in this article is drawn from work done in collaboration with them.

References

1. J. Blaizot and E. Iancu: "The quark-gluon plasma: Collective dynamics and hard thermal loops," to appear in Phys. Rep.; hep-ph/0101103
2. J.P. Blaizot: *"QCD at finite temperature"*, in "Probing the Standard Model of Particle Interactions", Les Houches, Session LXVIII, 1997, ed. by R. Gupta et al. (Elsevier, Amsterdam, 1999)
3. J. P. Blaizot: "The quark-gluon plasma and nuclear collisions at high energy," *Lecture given at Les Houches Summer School on Theoretical Physics, Session 66: Trends in Nuclear Physics, 100 Years Later, Les Houches, France, July 30 - August 30, 1996*
4. J.-P. Blaizot, E. Iancu, J.-Y. Ollitrault: in *Quark Gluon Plasma 2*, R.C. Hwa ed., World Scientific, Singapore (1995), p. 135
5. J.-P. Blaizot: in *Lecture Notes on the Workshop: Nuclear Equation of State*, Puri, India, Jan. 1994, A. Ansari and L. Satpathy eds. (World Scientific, Singapore 1996)
6. J-P. Blaizot: in Proceeding of the Fourth Summer School and Symposium on Nuclear Physics, *High Density and High Temperature Physics*, July 1-6, Namwon, Korea, Journ. Korean Phys. Soc., vol. 25
7. L. Kadanov, G. Baym: *Quantum Statistical Mechanics* (Benjamin/Cummings, London 1962)
8. A.A. Abrikosov, L.P. Gorkov, I.E. Dzyaloshinskii: *Methods of Quantum Field Theory in Statistical Physics*, (Dover, New-York, 1963)
9. A. Fetter, J.D. Walecka: *Quantum Theory of Many Particle Systems* (McGraw Hill, New-York, 1971)
10. J.P. Blaizot, G. Ripka: *Quantum Theory of Finite Systems* (MIT Press, Cambridge 1986)
11. J.I. Kapusta: *Finite Temperature Field Theory*, Cambridge Monographs in Mathematical Physics (Cambridge University Press, 1989)
12. M. Le Bellac: *Thermal Field Theory*, Cambridge Monographs in Mathematical Physics (Cambridge University Press, 1996)
13. J. P. Blaizot, E. Iancu, A. Rebhan: Phys. Rev. Lett. **83**, 2906 (1999) [hep-ph/9906340]
14. J. P. Blaizot, E. Iancu, A. Rebhan: Phys. Lett. **B470**, 181 (1999) [hep-ph/9910309]
15. J. P. Blaizot, E. Iancu, A. Rebhan: Phys. Rev. **D63**, 065003 (2001) [hep-ph/0005003]
16. K. Johnson, C.B. Thorn, A. Chodos, R.L. Jaffe, V.F. Weisskopf: Phys. Rev. **D9**, 3471 (1974); K. Johnson, A. Chodos, R.L. Jaffe, C.B. Thorn: Phys. Rev. **D10**, 2599 (1974)
17. P. Ginsparg: Nucl. Phys. **B170**, 388 (1980)

18. T. Appelquist, R.D. Pisarski: Phys. Rev. **D23**, 2305 (1981)
19. S. Nadkarni: Phys. Rev. **D27**, 917 (1983); Phys. Rev. **D38**, 3287 (1988)
20. N.P. Landsman: Nucl. Phys. **B322**, 498 (1989)
21. E. Braaten: Phys. Rev. Lett. **74**, 2164 (1995)
22. K. Kajantie, K. Rummukainen, M.E. Shaposhnikov: Nucl. Phys. **B407**, 356 (1993); K. Farakos, K. Kajantie, K. Rummukainen, M.E. Shaposhnikov: Nucl. Phys. **B425**, 67 (1994); *ibid.* **442**, 317 (1995)
23. A.D. Linde: Rep. Prog. Phys. **42**, 389 (1979)
24. A.D. Linde: Phys. Lett. **B96**, 289 (1980)
25. J.P. Blaizot, E. Iancu: Nucl. Phys. **B390**, 589 (1993)
26. J.P. Blaizot, E. Iancu: Phys. Rev. Lett. **70**, 3376 (1993); Nucl. Phys. **B417**, 608 (1994)
27. V.V. Klimov: Sov. J. Nucl. Phys. **33**, 934 (1981); Sov. Phys. JETP **55**, 199 (1982)
28. H.A. Weldon: Phys. Rev. **D26**, 1394 (1982)
29. H.A. Weldon: Phys. Rev. **D26**, 2789 (1982)
30. R. D. Pisarski: Phys. Rev. Lett. **63**, 1129 (1989); E. Braaten, R. D. Pisarski: Nucl. Phys. **B337**, 569 (1990); Nucl. Phys. **B339**, 310 (1990); Phys. Rev. Lett. **64**, 1338 (1990); Phys. Rev. **D42**, 2156 (1990)
31. J. Frenkel, J.C. Taylor: Nucl. Phys. **B334**, 199 (1990); J.C. Taylor, S.M.H. Wong: *ibid.* **B346**, 115 (1990)
32. R. Efraty, V.P. Nair: Phys. Rev. Lett. **68**, 2891 (1992); R. Jackiw, V.P. Nair: Phys. Rev. **D48**, 4991 (1993)
33. E.M. Lifshitz, L.P. Pitaevskii: *Physical Kinetics* (Pergamon Press, Oxford, 1981)
34. V.P. Silin: Sov. Phys. JETP **11**, 1136 (1960)
35. N.L. Balazs, B.K. Jennings: Phys. Rep. **104**, 347 (1984)
36. M. Hillery, R.F. O'Connell, M.O. Scully, E.P. Wigner: Phys. Rep. **106**, 121 (1984)
37. G. Baym, C.J. Pethick: *Landau Fermi-Liquid Theory: Concepts and Applications* (J. Wiley, New York, 1991)
38. S.K. Wong: Nuovo Cim. **65A**, 689 (1970)
39. U. Heinz: Phys. Rev. Lett. **51**, 351 (1983); J. Winter: J.Phys. (Paris) Suppl. **45**, C4-53 (1984); U. Heinz: Ann. Phys. **161**, 48 (1985)
40. H.-Th. Elze, U. Heinz: Phys. Rep. **183**, 81 (1989)
41. P.F. Kelly, Q. Liu, C. Lucchesi, C. Manuel: Phys. Rev. Lett. **72**, 3461 (1994); Phys. Rev. **D50**, 4209 (1994)
42. A. V. Selikhov, M. Gyulassy: Phys. Lett. **B316**, 373 (1993); Phys. Rev. **C49**, 1726 (1994)
43. Yu.A. Markov, M.A. Markova: Theor. Math. Phys. **103**, 444 (1995)
44. U. Kraemmer, A. Rebhan, H. Schulz: Ann. Phys. **238**, 286 (1995)
45. J.P. Blaizot, E. Iancu: Phys. Rev. Lett. **72**, 3317 (1994)
46. J.P. Blaizot, E. Iancu: Phys. Lett. **B326**, 138 (1994)
47. G. Baym, H. Monien, C.J. Pethick, D.G. Ravenhall: Phys. Rev. Lett. **64**, 1867 (1990)
48. P. Arnold, C. Zhai: Phys. Rev. **D50**, 7603 (1994); Phys. Rev. **D51**, 1906 (1995); C. Zhai, B. Kastening: Phys. Rev. **D52**, 7232 (1995). See also, E. Braaten: Phys. Rev. Lett. **74**, 2164 (1995); E. Braaten, A. Nieto: Phys. Rev. **D51**, 6990 (1995)

49. E. Braaten: Phys. Rev. Lett. **74**, 2164 (1995) [hep-ph/9409434]
50. E. Braaten, A. Nieto: Phys. Rev. Lett. **76**, 1417 (1996) [hep-ph/9508406]
51. E. Braaten, A. Nieto: Phys. Rev. **D53**, 3421 (1996) [hep-ph/9510408]
52. K. Kajantie, M. Laine, K. Rummukainen, Y. Schroder: Phys. Rev. Lett. **86**, 10 (2001) [hep-ph/0007109]
53. T. Hatsuda: Phys. Rev. **D56**, 8111 (1997) [hep-ph/9708257]
54. S. Chiku, T. Hatsuda: Phys. Rev. **D58**, 076001 (1998) [hep-ph/9803226]
55. R. R. Parwani: Phys. Rev. **D63**, 054014 (2001) [hep-ph/0010234]
56. R. R. Parwani: Phys. Rev. **D64**, 025002 (2001) [hep-ph/0010294]
57. A. Peshier, B. Kämpfer, O. P. Pavlenko, G. Soff: Phys. Rev. **D54**, 2399 (1996); A. Peshier: `hep-ph/9809379`
58. P. Lévai, U. Heinz: Phys. Rev. **C57**, 1879 (1998) and references therein
59. F. Karsch, A. Patkos, P. Petreczky: Phys. Lett. **B401**, 1997 (69)
60. J. O. Andersen, E. Braaten, M. Strickland: Phys. Rev. Lett. **83**, 2139 (1999), Phys. Rev. **D61**, 014017, 074016 (2000)
61. J. M. Luttinger, J. C. Ward: Phys. Rev. **118**, 1417 (1960); C. De Dominicis, P.C. Martin: J. Math. Phys. **5**, 14, 31 (1964)
62. G. Baym: Phys. Rev. **127**, 1391 (1962)
63. E. Riedel: Z. Phys. **210**, 403 (1968)
64. B. Vanderheyden, G. Baym: J. Stat. Phys. **93**, 843 (1998)
65. G. M. Carneiro, C. J. Pethick: Phys. Rev. **B11**, 1106 (1975)
66. I. T. Drummond, R. R. Horgan, P. V. Landshoff, A. Rebhan: Nucl. Phys. **B524**, 579 (1998) [hep-ph/9708426]
67. G. Boyd et al.: Nucl. Phys. **B469**, 419 (1996)

Thermal Gauge Field Theories

Anton Rebhan

Institut für Theoretische Physik, Technische Universität Wien
Wiedner Hauptstr. 8-10, A-1040 Vienna, Austria

1 Overview

The theoretical framework for describing ultrarelativistically hot and dense matter is quantum field theory at finite temperature and density. At sufficiently high temperatures and densities, asymptotic freedom should make it possible to describe even the fundamental theory of strong interactions, quantum chromodynamics (QCD), through analytical and mostly perturbative means. This article tries to cover both principal issues related to gauge freedom as well as specific problems of thermal perturbation theory in non-Abelian gauge theories.

After a brief review of the imaginary- and real-time formalisms of thermal field theory, the latter is extended to gauge theories. Aspects of different treatments of Faddeev-Popov ghosts and different gauge choices are discussed for general non-Abelian gauge theories, both in the context of path integrals and in covariant operator quantization. The dependence of the formalism on the gauge-fixing parameters introduced in perturbation theory is investigated in detail. Only the partition function and expectation values of gauge-invariant observables are entirely gauge independent. Beyond those it is shown that the location of singularities of gauge and matter propagators, which define screening behaviour and dispersion laws of the corresponding quasi-particle excitations, are gauge independent when calculated systematically.

At soft momentum scales it turns out to be necessary to reorganize perturbation theory such that (at least) the contribution of the so-called hard thermal (dense) loops (HTL/HDL) is resummed. The latter form a gauge-invariant effective action, and their gauge-fixing independence is verified. The existing results of such resummations on the modification of the spectrum of HTL quasi-particles at next-to-leading order (NLO) are reviewed, and a few cases discussed in more detail, with special attention given to gauge dependence questions.

It is shown how screening and damping phenomena become logarithmically sensitive to the strictly nonperturbative physics of the chromomagnetostatic sector, with the exception of the case of zero 3-momentum. In particular the definition of a non-Abelian Debye mass is discussed at length, also with respect to recent lattice results.

Real parts of the dispersion laws of quasiparticle excitations, on the other hand, are infrared-safe at NLO. However, as the additional collective modes of longitudinal plasmons and fermionic "plasminos" approach the light-cone, collinear singularities arise, signalling a qualitative change of the dispersion laws and requiring additional resummations. At high momenta only the regular modes of the fermions and the spatially transverse ones of the gauge bosons remain and they tend to asymptotic mass hyperboloids. The NLO corrections to the asymptotic thermal masses play an interesting role in a self-consistent reformulation of thermodynamics in terms of weakly interacting quasi-particles. In this application, the problem of poor convergence of resummed thermal perturbation theory resurfaces, but it is shown that it may be overcome through approximately self-consistent gap equations.

2 Basic Formulae

Before coming to quantum field theories and gauge theories in particular, let us begin by recalling some relevant formulae from quantum statistical mechanics.

We shall always consider the grand canonical ensemble, in which a system in equilibrium can exchange energy as well as particles with a reservoir such that only mean values of energy and other conserved quantities (baryon/lepton number, charge, etc.) are prescribed through Lagrange multipliers $\beta = 1/T$ and $\alpha_i = -\beta\mu_i$, respectively, where T is temperature and μ_i are the various chemical potentials associated with a set of commuting conserved observables $\hat{N}_i = \hat{N}_i^\dagger$ satisfying $[\hat{N}_i, \hat{N}_j] = 0$ and $[\hat{H}, \hat{N}_i] = 0$, where $\hat{H}$ is the Hamiltonian.

The statistical density matrix is given by

$$\hat{\varrho} = Z^{-1} \exp\left[-\alpha_i \hat{N}_i - \beta \hat{H}\right] \equiv Z^{-1} \exp\Big[-\beta\underbrace{(\hat{H} - \mu_i \hat{N}_i)}_{=:\,\bar{H}}\Big], \tag{1}$$

where Z is the partition function

$$Z = Z(V, T, \mu_i) = \mathrm{Tr}\, \mathrm{e}^{-\beta \bar{H}}, \tag{2}$$

and the thermal expectation value (ensemble average) of an operator $\hat{A}$ is given by

$$\langle \hat{A} \rangle = \mathrm{Tr}[\hat{\varrho}\hat{A}]\,. \tag{3}$$

When total energy and particle numbers are extensive quantities[1], i.e. proportional to the volume V, one also has $\ln Z \propto V$, and since we shall be interested in the limit $V \to \infty$, it is preferable to define intensive quantities. The most important one is the thermodynamic pressure

$$P = \frac{1}{\beta V} \ln Z, \tag{4}$$

[1] A notable exception occurs when general relativity has to be included.

which up to a sign is identical to the free energy density $F/V = -P$ (F is also referred to as *the* thermodynamic potential Ω).

Other thermodynamic (or should one say thermo-static?) quantities can be derived from P, such as particle/charge densities

$$n_i = \frac{1}{V}\langle \hat{N}_i \rangle = \frac{\partial P}{\partial \mu_i}, \tag{5}$$

energy density

$$\varepsilon = E/V = \frac{1}{V}\langle \hat{H} \rangle = -\frac{1}{V}\frac{\partial \ln Z}{\partial \beta} = -\frac{\partial(\beta P)}{\partial \beta} \tag{6}$$

(at fixed α_i), and entropy density (which will play a prominent role at the very end of this article)

$$\begin{aligned} s = S/V &= \frac{1}{V}\langle -\ln \hat{\varrho} \rangle = \frac{1}{V}\ln Z + \frac{\beta}{V}\langle \hat{H} - \mu_i \hat{N}_i \rangle \\ &= \frac{\partial P}{\partial T} = \beta(P + \varepsilon - \mu_i n_i). \end{aligned} \tag{7}$$

In the latter equations one recognizes the familiar Gibbs-Duhem relation

$$E = -PV + TS + \mu_i N_i, \tag{8}$$

which explains why P was defined as the (thermodynamic) pressure. A priori, the hydrodynamic pressure, which is defined through the spatial components of the energy-momentum tensor through $\frac{1}{3}\langle T^{ii} \rangle$, is a separate object. In equilibrium, it can be identified with the thermodynamic one through scaling arguments [92], which, however, do not allow for the possibility of scale (or "trace") anomalies that occur in all quantum field theories with non-zero β-function (such as QCD). In [51] it has been shown recently that the very presence of the trace anomaly can be used to prove the equivalence of the two in equilibrium.

All the above relations continue to hold in (special) relativistic situations, namely within the particular inertial frame in which the heat bath is at rest. In other inertial frames one has the additional quantity of heat-bath 4-velocity u^μ, and one can generalize the above formulae by replacing $V = \int d^3x \to \int_{\Sigma \perp u} d\Sigma_\mu u^\mu$, and the operators in (1) by

$$H \to \int d\Sigma_\mu T^{\mu\nu} u_\nu, \qquad N_i \to \int d\Sigma_\mu j_i^\mu. \tag{9}$$

α_i and β are Lorentz scalars (i.e., temperature is by definition measured in the rest frame of the heat bath). The partition function can then be written in covariant fashion as [69]

$$Z = \mathrm{Tr}\left[\exp \int d\Sigma_\mu \left(-\hat{T}^{\mu\nu}\beta_\nu - \hat{j}_i^\mu \alpha_i\right)\right], \tag{10}$$

where we have introduced an inverse-temperature 4-vector $\beta^\mu \equiv \beta u^\mu$.

However, in what follows we shall most of the time remain in the rest frame of the heat bath where $u^\mu = \delta_0^\mu$.

3 Complex Time Paths

With a complete set of states given by the eigenstates of an operator $\hat\varphi$, $\hat\varphi|\varphi\rangle = \varphi|\varphi\rangle$, we formally write

$$Z = \mathrm{Tr}\left[\mathrm{e}^{-\beta\hat H}\right] = \sum_\varphi \langle\varphi|\mathrm{e}^{-\beta\hat H}|\varphi\rangle. \tag{11}$$

In field theory, we shall boldly use the field operator $\hat\varphi = \hat\varphi(t,\mathbf{x})$ in the Heisenberg picture and write $|\varphi\rangle$ for its eigenstates at a particular time.

Using that the transition amplitude $\langle\varphi_1|\mathrm{e}^{-\mathrm{i}\hat H(t_1-t_0)}|\varphi_0\rangle$ [the overlap of states that have eigenvalue $\varphi_0(\mathbf{x})$ at time t_0 with states that have eigenvalue $\varphi_1(\mathbf{x})$ at time t_1] has the path integral representation

$$\begin{aligned}&\langle\varphi_1|\mathrm{e}^{-\mathrm{i}\hat H(t_1-t_0)}|\varphi_0\rangle\\ &= \mathcal{N}\int\limits_{\substack{\varphi(t_0,\mathbf{x})=\varphi_0\\ \varphi(t_1,\mathbf{x})=\varphi_1}} \mathcal{D}\varphi\, \exp\mathrm{i}\int_{t_0}^{t_1} dt \int d^3x\, \mathcal{L}(\varphi,\partial\varphi)\end{aligned} \tag{12}$$

we can give a path integral representation for Z that takes care of the density operator by setting $t_1 = t_0 - \mathrm{i}\beta$, and of the trace by integrating over all configurations with $\varphi_1 = \varphi_0$.

When there is a chemical potential $\mu \neq 0$, we have $\hat H \to \bar H = \hat H - \mu\hat N$ and this implies $L \to L + \mu N$ if there are no time derivatives in N, and we have

$$Z = \mathcal{N}\int \mathcal{D}\varphi\, \exp\mathrm{i}\int_{t_0}^{t_0-\mathrm{i}\beta} dt \int d^3x\, \bar{\mathcal{L}}, \tag{13}$$

where the path integral is over all configurations periodic in imaginary time, $\varphi(t_0,\mathbf{x}) = \varphi(t_0 - \mathrm{i}\beta, \mathbf{x})$.

In this formula, real time has apparently been fixed to t_0 and replaced by an imaginary time flow which is periodic with period β, the inverse temperature. In equilibrium thermodynamics, this seems only fitting as nothing depends on time in strict equilibrium.

However, we have not really required time to have a fixed real part. We have made the end point complex with the same real part, but the integration over t in (13) need not be a straight line with fixed t_0. Instead we shall consider a general complex time path, and this allows us to define Green functions by the path integral formula

$$\langle \mathrm{T}_c\hat\varphi_1\cdots\hat\varphi_n\rangle = \mathcal{N}\int \mathcal{D}\varphi\,\varphi_1\cdots\varphi_n \exp\mathrm{i}\int_C dt\int d^3x\,\bar{\mathcal{L}}, \tag{14}$$

where T_c now means contour ordering along the complex time path $\mathcal{C}$ from t_0 to $t_0 - \mathrm{i}\beta$ such that $t_i \in \mathcal{C}$, and $t_1 \succeq t_2 \succeq \cdots \succeq t_n$ with respect to a monotonically increasing contour parameter.

Through quantities like (14) we are not restricted to time-independent, thermo-static questions, but may also consider small perturbations of the equilibrium (response theory).

The time path introduced in (14) is in fact not completely arbitrary. Considering spectral representations in the energy representation leads to the conclusion [92] that $\mathcal{C}$ has to be such that the imaginary part of t is monotonically decreasing. This is a necessary condition for analyticity; in the limiting case of a constant imaginary part along (parts of) the contour, distributional quantities (generalized functions) arise.

Except for the periodic boundary conditions with regard to the end points of $\mathcal{C}$ (which become anti-periodic for fermionic field operators and Grassmann-valued "classical" fields), the path integral formula (14) is formally identical to the familiar one from $T = 0$ and $\mu = 0$.

Indeed, perturbation theory is set up in the usual fashion. Using the interaction-picture representation one can derive

$$\langle \mathrm{T}_c \mathcal{O}_1 \cdots \mathcal{O}_n \rangle = \frac{Z_0}{Z} \langle \mathrm{T}_c \mathcal{O}_1 \cdots \mathcal{O}_n \, \mathrm{e}^{\mathrm{i} \int_C \mathcal{L}_I} \rangle_0, \tag{15}$$

where $\mathcal{L}_I$ is the interaction part of $\mathcal{L}$, and the correlators on the right-hand-side can be evaluated by a Wick(-Bloch-DeDominicis) theorem:

$$\langle \mathrm{T}_c \mathrm{e}^{\mathrm{i} \int_C d^4x\, j\varphi} \rangle_0 = \exp\{-\frac{1}{2} \int_C \int_C d^4x\, d^4x' j(x) D^c(x - x') j(x')\}, \tag{16}$$

where D^c is the 2-point function and this is the only building block of Feynman graphs with an explicit T and μ dependence. It satisfies the KMS (Kubo-Martin-Schwinger) condition

$$D^c(t - \mathrm{i}\beta) = \pm \mathrm{e}^{-\mu\beta} D^c(t), \tag{17}$$

stating that $\mathrm{e}^{\mathrm{i}\mu t} D(t)$ is periodic (antiperiodic) for bosons (fermions).

3.1 Imaginary-Time (Matsubara) Formalism

The simplest possibility for choosing the complex time path is the straight line from t_0 to $t_0 - \mathrm{i}\beta$, which is named after Matsubara [97] who first formulated perturbation theory based on this contour. It is also referred to as imaginary-time formalism (ITF), because for $t_0 = 0$ one is exclusively dealing with imaginary times.

Because of the (quasi-)periodicity (17), the propagator is given by a Fourier series

$$D^c(t) = \frac{1}{-\mathrm{i}\beta} \sum_\nu \tilde{D}(z_\nu) \mathrm{e}^{-\mathrm{i}z_\nu t}, \quad \tilde{D}(z_\nu) = \int_0^{-\mathrm{i}\beta} dt\, D^c(t) \mathrm{e}^{\mathrm{i}z_\nu t} \tag{18}$$

with discrete complex (Matsubara) frequencies

$$z_\nu = 2\pi i\nu/\beta + \mu, \quad \nu \in \begin{cases} \mathbb{Z} & \text{bos.} \\ \mathbb{Z} - \frac{1}{2} & \text{ferm.} \end{cases} \tag{19}$$

The transition to Fourier space turns the integrands of Feynman diagrams from convolutions to products as usually, with the difference that there is no longer an integral but a discrete sum over the frequencies, and compared to standard momentum-space Feynman rules one has

$$\int \frac{d^4k}{i(2\pi)^4} \to \beta^{-1} \sum_\nu \int \frac{d^3k}{i(2\pi)^3}, \qquad i(2\pi)^4\delta^4(k) \to \beta(2\pi)^3\delta_{\nu,0}\delta^3(k). \tag{20}$$

However, all Green functions that one can calculate in this formalism are initially defined only for times on $\mathcal{C}$, so all time arguments have the same real part. The analytic continuation to different times on the real axis is, however, frequently a highly non-trivial task [92], so that it can be advantageous to use a formalism that supports real time arguments from the start.

3.2 Real-Time (Schwinger-Keldysh) Formalism

In the so-called real-time formalisms, the complex time path $\mathcal{C}$ is chosen such as to include the real-time axis from an initial time t_0 to a final time t_f. Since we have to end up at $t_0 - i\beta$, this requires a further part of the contour to run backward in real time [118,19] and to finally pick up the imaginary time $-i\beta$. There are a couple of paths $\mathcal{C}$ that have been proposed in the literature. The oldest one due to Keldysh [78] is shown in Fig. 1, and this is also (again) the most popular one.

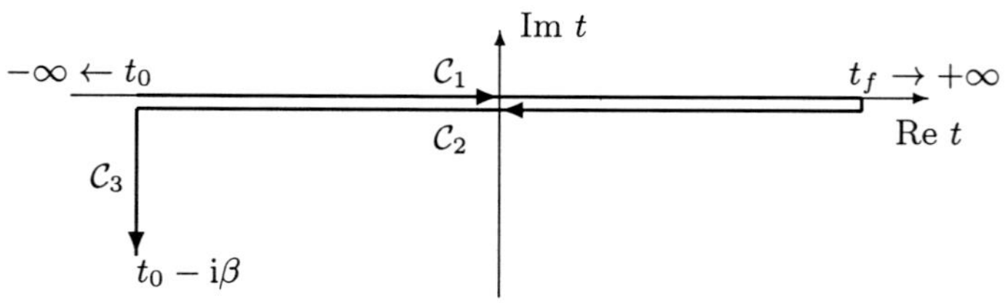

Fig. 1. Complex time path in the Schwinger-Keldysh real-time formalism

Clearly, if none of the field operators in (14) has a time argument on $\mathcal{C}_1$ or $\mathcal{C}_2$, the contributions from these parts of the contour simply cancel and one is back to the ITF.

On the other hand, if $t_0 \to -\infty$, and all operators have finite real time arguments, the contribution from contour $\mathcal{C}_3$ decouples because from the spectral representation one has for instance for the propagator connecting contour $\mathcal{C}_1$ and $\mathcal{C}_3$

$$D^{13}(k, t - (t_0 - \mathrm{i}\lambda)) = \int_{-\infty}^{\infty} d\omega\, \mathrm{e}^{-\mathrm{i}\omega(t-t_0)} \frac{\sigma \mathrm{e}^{\lambda\omega}}{\mathrm{e}^{\beta\bar{\omega}} - 1} \varrho(k, \omega) \stackrel{t_0 \to -\infty}{\longrightarrow} 0 \tag{21}$$

for $\lambda \in (0, \beta)$ by Riemann-Lebesgue [92].[2]

With only $\mathcal{C}_1$ and $\mathcal{C}_2$ contributing, the action in the path integral decomposes according to

$$\int_{\mathcal{C}_1 \cup \mathcal{C}_2} \mathcal{L}(\varphi) = \int_{-\infty}^{\infty} dt\, \mathcal{L}(\varphi^{(1)}) - \int_{-\infty}^{\infty} dt\, \mathcal{L}(\varphi^{(2)}, \tag{22}$$

where we have to distinguish between fields of type 1 (those from contour $\mathcal{C}_1$) and of type 2 (those from contour $\mathcal{C}_2$) because of the prescription of contour ordering in (14).[3] From (22) it follows that type-1 fields have vertices only among themselves, and the same holds true for the type-2 fields. However, the two types of fields are coupled through the propagator, which is a 2×2 matrix with non-vanishing off-diagonal elements:

$$\mathbf{D}^c(t, t') = \begin{pmatrix} \langle \mathrm{T}\varphi(t)\varphi(t')\rangle & \sigma\langle\varphi(t')\varphi(t)\rangle \\ \langle\varphi(t)\varphi(t')\rangle & \langle\tilde{\mathrm{T}}\varphi(t)\varphi(t')\rangle \end{pmatrix}. \tag{23}$$

Here $\tilde{\mathrm{T}}$ denotes anti-timeordering for the 2-2 propagator and σ is a sign which is positive for bosons and negative for fermions. The off-diagonal elements do not need a time-ordering symbol because type-2 is by definition always later (on the contour) than type-1.

In particular, for a massive scalar field one obtains

$$\mathbf{D}^c(k) = \begin{pmatrix} \frac{\mathrm{i}}{k^2 + m^2 + \mathrm{i}\varepsilon} & 2\pi\delta^-(k^2 - m^2) \\ 2\pi\delta^+(k^2 - m^2) & \frac{-\mathrm{i}}{k^2 + m^2 - \mathrm{i}\varepsilon} \end{pmatrix}$$
$$+ 2\pi\delta(k^2 - m^2)\frac{1}{\mathrm{e}^{\beta|k_0|} - 1}\begin{pmatrix} 1 & 1 \\ 1 & 1 \end{pmatrix}, \tag{24}$$

where $\delta^{\pm}(k^2 - m^2) = \theta(\pm k_0)\delta(k^2 - m^2)$. The specifically thermal contribution is that of the second line. Mathematically, it is a homogeneous Green function, as it should be, because it is proportional to $\delta(k^2 - m^2)$. Physically, this part corresponds to Bose-Einstein-distributed, real particles on mass-shell.

[2] There are cases where this line of reasoning breaks down. The decoupling of the vertical part of the contour in RTF does, however, take place provided the statistical distribution function in the free RTF propagator defined below in (24) does have as its argument $|k_0|$ and not the seemingly equivalent $\sqrt{\mathbf{k}^2 + m^2}$ [104,62].

[3] Type-2 fields are sometimes called "thermal ghosts", which misleadingly suggests that type-1 fields are physical and type-2 fields unphysical. In fact, they differ only with respect to the time-ordering prescriptions they give rise to.

The matrix propagator (24) can also be written in a diagonalized form [98,99]

$$\mathbf{D}^c(k) = \mathbf{M}(k_0)\begin{pmatrix} iG_F & 0 \\ 0 & -\mathrm{i}G_F^* \end{pmatrix}\mathbf{M}(k_0) \tag{25}$$

with $G_F \equiv 1/(k^2 + m^2 + \mathrm{i}\varepsilon)$ and

$$\mathbf{M}(k_0) = \frac{1}{\sqrt{\mathrm{e}^{\beta|k_0|} - 1}}\begin{pmatrix} \mathrm{e}^{\frac{1}{2}\beta|k_0|} & \mathrm{e}^{-\frac{1}{2}\beta k_0} \\ \mathrm{e}^{\frac{1}{2}\beta k_0} & \mathrm{e}^{\frac{1}{2}\beta|k_0|} \end{pmatrix}. \tag{26}$$

In the $T \to 0$ limit one has

$$\mathbf{M}(k_0) \stackrel{\beta\to\infty}{\longrightarrow} \mathbf{M}_0(k_0) = \begin{pmatrix} 1 & \theta(-k_0) \\ \theta(k_0) & 1 \end{pmatrix} \tag{27}$$

so that one still has propagators connecting fields of different type. However, if all the external lines of a diagram are of the same type, then also all the internal lines are, because $\prod_i \theta(k^0_{(i)}) = 0$ when $\sum_i k^0_{(i)} = 0$ and any connected region of the other field-type leads to a factor of zero.

4 Gauge Theories – Feynman Rules

As a simple application of the formalism developed above and as a demonstration of the need for more formalism for gauge theories, let us try to calculate the thermodynamic pressure of photons in the imaginary-time formalism and in a covariant gauge. (There is no need for the real-time formalism here, because we are not considering Green functions external lines.) The simplest gauge to perform calculations is usually Feynman's gauge, which simplifies the Lagrangian of the electromagnetic fields according to $\mathcal{L} = -\frac{1}{4}F_{\mu\nu}F^{\mu\nu} \to \frac{1}{2}A_\nu \Box A^\nu$.

One would therefore expect the partition function to be given by

$$\int_{\text{periodic}} \mathcal{D}A \,\exp \mathrm{i}\int_0^{-\mathrm{i}\beta} dt\, d^3x\, \mathcal{L} = \text{const.}\times(\det\Box)^{-\frac{1}{2}\times 4}_{\text{periodic}},$$

and the thermal pressure would be calculated from

$$\begin{aligned}\ln Z &= -4 \times \frac{1}{2}\mathrm{Tr}\ln\Box + \text{const} \\ &= -4 \times \frac{1}{2}V\sum_n \int \frac{d^3k}{(2\pi)^3}\ln\left(\omega_n^2 + k^2\right) + \text{const} \\ &= 4V\int\frac{d^3k}{(2\pi)^3}\left[-\frac{\beta k}{2} - \ln(1 - \mathrm{e}^{-\beta k})\right] + \text{const}\end{aligned}$$

as

$$P(T) - P(0) = \frac{1}{\beta V}\ln Z = 4T\int\frac{d^3k}{(2\pi)^3}\ln(1 - \mathrm{e}^{-\beta k})^{-1} = 2 \times \frac{\pi^2 T^4}{45}$$

giving *twice* the correct result of Planck for black-body radiation.

The error we made is that we have not calculated $\mathrm{Tr} e^{-\beta \bar{H}}$ in a physical Hilbert space (in fact in no Hilbert space at all, because there are negative-norm states). Instead of only two physical (transverse) degrees of freedom, we have added up the contributions from four. The standard way to get rid of the unphysical degrees of freedom is to cancel their contribution by ghost contributions, which evidently are required already in the Abelian case.

4.1 Path Integral – Faddeev-Popov Trick

Because of gauge invariance under $\delta A_\mu^a = D_\mu^{ab}(A)\omega^b$ (where a, b are possible color indices), there is a redundancy in the path integral that leads to zero modes in the kinetic kernels, making them non-invertible. Thus, in order to be able to write down propagators and do perturbation theory, one needs to remove this redundancy. Using a suitable gauge fixing function $F^a[A](x)$, this can be done by inserting

$$\prod_{a,x} \delta\left(F^a[A](x)\right) \cdot \det \frac{\partial F^a}{\partial \omega^b} \tag{28}$$

into the measure of the path integral, selecting only one gauge field configuration per gauge orbit.

One can equally well use $F^a[A](x) - \zeta^a(x)$ in place of $F^a[A](x)$ with arbitrary functions ζ^a and perform a Gaussian average

$$\int \mathcal{D}\zeta \, e^{\frac{i}{2\alpha}\zeta^2} \ldots$$

over the latter. This gives so-called general or inhomogeneous gauge breaking terms

$$\mathcal{L} \to \mathcal{L} + \frac{1}{2\alpha} F^a[A]^2 \tag{29}$$

with a gauge fixing parameter α.

In covariant gauges $F^a[A](x) = \partial^\mu A_\mu^a(x)$ and Abelian electromagnetism, where a takes only one value and $D_\mu^{ab}(A) \to \partial_\mu$, the determinant in (28) is

$$\det \frac{\partial F}{\partial \omega} = \det \Box = (\det \Box)^{+\frac{1}{2}\times 2},$$

so this indeed compensates for the two unphysical degrees of freedom in the above miscalculation of black-body radiation.

Usually, this "Faddeev-Popov determinant" does not play a role in QED because it is field independent. For calculating thermodynamic potentials, it does, because this determinant depends on temperature through the boundary conditions.

In non-Abelian gauge theories, $\det \frac{\partial F^a}{\partial \omega^b} = \det\left(\frac{\partial F^a}{\partial A^b_\mu} D^{bc}_\mu(A)\right)$ is field dependent, and it is convenient to introduce anticommuting and real[4] Faddeev-Popov ghost fields

$$\int \mathcal{D}c\mathcal{D}\bar{c}\,\exp \mathrm{i}\int_C \bar{c}^a \frac{\partial F^a}{\partial A^b_\mu} D^{bc}_\mu(A) c^c = \text{const.} \times \det\left(\frac{\partial F^a}{\partial A^b_\mu} D^{bc}_\mu(A)\right). \tag{30}$$

The correct boundary conditions are clearly those of the gauge potentials and thus are periodic in imaginary time despite the fact that the ghosts are anti-commuting and thus behave like fermions with regard to combinatorial factors in front of Feynman diagrams [20].

4.2 Covariant Operator Quantization

While the path integral makes it evident how to treat ghosts at finite temperature, one can arrive at the same conclusion without recourse to path integrals in covariant (BRS) operator quantization [86,87], where at first it is somewhat surprising that anticommuting fields should end up having periodic rather than antiperiodic boundary conditions in imaginary time.

BRS quantization is preferably done with Lagrange multiplier fields B and a gauge-fixed Lagrangian (in general covariant gauges)

$$\mathcal{L} = \mathcal{L}_{\text{inv}} - A^a_\mu \partial^\mu B^a + \frac{\alpha}{2} B^a B^a - \mathrm{i}(\partial^\mu \bar{c}^a) D_\mu c \tag{31}$$

with c and $\bar{c}$ being anticommuting field operators.

The gauge-fixed Lagrangian possesses a global fermionic (BRS) symmetry, which in any gauge reads

$$\begin{aligned} &[\mathrm{i}Q_{\text{BRS}}, A_\mu] = D_\mu c, \qquad &&\{\mathrm{i}Q_{\text{BRS}}, c\} = -\tfrac{g}{2} c \times c, \\ &[\mathrm{i}Q_{\text{BRS}}, B] = 0, &&\{\mathrm{i}Q_{\text{BRS}}, \bar{c}\} = iB, \end{aligned} \tag{32}$$

where we used a vectorial notation to write for instance $D_\mu c = (\partial_\mu + g A_\mu \times) c$. In covariant gauges for instance this is generated by

$$Q_{\text{BRS}} = \int d^3x \left[B \cdot D_0 c - c \cdot \partial_0 B + \frac{\mathrm{i}g}{2} (\partial_0 \bar{c}) \cdot (c \times c) \right]. \tag{33}$$

The BRS operator is nilpotent, $Q^2_{\text{BRS}} = 0$ and commutes with Lagrangian and Hamiltonian, $[\mathrm{i}Q_{\text{BRS}}, \mathcal{L}] = 0 = [\mathrm{i}Q_{\text{BRS}}, \mathcal{H}]$.

There is one further global symmetry, ghost number, with conserved charge

$$N_c = \int d^3x \, [\partial_0 \bar{c} \cdot c - \bar{c} \cdot D_0 c] \tag{34}$$

[4] $\bar{c}$ is not the conjugate of c, but an independent field.

satisfying

$$[N_c, c] = c, \quad [N_c, \bar{c}] = -\bar{c}, \quad [N_c, A_\mu] = 0, \quad [N_c, B] = 0. \tag{35}$$

N_c is anti-Hermitian, $N_c = -N_c^\dagger$, although it has real eigenvalues $n_{\rm gh}$, which is possible because our arena is a non-Hilbert space $\mathcal{V}$ containing negative-norm states.

The negative-norm states can be eliminated by the linear condition

$$\mathcal{V} \to \mathcal{V}_{\rm phys} : Q_{\rm BRS}|{\rm phys}\rangle = 0, \tag{36}$$

and the true physical Hilbert space is finally obtained by modding out zero-norm states,

$$\mathcal{H}_{\rm phys} = \overline{\mathcal{V}_{\rm phys}/\mathcal{V}_0}\,. \tag{37}$$

The corresponding projection operator $\mathcal{P}$ in $\mathcal{H}_{\rm phys} = \mathcal{P}\mathcal{V}$ can be shown [86,87] to have a complement that is "BRS exact", meaning

$$\mathcal{P} + \{Q_{\rm BRS}, \mathcal{R}\} = \mathbf{1}, \tag{38}$$

but the actual construction of these operators is rather complicated.

However, we apparently need them to be able to define the trace restricted to the physical Hilbert space that appears in $Z = {\rm Tr}\big|_{\mathcal{H}_{\rm phys}} {\rm e}^{-\beta H} = {\rm Tr}\,\mathcal{P}\,{\rm e}^{-\beta H}$ or in expectation values of observables.

Hata-Kugo Trick. There is however an elegant trick that avoids the explicit construction of $\mathcal{P}$ [66]: From $[N_c, Q_{\rm BRS}] = Q_{\rm BRS}$ it follows that $N_c^n Q_{\rm BRS} = Q_{\rm BRS}(N_c+1)^n$ and therefore ${\rm e}^{{\rm i}\pi N_c} Q_{\rm BRS} = Q_{\rm BRS}{\rm e}^{{\rm i}\pi(N_c+1)} = -Q_{\rm BRS}{\rm e}^{{\rm i}\pi N_c}$, so

$$\{{\rm e}^{{\rm i}\pi N_c}, Q_{\rm BRS}\} = 0. \tag{39}$$

This, together with $N_c|\psi\rangle = 0$ for $|\psi\rangle \in \mathcal{H}_{\rm phys}$ can be used to write

$$\begin{aligned} Z &= {\rm Tr}\,\mathcal{P}\,{\rm e}^{-\beta H} = {\rm Tr}\,\mathcal{P}\,{\rm e}^{{\rm i}\pi N_c}{\rm e}^{-\beta H} \\ &= {\rm Tr}\,{\rm e}^{{\rm i}\pi N_c}{\rm e}^{-\beta H} - \underbrace{{\rm Tr}\,\{Q_{\rm BRS}, \mathcal{R}\}\,{\rm e}^{{\rm i}\pi N_c}{\rm e}^{-\beta H}}_{{\rm Tr}\,\mathcal{R}\underbrace{\{{\rm e}^{{\rm i}\pi N_c}, Q_{\rm BRS}\}}_{0}{\rm e}^{-\beta H}} \end{aligned} \tag{40}$$

since $[Q_{\rm BRS}, H] = 0$.

So the comparatively simple operator ${\rm e}^{{\rm i}\pi N_c}$ can be used in place of the complicated $\mathcal{P}$ to express Z, and similarly thermal expectation values of gauge-invariant observables $\mathcal{O}$, through a trace in unrestricted $\mathcal{V}$ containing ghosts and other unphysical degrees of freedom,

$$Z = {\rm Tr}\left[{\rm e}^{{\rm i}\pi N_c}{\rm e}^{-\beta H}\right], \quad \langle\mathcal{O}\rangle = Z^{-1}{\rm Tr}\left[{\rm e}^{{\rm i}\pi N_c}{\rm e}^{-\beta H}\mathcal{O}\right]. \tag{41}$$

This result shows that the anticommuting ghosts, which naturally are subject to antiperiodic boundary conditions, acquire a purely imaginary chemical potential

$$\mu_c = \mathrm{i}\pi/\beta. \tag{42}$$

In the ITF, the Matsubara frequencies of the ghosts are therefore

$$z_\nu = 2\pi\mathrm{i}(n-\frac{1}{2})/\beta + \mu_c = 2\pi\mathrm{i}n/\beta, \quad n \in \mathbb{Z} \tag{43}$$

like those of ordinary bosons. Thus, while they do have fermionic combinatorics in Wick contractions and the like, thermodynamically they behave like bosons.

In the RTF, where the matrix-valued propagator (24) can be written as

$$-\mathrm{i}\mathbf{D} = \begin{pmatrix} G_F & 0 \\ 0 & -G_F^* \end{pmatrix} + (G_F - G_F^*) \times$$
$$\times \sigma \begin{pmatrix} \theta(\bar{k}_0)n(\bar{k}_0) + \theta(-\bar{k}_0)n(-\bar{k}_0) & \mathrm{sgn}(\bar{k}_0)n(\bar{k}_0) \\ \mathrm{sgn}(\bar{k}_0)(\sigma + n(\bar{k}_0)) & \theta(\bar{k}_0)n(\bar{k}_0) + \theta(-\bar{k}_0)n(-\bar{k}_0) \end{pmatrix} \tag{44}$$

with $\sigma = \pm$ for bosons/fermions, $n(x) = \frac{1}{\mathrm{e}^{\beta x} - \sigma}$, and $\bar{k}_0 = k_0 - \mu$, we have $\sigma = -$ for the Faddeev-Popov ghosts, but

$$n_{\mathrm{FD}}(k_0 - \mathrm{i}\pi/\beta) = \frac{1}{\mathrm{e}^{\beta k_0 - \mathrm{i}\pi} + 1} = -n_{\mathrm{BE}}(k_0), \tag{45}$$

so the imaginary chemical potential (42) in effect negates σ and replaces Fermi-Dirac by Bose-Einstein statistics.

Only the signs arising in Wick contractions are those of fermions, which shows that the Faddeev-Popov ghost propagators have indeed the right form to be able to compensate for unphysical degrees of freedom contained in the gauge boson propagator, which naturally has $\sigma = +$ and Bose-Einstein statistical factors.

Compared to (44), the gauge boson propagator also has a factor $\mathcal{G}_{\mu\nu} = (-g_{\mu\nu} + (1-\alpha)\frac{k_\mu k_\nu}{k^2})$ in covariant gauges. We shall also consider other gauges in what follows, which can be characterized by

$$g_{\mu\nu} \to \mathcal{G}_{\mu\nu} = g_{\mu\nu} - \frac{k^\mu \tilde{f}^\nu + \tilde{f}^\mu k^\nu}{\tilde{f}\cdot k} + (\tilde{f}^2 - \alpha k^2)\frac{k^\mu k^\nu}{(\tilde{f}\cdot k)^2}, \tag{46}$$

where $\tilde{f}$ is the momentum-space version of f in $F^a[A] = f^\mu A^a_\mu$.

Popular gauge choices besides the familiar covariant gauges include axial gauges ($F^a = n^\mu A^a_\mu$, n^μ const.) and Coulomb gauge(s) ($F^a = \partial^i A^a_i$).

In axial gauges, ghosts decouple completely, because the Faddeev-Popov determinant $\det(n\cdot\partial)$ is field- and temperature-independent, however they are fraught with technical difficulties, already at zero temperature. The particularly attractive "temporal" gauge $n^\mu = \delta^\mu_0$, which does not cause *additional* breaking of Lorentz symmetry, is unfortunately inconsistent with periodic boundary conditions. Relaxing those leads to rather complicated Feynman rules in the ITF [71], while the RTF version seems more tractable [70], at least it does not appear to be more problematic than at zero temperature.

Coulomb gauge is in fact widely used at finite temperature, because it also does not lead to additional Lorentz symmetry breaking. However, it does have less simple Slavnov-Taylor identities [48] because ghosts do not decouple, although they frequently do not contribute, since their (RTF) propagator does not contain statistical distribution functions.

4.3 Frozen Ghosts

In [90], alternative Feynman rules have been proposed that avoid thermalized ghosts even in covariant gauges. In the RTF one can switch off the interactions as $t_0 \to -\infty$, and define the physical Hilbert space in terms of Abelianized in-states. Without additional Lorentz symmetry breaking, physical in-states can then be chosen as

$$|\mathrm{phys,in}\rangle = |A_{\mathrm{phys.}}\text{-quanta}\rangle|0 \text{ w.r.t. } A_{\mathrm{unphys.}}, B, \bar{c}, c\rangle \tag{47}$$

with $A^{a\mu}_{\mathrm{phys.}}(k) = \mathcal{A}^{\mu\nu}(k)A^a_\nu(k)$ and

$$\mathcal{A}^{0\mu} = 0, \quad \mathcal{A}^{ij} = -(\delta^{ij} - \frac{k^i k^j}{\mathbf{k}^2}). \tag{48}$$

The unphysical states correspond to

$$A^{a\mu}_{\mathrm{unphys.}}(k) = (g^{\mu\nu} - \mathcal{A}^{\mu\nu}(k))\,A^a_\nu(k), \quad B^a, \quad \bar{c}^a, \text{ and } c^a.$$

The interaction-picture free Hamiltonian then separates into two commuting parts

$$H_{0I} = H_{0I}^{\mathrm{phys}} + H_{0I}^{\mathrm{unphys}}, \quad [H_{0I}^{\mathrm{phys}}, H_{0I}^{\mathrm{unphys}}] = 0, \tag{49}$$

and, because $H_{0I}^{\mathrm{unphys}}|\mathrm{phys,in}\rangle = 0$, thermal averages factorize:

$$\sum_{\mathrm{phys}} \langle \mathrm{phys,in}|\mathrm{e}^{-\beta H_{0I}} \cdots A_{\mathrm{phys.}} \cdots A_{\mathrm{unphys.}} \cdots \bar{c} \cdots c \cdots |\mathrm{phys,in}\rangle =$$

$$\sum_{\mathrm{phys}} \langle A_{\mathrm{phys.}}|\mathrm{e}^{-\beta H_{0I}^{\mathrm{phys.}}} \cdots A_{\mathrm{phys.}} \cdots |A_{\mathrm{phys.}}\rangle\langle 0|\cdots A_{\mathrm{unphys.}} \cdots \bar{c} \cdots c \cdots |0\rangle$$

with the thermal Wick theorem applying to the first factor under the latter sum, and the $T = 0$ Wick theorem to the second one.

This leads to alternative Feynman rules for gauge theories in RTF in which only the transverse projection of the gauge bosons have thermal (matrix) propagators, and all other fields have only the $T = 0$ limits of those. The rest of the Feynman rules is as usual in RTF.

E.g., in Feynman gauge ($\alpha = 1$) the gauge boson and ghost propagators now read

$$\mathbf{D}_{\mu\nu} = -\mathrm{i}\mathcal{A}_{\mu\nu}\mathbf{M}\begin{pmatrix} G_F & 0 \\ 0 & -G_F^* \end{pmatrix}\mathbf{M} - \mathrm{i}(g_{\mu\nu} - \mathcal{A}_{\mu\nu})\mathbf{M}_0\begin{pmatrix} G_F & 0 \\ 0 & -G_F^* \end{pmatrix}\mathbf{M}_0$$
$$= -g_{\mu\nu}\begin{pmatrix} \frac{\mathrm{i}}{k^2+\mathrm{i}\varepsilon} & 2\pi\delta^-(k^2) \\ 2\pi\delta^+(k^2) & \frac{-\mathrm{i}}{k^2-\mathrm{i}\varepsilon} \end{pmatrix} - 2\pi\delta(k^2)\mathcal{A}_{\mu\nu}n(|k_0|)\begin{pmatrix} 1 & 1 \\ 1 & 1 \end{pmatrix}, \quad (50)$$

$$\mathbf{D}^{\mathrm{gh}} = \begin{pmatrix} \frac{\mathrm{i}}{k^2+\mathrm{i}\varepsilon} & 2\pi\delta^-(k^2) \\ 2\pi\delta^+(k^2) & \frac{-\mathrm{i}}{k^2-\mathrm{i}\varepsilon} \end{pmatrix}. \quad (51)$$

In general linear gauges one has to replace $g_{\mu\nu}$ in the vacuum part according to (46); the thermal part remains unchanged.

Using these Feynman rules simplifies certain calculations in covariant and other gauges [90], because, although ghosts are present, they do not carry statistical distribution functions, but are "frozen". On the other hand, the usual cancellation of pinch singularities in the RTF (absence of "pathologies"), turns out be more complicated, and occurs in general only upon Dyson resummations [91].

5 Gauge Dependence Identities

As we have seen, perturbation theory and its Feynman rules require the introduction of gauge fixing terms. Clearly, physical results have to come out independent of those. We shall therefore now study in detail to what extent perturbative calculations will exhibit dependences on the gauge fixing parameters by considering

$$F^a[A] \to F^a[A] + \delta F^a[A]. \quad (52)$$

With $\delta F^a[A] \propto F^a[A]$, this also comprises a possible variation of the gauge parameter α in (29) or (31).

5.1 Gauge Independence of the Partition Function

Path Integral. The introduction of the gauge fixing term together with the Faddeev-Popov determinant according to (28) was done in a way that picks one representative field configuration from each gauge orbit.[5] So by construction the partition function or averages of gauge-invariant operators are independent of the gauge fixing terms.

[5] At least perturbatively this is guaranteed by the existence of $\partial F/\partial\omega$; non-perturbatively there may be obstructions to worry about.

This can be checked explicitly by noting that the variation (52) of the gauge fixing function can be written as

$$\delta F[A] = \frac{\partial F[A]}{\partial A_\mu} D_\mu[A] \underbrace{\left[\frac{\partial F[A]}{\partial A} \cdot D[A]\right]^{-1} \delta F[A]}_{\tilde{\delta}\xi[A]}. \tag{53}$$

A corresponding change of the gauge breaking term $\frac{1}{2\alpha}(F^a)^2$ is thus equivalent to a gauge transformation $\tilde{\delta} A_\mu = D_\mu[A]\tilde{\delta}\xi$ with the above non-local, field-dependent parameter $\tilde{\delta}\xi[A]$.

The invariant part of the action is, of course, invariant under $A \to A + \tilde{\delta} A$, as are any gauge invariant operators that might have been inserted, so it remains to check that the path integral measure $\mathcal{D}A$ together with the Faddeev-Popov determinant is invariant, too. This can indeed be verified by writing $\tilde{\delta}\mathcal{D}A$ as $\mathrm{tr}\partial\tilde{\delta}A/\partial A$, and $\tilde{\delta}\det[\frac{\partial F}{\partial A}\cdot D] = \det[\frac{\partial F}{\partial A}\cdot D]\times\tilde{\delta}(\mathrm{tr}\log[\frac{\partial F}{\partial A}\cdot D]) = \det[\frac{\partial F}{\partial A}\cdot D]\times\mathrm{tr}([\frac{\partial F}{\partial A}\cdot D]^{-1}\tilde{\delta}[\frac{\partial F}{\partial A}\cdot D])$, and then using that gauge transformations form a group [47,83].

Covariant Operator Formalism. In the covariant operator formalism the Lagrangian in a general gauge F can be written as

$$\mathcal{L} = \mathcal{L}_{\mathrm{inv}} + B \cdot F + \frac{\alpha}{2} B \cdot B - \bar{c} \cdot [Q_{\mathrm{BRS}}, F] \tag{54}$$

and variations of F correspond to BRS transformations with parameter $\bar{c}\cdot\delta F$:

$$\delta\mathcal{L} = B \cdot \delta F - \bar{c} \cdot [Q_{\mathrm{BRS}}, \delta F] = \{Q_{\mathrm{BRS}}, \bar{c} \cdot \delta F\}. \tag{55}$$

It is always possible to construct an operator δG such that $\delta H = \{Q_{\mathrm{BRS}}, \delta G\}$ (if no time derivatives are involved, one simply has $\delta G = -\int d^3x\, \bar{c}\cdot\delta F$).

Gauge independence of the partition function and of thermal averages of gauge invariant operators defined by (41) can be verified as follows [66]: Variations of exponentiated operators can be expressed as

$$\mathrm{e}^{A+\delta B} - \mathrm{e}^{A} = \int_0^\lambda d\lambda\, \mathrm{e}^{\lambda A}\, \delta B\, \mathrm{e}^{(1-\lambda)A} + O(\delta^2) \tag{56}$$

and using this one finds

$$\begin{aligned} \delta \mathrm{Tr}[\mathrm{e}^{-\beta H + \mathrm{i}\pi N_c}\mathcal{O}] &= -\int_0^\beta d\lambda\, \mathrm{Tr}[\mathrm{e}^{-\lambda H}\{Q_{\mathrm{BRS}}, \delta G\}\mathrm{e}^{\lambda H}\mathrm{e}^{-\beta H}\mathrm{e}^{\mathrm{i}\pi N_c}\mathcal{O}] \\ &= -\int_0^\beta d\lambda\, \mathrm{Tr}[\mathrm{e}^{-\lambda H}\,\delta G\, \mathrm{e}^{\lambda H}\underbrace{\{\mathrm{e}^{\mathrm{i}\pi N_c}, Q_{\mathrm{BRS}}\}}_{0}\mathrm{e}^{-\beta H}\mathcal{O}] \end{aligned} \tag{57}$$

because of cyclic invariance of the trace and $[Q_{\mathrm{BRS}}, \mathcal{O}] = 0$ for a gauge-invariant operator $\mathcal{O}$.

5.2 Gauge Dependence of Green Functions

Green functions, i.e. thermal averages of products of field operators like $\langle \mathrm{T}_c \cdots A_\mu \cdots \psi \cdots \bar{\psi} \cdots \rangle$, are, however, gauge-variant objects, and will therefore in general depend on gauge fixing parameters.[6]

In particular, the propagators of gauge and matter fields will contain all sorts of gauge parameter dependences. Yet, they are among the prime objects of linear response theory as they are used to derive the properties of quasiparticles.

Historically, a stimulating failure was the attempt to extract the damping constant of long-wavelength plasmons in a gluon plasma from one-loop thermal perturbation theory. Some of the results that were accumulated in the 80's are summarized in Table 1. These turned out to be gauge independent in algebraic gauges but gauge-parameter dependent in covariant ones. Moreover, in the latter the damping constant came out with the wrong sign which some took as signal of an instability of the perturbative ground state [65,102].

Table 1. Bare one-loop gluonic plasmon damping constant $\gamma(|\mathbf{k}| \to 0)$

$\gamma_{\mathrm{pl.}}/[\frac{g^2TN}{24\pi}]$	gauge	published
$-[\frac{11}{4}+(\frac{\alpha}{2}-2)^2]$	covariant gauges	1980 [75]
$+1$	temporal gauge	1985 [73]
$+1$	Coulomb gauge	1987 [67]
$-[11+\frac{1}{4}(1-\alpha)^2]$	background covariant gauges [3]	1987 [65]
$-\frac{45}{4}$	gauge-independent effective action [123,112]	1987 [65]
-11	gauge-independent pinch technique [44]	1988 [102]
$\vdots$		

It was in particular Pisarski [108] who argued that these results were simply incomplete, and who together with Braaten [38,39,36] devised an appropriate resummation scheme. However, since explicit calculations can only be performed in practice with a rather limited choice of gauge parameters, it is important to investigate more generally how gauge dependent the full propagators are and whether they contain gauge-independent information at all. To this end, we shall first derive rather general "gauge dependence identi-

[6] Notice that gauge invariance and gauge-fixing parameter independence are separate issues: a functional of fields can be gauge invariant and yet depend parametrically on the gauge-fixing function; conversely, independence of gauge fixing parameters (within a class of gauges) does not imply that a particular functional (e.g. of mean fields) is a gauge-invariant one.

ties" and study their consequences for the thermal Green functions of interest [82,83].

Primary Diagrams. In order to unclutter the relevant relations, we shall temporarily switch to the compact notation of DeWitt [47], where a single index i comprises all discrete and continuous field labels (e.g. $i = (A, \mu, a, x)$) and Einstein's summation convention is extended to include integration over all of space and time (the latter along the contour $\mathcal{C}$). This way, φ^i will represent an arbitrary gauge or matter field $\varphi^i = \{A^{a_i}_{\mu_i}(x_i), \psi^{a_i}_{\sigma_i}(x_i), \ldots\}$; only the Faddeev-Popov ghosts fields will be treated separately.

The generating functional of Green functions reads

$$Z[J] = \langle e^{iJ_i\varphi^i} \rangle \quad \text{with} \quad J_i\varphi^i = \int_{\mathcal{C}} d^4x [J_{(A)\mu a}(x) A^{\mu a}(x) + \ldots] \tag{58}$$

and depends implicitly on a gauge fixing functional $F^\alpha[\varphi]$, where $\alpha = (a_\alpha, x_\alpha)$ comprises both a group and a space-time index.

Information on the dependence on F^α can be obtained either by using BRS techniques or equivalently by employing the non-local gauge transformation of (53), which in compact notation reads

$$\delta\varphi^i = D^i_\alpha[\varphi]\delta\xi^\alpha \quad \text{with} \quad \delta\xi^\alpha = \delta\xi^\alpha[\varphi] = -\mathcal{G}^\alpha{}_\beta[\varphi]\, \delta F^\beta[\varphi], \tag{59}$$

where D^i_α is a generalized function containing the gauge generators, and $\mathcal{G}^\alpha{}_\beta[\varphi] = -(F^\beta_{,i} D^i_\alpha)^{-1}[\varphi]$ is the Faddeev-Popov ghost propagator in a background field φ. This immediately gives

$$\delta \ln Z[J] = iJ_j \left\langle D^j_\alpha[\varphi] \mathcal{G}^\alpha{}_\beta[\varphi]\, \delta F^\beta[\varphi] \right\rangle [J] \quad \text{under } F^\alpha \to F^\alpha + \delta F^\alpha. \tag{60}$$

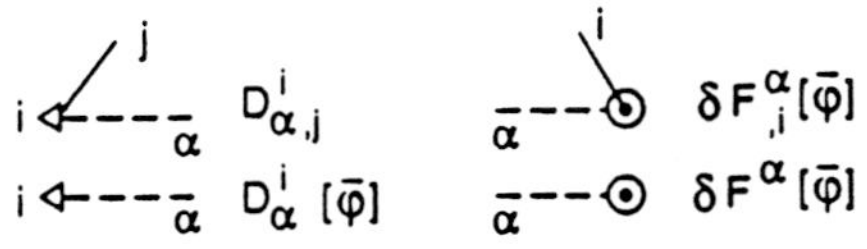

Fig. 2. Additional Feynman rules for $\delta X^j[\bar{\varphi}]$ in the case of linear gauge generators D and linear gauge fixing F

The diagrammatic content of (60) is more conveniently investigated after a Legendre transformation of $W[J] \equiv -i \ln Z[J]$ to the effective action

$$\Gamma[\bar{\varphi}] = W[J] - J_j\bar{\varphi}^j, \qquad \bar{\varphi}^j = \frac{\delta W[J]}{\delta J_j}, \tag{61}$$

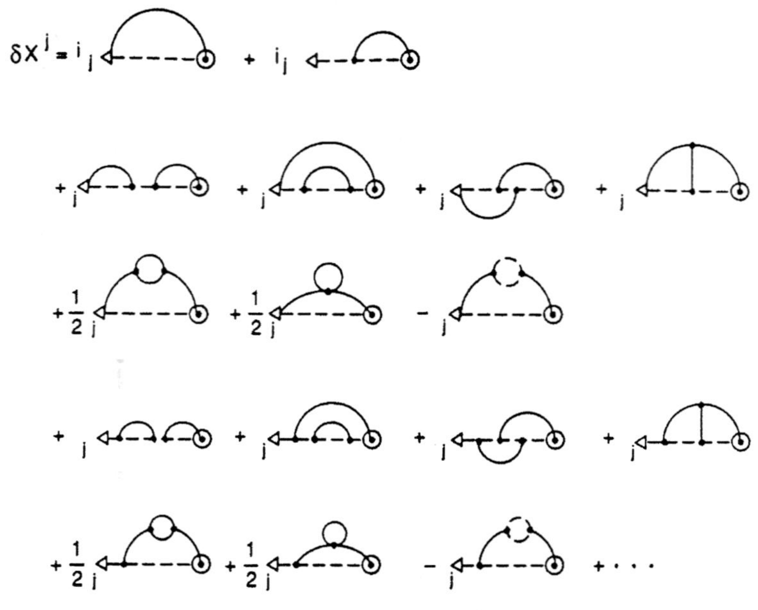

Fig. 3. Primary diagram expansion of $\delta X^j[\bar\varphi]$ through 2-loop order. Contributions involving an undifferentiated $\delta F^\alpha[\bar\varphi]$ have been dropped, which corresponds to omitting the trivial tree-level gauge dependence [83]

which is the generating functional of one-particle-irreducible (1-p-i) diagrams. Equation (60) then becomes

$$\delta\Gamma[\bar\varphi] = \frac{\delta\Gamma[\bar\varphi]}{\delta\bar\varphi^j} \left\langle D^j_\alpha[\varphi]\mathcal{G}^\alpha{}_\beta[\varphi]\,\delta F^\beta[\varphi]\right\rangle[\bar\varphi] \equiv \Gamma_{,j}[\bar\varphi]\,\delta X^j[\bar\varphi]\,. \qquad (62)$$

Diagrammatically, $\Gamma_{,j}[\bar\varphi]$ is the sum of all mean-field dependent (primary) 1-p-i one-point diagrams, while $\delta X^j[\bar\varphi]$ is given by primary diagrams which involve the additional vertices introduced in Fig. 2 and which are 1-p-i except for the basic ghost line attached to δF^β, as shown in Fig. 3.

From these relations one can derive gauge dependence identities for 1-p-i vertex functions by differentiation with respect to $\bar\varphi$. For example, the gauge dependences of the 2-point vertex function (self-energy) $\Gamma_{,ij}[0]$ are determined by the diagrams shown in Fig. 4.

QED. As a first application let us consider the case of an Abelian theory such as QED. The additional Feynman rules of Fig. 2 involve the ghost propagator, but there are no further ghost vertices in the theory (for linear gauge fixing), so only the very first diagram in Fig. 4 arises.

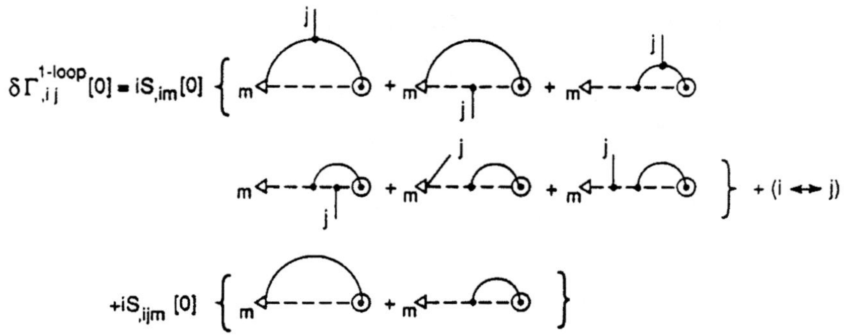

Fig. 4. Gauge dependence of the 2-point vertex function $\Gamma_{ij}[0]$ assuming that all one-point functions vanish at $\bar{\varphi} = 0$

Furthermore, the structure of the gauge generator is such that

$$\text{i ◁- - -} = \begin{cases} \partial^{(x_i)}_{\mu_i} & \text{if } i \leftrightarrow A_{\mu_i}(x_i) \\ 0 & \text{else} \end{cases} \qquad \text{i ◁- - - (with leg j)} = \begin{cases} 0 & \text{if } i,j \leftrightarrow A_\mu \\ \pm ie & \text{if } i,j \leftrightarrow \overset{(-)}{\psi} \end{cases} \tag{63}$$

If the external indices i, j of $\delta\Gamma_{,ij}$ correspond to photons, one finds that one cannot even build the one remaining diagram of Fig. 4, so $\delta\Gamma_{A^\mu A^\nu}$ proves to be completely gauge-fixing independent. This is in fact a well-known result which can be understood also by the gauge invariance of the electromagnetic current operator.

On the other hand, if the external lines are fermionic, there is a non-trivial right-hand-side to Fig. 4, as shown in Fig. 5, so the fermion self-energy is a gauge fixing dependent quantity, already in QED.

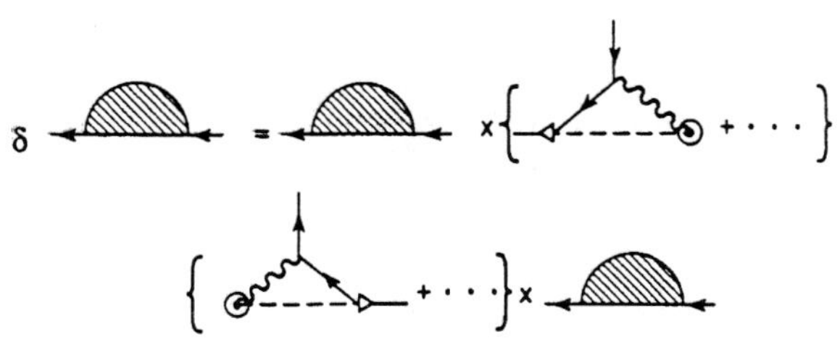

Fig. 5. Gauge dependence identity for the fermion self-energy

Hard Thermal/Dense Loops. In the high-temperature (large-chemical-potential) limit of QED and QCD, it turns out that the leading contributions to the 1-loop vertex functions, the so-called "hard thermal (dense) loops" (HTL/HDL) obey tree-level-type ghost-free Ward identities and appear to be gauge-fixing independent [59,39,37]. This gauge independence, however, does not arise in an obvious way and involves non-trivial cancellations in the various gauges that have been considered.

The above gauge dependence identities can be used to verify the gauge independence of the HTL's in a rather simple manner. The only further ingredients needed are the temperature power-counting rules given in [39], which, roughly, read as follows: in a Feynman diagram, explicit loop momenta in the numerator give a factor T, each propagator counts as T^{-1}, and the sum-integral over the loop momentum contributes T^3 unless there are two or more propagators with the same statistics, in which case the sum-integral counts as T^2.

By this, the leading temperature contributions to a 1-loop vertex function are found to be proportional to T^2, such that an N-point gluon vertex function scales as $\Gamma_{,(N)} \sim g^N T^2 k^{2-N}$ (where k represents generically components of external momenta). If two external lines are fermionic, we have $\Gamma_{,(N)} \sim g^N T^2 k^{1-N}$, while vertex functions with more than two external fermion lines do not contribute terms $\propto T^2$.

Considering e.g. vertex functions with only external gluons, all of the potential gauge dependences of the HTL's are contained in the 1-loop contributions to $\delta\Gamma_{,(N)} = \sum_{M=0 \atop perms.}^{N} \Gamma_{,(N+1-M)} \delta X^{\cdot}_{,(M)}$ with the diagrammatic structure of $\delta X^{\cdot}_{,(M)}$ as given by

$$\delta X^{\cdot}_{,(M)} = \sum_{\text{perm.}} \Bigg\{$$
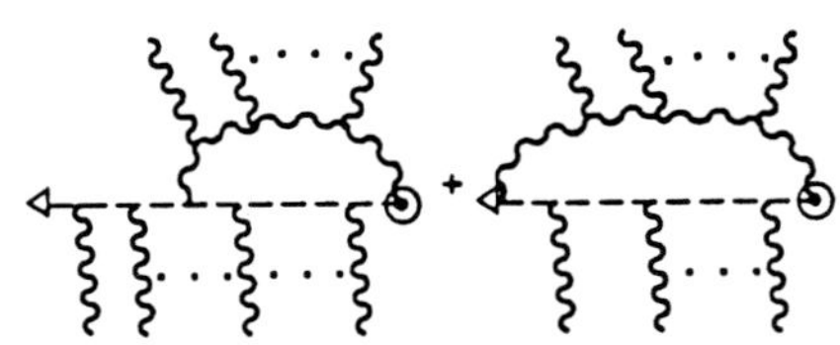
$$\Bigg\}. \qquad (64)$$

The above temperature power-counting rules are modified only by the additional vertex $\delta F^{\alpha}_{,i}[0]$ which may bring in one power of loop momentum and thus one power of T in derivative gauges, or none in algebraic ones. Adding up, one finds that the right-hand-side of (64) is proportional to $T^{0\ldots1}$. Hence, gauge dependences of one-loop vertex functions can occur only at subleading order $\propto T$. HTL(HDL)'s on the other hand are found to be completely gauge-fixing independent.

5.3 Gauge Independence of Propagator Singularities

In non-Abelian gauge theories, all the matter and gauge field vertex functions, self-energies, and propagators contain highly nontrivial gauge dependences, which raises the question whether there is any gauge-independent and therefore potentially physical information in those at all. We shall give an affirmative answer by showing that (the locations of) certain singularities of the full propagators are indeed gauge independent [82].

Non-Abelian Gauge-Boson Propagator. We begin by analysing the Lorentz structure of the gluon propagator in the case of a general gauge that preserves the rotational symmetry. Moreover, we shall simplify things by dropping any color indices, which presupposes the absence of color symmetry breaking.[7]

In momentum space one can define a transverse projection of the 4-velocity of the heat bath, $\tilde{n}^\mu = (g^{\mu\sigma} - \frac{k^\mu k^\sigma}{k^2})u_\sigma$, and use it to write the general structure of the gauge-boson propagator as

$$G^{\mu\nu}(k) = \Delta_A \mathcal{A}^{\mu\nu} + \Delta_B \mathcal{B}^{\mu\nu} + \Delta_C \mathcal{C}^{\mu\nu} + \Delta_D \mathcal{D}^{\mu\nu} \tag{65}$$

with

$$\begin{aligned} \mathcal{A}^{\mu\nu}(k) &= [g^{\mu\nu} - \frac{k^\mu k^\nu}{k^2}] - \frac{\tilde{n}^\mu \tilde{n}^\nu}{\tilde{n}^2}, & \mathcal{B}^{\mu\nu}(k) &= \frac{\tilde{n}^\mu \tilde{n}^\nu}{\tilde{n}^2}, \\ \mathcal{C}^{\mu\nu}(k) &= \frac{1}{|\mathbf{k}|}\{\tilde{n}^\mu k^\nu + k^\mu \tilde{n}^\nu\}, & \mathcal{D}^{\mu\nu}(k) &= \frac{k^\mu k^\nu}{k^2}. \end{aligned} \tag{66}$$

Here $\mathcal{A}^{\mu\nu}$ is the spatially transverse projector introduced already in (48), and $\mathcal{B}^{\mu\nu}$ is a second, independent tensor that is likewise transverse with respect to 4-momentum, but longitudinal with respect to 3-momentum. $\mathcal{C}^{\mu\nu}$ and $\mathcal{D}^{\mu\nu}$ complete the basis of symmetric tensors, with $\mathcal{C}^{\mu\nu}$ chosen such that $k_\mu \mathcal{C}^{\mu\nu} k_\nu = 0$, and $\mathcal{D}^{\mu\nu}$ longitudinal with respect to 4-momentum.

$\mathcal{A}$, $\mathcal{B}$, and $\mathcal{D}$ are idempotent, whereas $\mathcal{C}^2 = -(\mathcal{B} + \mathcal{D})$. Under a Lorentz trace, products of one such tensor with a different one vanish; without trace, $\mathcal{A}$ is orthogonal to all the others, but among the rest one only has $\mathcal{B} \perp D$.

Similarly, we shall decompose the gluon self-energy $\Pi^{\mu\nu} = G^{-1\mu\nu} - G_0^{-1\mu\nu}$ according to

$$\Pi^{\mu\nu} = -\Pi_A \mathcal{A}^{\mu\nu} - \Pi_B \mathcal{B}^{\mu\nu} - \Pi_C \mathcal{C}^{\mu\nu} - \Pi_D \mathcal{D}^{\mu\nu} . \tag{67}$$

At momentum scales $\omega, k \ll T, \mu$, the leading-order term in the one-loop polarisation tensor $\Pi^{\mu\nu}$ is given by the HTL (HDL) $\sim \max(T^2, \mu^2)$, which has only 4-d-transverse contributions

$$\Pi_A^{\rm HTL} = \frac{1}{2}(\Pi^{\rm HTL}{}_\mu{}^\mu - \Pi_B^{\rm HTL}), \tag{68a}$$

$$\Pi_B^{\rm HTL} = -\frac{k^2}{\mathbf{k}^2}\Pi_{00}^{\rm HTL}, \tag{68b}$$

$$\Pi_C^{\rm HTL} = 0, \tag{68c}$$

$$\Pi_D^{\rm HTL} = 0, \tag{68d}$$

[7] The extension of the following results to color superconducting situations has not yet been worked out, but would be of great interest in view of the gauge dependence issues there [111].

where, at high T,

$$\Pi^{\mathrm{HTL}}{}_{\mu}{}^{\mu} = \frac{e^2T^2}{3}, \tag{69}$$

$$\Pi^{\mathrm{HTL}}_{00} = \frac{e^2T^2}{3}\left(1 - \frac{k^0}{2|\mathbf{k}|}\ln\frac{k^0+|\mathbf{k}|}{k^0-|\mathbf{k}|}\right). \tag{70}$$

As a function of frequency and 3-momentum, the result is identical in QED and QCD, if for the latter we define $e^2 := g^2(N+N_f/2)$ (for SU(N) with N_f flavors) [75,124]. If there is also a nonvanishing chemical potential μ, a similar result holds where $T^2 \to T^2 + \frac{3}{\pi^2}\mu^2$ in terms $\propto N_f$ (all of them in QED)—with obvious generalization to the case of different chemical potentials μ_i for different flavors i.

The HTL-dressed propagator $G^{\mathrm{HTL}} = (G_0^{-1} + \Pi^{\mathrm{HTL}})^{-1}$ has poles off the usual light-cone, which come in two branches determined by

$$\Delta_A^{\mathrm{HTL}-1} = k^2 - \Pi_A^{\mathrm{HTL}} = 0, \tag{71a}$$

$$\Delta_B^{\mathrm{HTL}-1} = k^2 - \Pi_B^{\mathrm{HTL}} = 0. \tag{71b}$$

Since, as we have seen above, the HTL contribution is completely gauge independent and the gauge fixing parameters contained in G_0^{-1} do not appear in (71a,71b), the A- as well as the B-part of the HTL propagator is completely gauge independent.

The physical interpretation of the A- and B-branch of propagator poles (displayed in Fig. 6) is that the former represents quasi-particles which are in-medium versions of the physical polarisation of the gauge bosons, while the appearance of the B-branch is a purely collective phenomenon corresponding to charge density oscillations (plasmons) above the plasma frequency and to charge screening below.

Beyond the HTL approximation and in non-Abelian theories, however, one has gauge parameter dependences within Π, and also $\Pi_{\mu\nu}k^\mu \neq 0$ so that $\Pi_C \neq 0, \Pi_D \neq 0$.

Considering a general, rotationally invariant gauge $\tilde{f}^\mu(k)\tilde{A}_\mu(k)$ as in (46), this can be parameterized as

$$\tilde{f}^\mu(k) = \tilde{\beta}(k)k^\mu + \tilde{\gamma}(k)\tilde{n}^\mu. \tag{72}$$

Covariant gauges then correspond to $\tilde{\beta} = 1, \tilde{\gamma} = 0$, Coulomb gauges to $\tilde{\beta} = \tilde{n}^2, \tilde{\gamma} = -k^0$, and temporal axial gauge to $\tilde{\beta} = k^0/k^2, \tilde{\gamma} = 1$.

The structure functions of the gauge propagator become more complicated, to wit,

$$\Delta_A = [k^2 - \Pi_A]^{-1} \tag{73a}$$

$$\Delta_B = [k^2 - \Pi_B - \frac{2\tilde{\beta}\tilde{\gamma}|\mathbf{k}|\Pi_C - \alpha\Pi_C^2 + \tilde{\gamma}^2\tilde{n}^2\Pi_D}{\tilde{\beta}^2k^2 - \alpha\Pi_D}]^{-1} \tag{73b}$$

$$\Delta_C = -\frac{\tilde{\beta}\tilde{\gamma}|\mathbf{k}| - \alpha\Pi_C}{\tilde{\beta}^2k^2 - \alpha\Pi_D} \tag{73c}$$

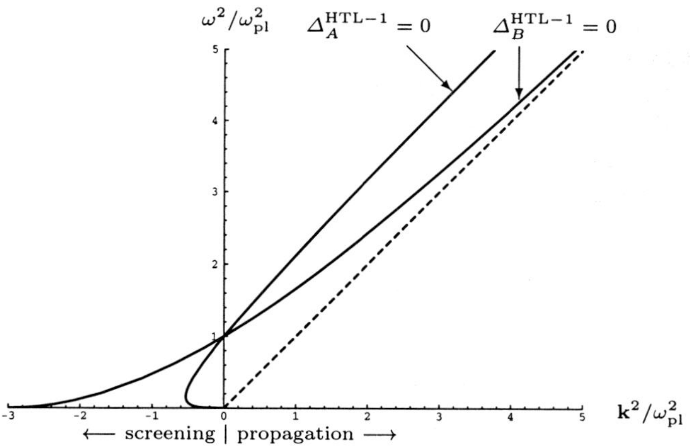

Fig. 6. The location of the zeros of $\Delta_A^{\rm HTL-1}$ (transverse gluons) and of $\Delta_B^{\rm HTL-1}$ (longitudinal plasmons) in quadratic scales such as to show propagating modes and screening phenomena on one plot. Above a common plasma frequency $\omega_{\rm pl.}$ there are propagating quasi-particle modes, which for large momenta in branch A tend to a mass hyperboloid with asymptotic mass $m_\infty^2 = \frac{3}{2}\omega_{\rm pl.}^2$, and in branch B approach the light-cone exponentially with exponentially vanishing residue. For $\omega < \omega_{\rm pl.}$, $|\mathbf{k}|$ is the inverse screening length, which in the static limit vanishes for mode A (absence of magnetostatic screening), but reaches the Debye mass, $\hat{m}_D^2 = 3\omega_{\rm pl.}^2$, for mode B (electrostatic screening)

$$\Delta_D = \frac{\tilde{\gamma}^2\tilde{n}^2 + \alpha(k^2 - \Pi_B)}{\tilde{\beta}^2 k^2 - \alpha\Pi_D}\Delta_B \tag{73d}$$

and there are gauge parameters everywhere, both explicitly and also within the structure functions of Π.

These gauge dependences are controlled by the gauge dependence identities discussed above, and, in compact notation, they have the form

$$\delta\Delta^{ij}\Big|_{J=0} = -(\Delta^{im}\delta X^j_{,m} + \delta X^i_{,m}\Delta^{mj})\Big|_{J=0} \tag{74}$$

for the full propagator. Specialized to the thermal gauge-boson propagator in f^μ-gauge, one finds [82,83]

$$\delta\Delta_A^{-1} = \Delta_A^{-1}\Big[-A^\mu_\nu(k)\delta X^\nu_{,\mu}(k)\Big] \equiv \Delta_A^{-1}\delta Y \tag{75a}$$

$$\delta\Delta_B^{-1} = \Delta_B^{-1}\left[-\frac{\tilde{n}^\mu}{\tilde{n}^2} + \frac{\tilde{\gamma}\tilde{\beta} - \alpha\Pi_C/|\mathbf{k}|}{\tilde{\beta}^2 k^2 - \alpha\Pi_D}k^\mu\right] 2\tilde{n}_\nu\delta X^\nu_{,\mu} \equiv \Delta_B^{-1}\delta Z \tag{75b}$$

but no such relations for Δ_C and Δ_D.

If δY and δZ are regular on the two "mass-shells" defined by $\Delta_A^{-1} = 0$ and $\Delta_B^{-1} = 0$, the relations (75a,75b) imply that the locations of these particular singularities of the gluon propagator are gauge fixing independent, for if $\Delta_A^{-1} = 0 = \Delta_B^{-1}$ then also $\Delta_A^{-1} + \delta\Delta_A^{-1} = 0 = \Delta_B^{-1} + \delta\Delta_B^{-1}$.

So everything depends on whether the possible singularities of $\delta X^{\nu}_{,\mu}$ could coincide with the expectedly physical dispersion laws $\Delta_A^{-1} = 0$ and $\Delta_B^{-1} = 0$. Diagrammatically, $\delta X^{\nu}_{,\mu}$ is obtained from the primary diagrams of Fig. 3 by inserting one additional vertex in all possible ways (and omitting all resulting tadpole-like diagrams in the case of no spontaneous symmetry breaking). Since the primary diagrams are 1-particle reducible with respect to the basic ghost line attached to δF_i^{α}, $\delta X^{\nu}_{,\mu}$ will have singularities like the (full) ghost propagator. These singularities are however generically different from those that define the spatially transverse and longitudinal gauge-boson quasiparticles. Indeed, in leading-order thermal perturbation theory, the temperature power-counting rules referred to in Sect. 5.2 imply that the ghost propagator does not receive contributions $\sim e^2T^2$ and thus will have completely different dispersion laws.

The other parts of the diagrams making up $\delta X^{\nu}_{,\mu}$ are 1-p-i and may develop singularities for other reasons, namely when one line of such an 1-p-i diagram is of the same type as the external one, and the remaining ones are massless. This may potentially give rise to infrared or mass-shell singularities. However, these singularities will be absent as soon as an overall infrared cut-off is introduced, for example by considering first a finite volume. In every finite volume, this obstruction to the gauge-independence proof is then avoided, and $\Delta_A^{-1} = 0$ and $\Delta_B^{-1} = 0$ define gauge-independent dispersion laws if the infinite volume limit is taken last of all [114].

This reasoning leads to the conclusion that the positions of all the singularities of Δ_A are gauge-fixing independent, though not necessarily their type or e.g. their residues if they are simple poles. In the case of Δ_B, there is a slight complication by the contents of the square bracket in (75b). There is a kinematical pole $1/k^2$ hidden in the $\tilde{n}$'s, and there is a contribution from the obviously gauge-dependent Δ_D (cf. (73d)). These gauge artefacts have to be excluded, but they are gauge dependent already at tree level and thus easy to identify. For example, Δ_B as defined above has a factor of k^2 which cancels in the Coulomb gauge propagator but not in that of covariant gauges, so this massless mode is a gauge mode and thus unphysical.

The gauge-(in)dependence identities (75a,75b) also explain the gauge dependences found in the one-loop calculation of the plasmon damping constant mentioned above. Truncating e.g. (75a) at one-loop order gives

$$\delta\Delta_A^{-1(1)}(k) = \Delta_A^{-1(0)}\delta Y^{(1)}, \tag{76}$$

with superscripts referring to bare loop order and using that $\delta Y^{(0)} \equiv 0$. However, the HTL plasma dispersion law is derived from $\Delta_A^{-1(0)} + \Delta_A^{-1(1)} = 0$, and the "correction" $\Delta_A^{-1(1)} \sim \Pi^{\mathrm{HTL}} \sim g^2T^2 \sim \omega_{\mathrm{pl.}}^2$ is not small but sets the scale for everything. The temperature-power-counting rules of Sect. 5.2 give $\delta Y^{(1)} \sim g^2T/\omega_{\mathrm{pl.}}$, so the right-hand side of (76) does not vanish at the order of the damping contribution $\gamma \times \omega_{\mathrm{pl.}} \sim g\omega_{\mathrm{pl.}}^2$.

On the other hand, if one does have a good expansion parameter (which bare loop order obviously is not), then the identities (75a,75b) imply order-by-order gauge independence.

As will be discussed further below, HTL perturbation theory [38,39] claims to be a systematic framework, although not up to arbitrarily high orders, and the expansion parameter is essentially $\sqrt{g^2}$. In [36], the long-wavelength plasmon damping constant has been calculated by Braaten and Pisarski with the result $\gamma(|\mathbf{k}|=0)/[\frac{g^2TN}{24\pi}] = +6.635\ldots$ and formal checks as to its gauge independence were positive.

More explicit calculations by Baier et al., however, revealed that, in covariant gauges and on plasmon-mass-shell, HTL-resummed perturbation theory still leads to explicit gauge dependent contributions to the damping of fermionic [17] as well as gluonic [16] quasi-particles. But, as was pointed out in [114], these apparent gauge dependences are avoided if the quasi-particle mass-shell is approached with a general infrared cut-off such as finite volume, and this cut-off lifted only in the end. This procedure defines gauge-independent dispersion laws and the gauge dependent parts are found to pertain to the residue, which at finite temperature happens to be linearly infrared singular in covariant gauges, rather than only logarithmically as at zero temperature, due to Bose enhancement.

Extension to Fermions. The fermion propagator at non-zero temperature or density has one more structure function than usually. In the ultrarelativistic limit where masses can be neglected, the fermion self-energy can be parametrized according to

$$\Sigma(k_0,\mathbf{k}) \;=\; a(k_0,k)\,\gamma^0 \;+\; b(k_0,k)\hat{\mathbf{k}}\cdot\boldsymbol{\gamma} \tag{77}$$

with $\hat{\mathbf{k}} = \mathbf{k}/|\mathbf{k}|$ (again neglecting the possibility of color superconductivity).

Defining $\Sigma_\pm(k_0,k) \equiv b(k_0,k) \pm a(k_0,k)$, a natural decomposition of the fermion self-energy and propagator $S^{-1} = -\not{k} + \Sigma$ is given by

$$\gamma_0\Sigma(k_0,\mathbf{k}) = \Sigma_+(k_0,k)\Lambda_+(\hat{\mathbf{k}}) \;-\; \Sigma_-(k_0,k)\Lambda_-(\hat{\mathbf{k}}), \tag{78}$$

$$\gamma_0 S^{-1}(k_0,\mathbf{k}) = \Delta_+^{-1}(k_0,k)\Lambda_+(\hat{\mathbf{k}}) \;+\; \Delta_-^{-1}(k_0,k)\Lambda_-(\hat{\mathbf{k}}) \tag{79}$$

with $\Delta_\pm^{-1} \equiv -[k_0 \mp (k+\Sigma_\pm)]$ and spin matrices

$$\Lambda_\pm(\hat{\mathbf{k}}) \equiv \frac{1\pm\gamma^0\boldsymbol{\gamma}\cdot\hat{\mathbf{k}}}{2}, \qquad \Lambda_+ + \Lambda_- = 1, \tag{80}$$

$$\Lambda_\pm^2 = \Lambda_\pm, \qquad \Lambda_+\Lambda_- = \Lambda_-\Lambda_+ = 0, \qquad \mathrm{Tr}\Lambda_\pm = 2, \tag{81}$$

projecting onto spinors whose chirality is equal (Λ_+), or opposite (Λ_-), to their helicity.

In the HTL approximation where $|k_0|, |\mathbf{k}| \ll \max(T, \mu)$, the fermion self-energy has been first calculated by Klimov [80] as

$$\Sigma_{\pm}^{\mathrm{HTL}}(k_0, k) = \frac{\hat{M}^2}{k}\left(1 - \frac{k_0 \mp k}{2k} \log \frac{k_0 + k}{k_0 - k}\right), \tag{82}$$

where $\hat{M}^2$ is the plasma frequency for fermions, i.e., the frequency of long-wavelength ($k \to 0$) fermionic excitations

$$\hat{M}^2 = \frac{g^2 C_f}{8}\left(T^2 + \frac{\mu^2}{\pi^2}\right). \tag{83}$$

($C_f = (N^2 - 1)/2N$ in SU(N) gauge theory, and $g^2 C_f \to e^2$ in QED.)

This leads to two separate branches of dispersion laws of fermionic quasi-particles $\Delta_{\pm}(\omega, k)^{-1} = 0$ carrying particle and hole quantum numbers respectively [125,107]. As shown in Fig. 7, the $(-)$-branch, which is occasionally nicknamed "plasmino", exhibits a curious dip reminiscent of the dispersion law of rotons in liquid helium.

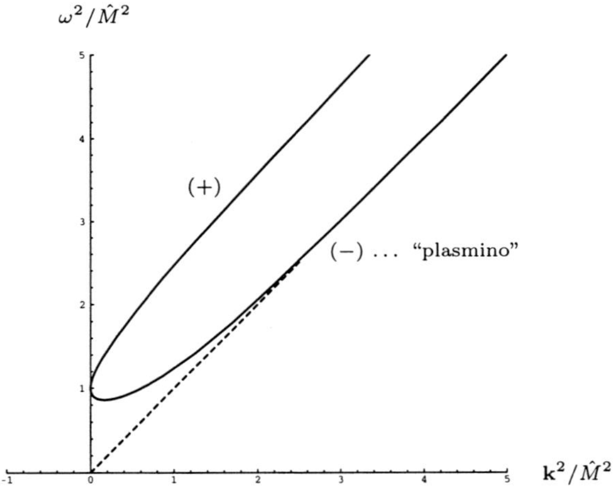

Fig. 7. The location of the zeros of $\Delta_{\pm}^{-1}$ in the HTL approximation in quadratic scales. The additional collective modes of branch $(-)$ ("plasminos") has a minimum of ω at $\omega/\hat{M} \approx 0.93$ and $|\mathbf{k}|/\hat{M} \approx 0.41$ and approaches the light-cone for large momenta, but with exponentially vanishing residue. The regular branch approaches a mass hyperboloid (here a straight line parallel to the diagonal) with asymptotic mass $\sqrt{2}\hat{M}$

As we have seen in Sect. 5.2, gauge dependences start at order T in the high-temperature expansion. While the HTL result (82) is completely gauge independent, gauge parameter dependences enter at subleading order. The gauge dependence identity for the fermion propagator is, however, somewhat simpler than that for the gluon propagator. The singularities in the fermion propagator can be summarized by

$$\det\left(S^{-1}_{\sigma\bar{\varrho}}\right) = 0, \tag{84}$$

where $\sigma, \bar{\varrho}$ are spinor indices. The gauge dependence identity thus takes the form

$$\delta \det\left(S^{-1}_{\sigma\bar{\varrho}}\right)(k) = -\det\left(S^{-1}_{\sigma\bar{\varrho}}\right)(k)\left[\delta X^{\tau}_{,\tau}(k) + \delta X^{\bar{\tau}}_{,\bar{\tau}}(k)\right], \tag{85}$$

where the two expressions within the square brackets are Dirac traces of the diagrams appearing within the braces in Fig. 5.

The conclusion of gauge independence of the solutions of (84) can now be reached by essentially the same arguments as those for the gauge boson propagator (with the exception that now there are no additional, gauge-dependent kinematical poles like those arising in the projection onto mode B in (75b)). The only obstruction to gauge independence comes from singularities of $\delta X^{\tau}_{,\tau}(k)$ and $\delta X^{\bar{\tau}}_{,\bar{\tau}}(k)$. If mass-shell singularities from massless gauge modes are avoided by infrared regularization or finite volume, only the singularities of the ghost propagator need to be considered. Again, the latter are generically different from those leading to the now fermionic quasi-particles since ghosts do not have HTL self-energies $\sim T^2$.

We have thus seen that all the singularities of the fermion propagator as well as those of the A- and B-branch of the gluon propagator (with some exceptions for the latter) are gauge-fixing independent. On the other hand, residues (if those singularities are simple poles at all) are not protected and may be gauge dependent; even the nature of the singularity may be different from gauge to gauge, as is well known to be the case already for the electron propagator in zero-temperature QED [31].

6 Quasiparticles in HTL Perturbation Theory

We have already seen that loop order in bare perturbation theory is not a good expansion parameter for calculating corrections to quasi-particle properties at soft scales $\sim gT$. Technically what happens is that the HTL contributions to one-loop vertex functions are of the same order of magnitude as their tree-level counterparts for external momenta $k \sim gT$:

$$\Gamma^{\text{HTL}}_{,N} \sim g^N T^2 k^{2-N} \sim g^{N-2} k^{4-N} \sim \left.\frac{\partial^N \mathcal{L}}{\partial A^N}\right|_{k\sim gT}. \tag{86}$$

Therefore all HTL contributions need to be resummed in Feynman diagrams that are sensitive to the soft regime $k \sim gT$.

Since HTL's are the leading contributions from hard momenta $k \sim T$, this can be understood as the transition from the bare Lagrangian to an effective, Wilson-renormalized one for $k \sim gT$, $\mathcal{L} \to \mathcal{L} + \mathcal{L}^{\text{HTL}}$. $\mathcal{L}^{\text{HTL}}$ is the effective

Lagrangian containing all the HTL diagrams and arises from integrating out all hard modes.

Soon after the identification of all the HTL's of QCD in [59,39], it has been found that, formally, $\mathcal{L}^{\mathrm{HTL}}$ has a comparatively simple and manifestly gauge-invariant integral representation [120,41,60]

$$\begin{aligned} \mathcal{L}^{\mathrm{HTL}} &= \hat{M}^2 \int \frac{d\Omega_v}{4\pi} \bar{\psi} \gamma^\mu \frac{v_\mu}{v \cdot D(A)} \psi \\ &- \frac{3}{2} \omega_{pl.}^2 \operatorname{tr} \int \frac{d\Omega_v}{4\pi} F^{\mu\alpha} \frac{v_\alpha v^\beta}{(v \cdot D_{adj.}(A))^2} F_{\mu\beta} \end{aligned} \tag{87}$$

with $\hat{M}^2$ the fermionic plasma frequency given in (83) and $\omega_{pl.}^2 = \frac{1}{3} \Pi_\mu^{\mathrm{HTL}\mu}$ the more familiar one of the gauge bosons (cf. (69)). In this integral representation $v = (1, \mathbf{v})$ is a light-like 4-vector, i.e. with $\mathbf{v}^2 = 1$, and its spatial components are averaged over by $d\Omega_v$. v is the remnant of the hard plasma constituents' momenta $p^\mu \sim T v^\mu$, namely their light-like 4-velocity, and the overall scale T has combined with the coupling constant to form the scale of thermal masses, $\hat{M}, \omega_{pl.} \sim gT$.

The HTL effective Lagrangian (87) is manifestly gauge invariant and moreover gauge independent ($\hat{M}$ and $\omega_{pl.}$ do not depend on the gauge fixing parameters used to integrate out the hard modes). It is non-local and Hermitian only in a Euclidean form, i.e. prior to analytic continuation to real time/frequencies. It has cuts which physically correspond to the phenomenon of Landau damping. The equations of motions associated with (87) can also be obtained from kinetic theory, which is extremely useful to gain further physical insight [21,22,79,26]. There is also a noteworthy connection to Chern-Simons theory [52,53,103].

Using (87) as an effective theory for soft scales $\sim gT$ means that the bare propagators are to be replaced by those of HTL quasi-particles, and these have infinitely many nonlocal vertices. E.g., the three-gluon vertex becomes

$$\begin{aligned} \Gamma_{\mu\nu\varrho}^{abc\;\mathrm{cl+HTL}}(k,q,r) &= \mathrm{i} g f^{abc} \Big\{ g_{\mu\nu}(k-q)_\varrho + \mathrm{cycl.} \\ &+ 3\omega_{\mathrm{pl.}}^2 \int \frac{d\Omega_v}{4\pi} v_\mu v_\nu v_\varrho \left[\frac{r_0}{k \cdot v\, r \cdot v} - \frac{q_0}{k \cdot v\, q \cdot v} \right] \Big\}. \end{aligned} \tag{88}$$

In QCD, there are HTL vertices for any number of external gluons and up to two quark lines, whereas in QED, where $v \cdot D_{\mathrm{adj.}}(A)) \to v \cdot \partial$ in (87), there is "only" an HTL photon self-energy $\Pi_{\mu\nu}$, an HTL fermion self-energy Σ, and vertices involving two fermions and an arbitrary number of photons.

While the effective Lagrangian (87) is gauge invariant and gauge independent in its entirety, NLO corrections won't be so. However, as we have seen in the previous section, the positions of singularities of the effective (quasi-particle) propagators are protected against gauge dependences by the identities (75a), (75b), and (85).

6.1 Long-Wavelength Plasmon Damping

The first such correction to be calculated by means of the HTL-resummed perturbation theory was the damping rate of long-wavelength plasmons[8] from the shift of the pole of the gluon propagator at $\mathbf{k} = 0$ from $\omega = \omega_{\rm pl.}^{\rm HTL} \to \omega = \omega_{\rm pl.} - {\rm i}\gamma(\mathbf{k} = 0)$ with the result [36]

$$\gamma(\mathbf{k} = 0) = +6.635\ldots \frac{g^2 NT}{24\pi} = 0.264\sqrt{N} g\, \omega_{\rm pl.}^{\rm HTL}, \tag{89}$$

implying the existence of weakly damped plasmons for $g \ll 1$.

In QCD, where one is interested in the range $g \sim 1$, one finds that the existence of plasmons as quasi-particles requires that g is significantly less than 2.2, so real QCD is on the borderline of having identifiable long-wavelength quasi-particles.

The corresponding quantity for fermionic quasi-particles has been calculated in [81,40] with a comparable result: weakly damped long-wavelength fermionic quasi-particles in 2- or 3-flavor QCD require that g is significantly less than 2.7.

6.2 NLO Correction to Gluonic Plasma Frequency

In [117], Schulz has calculated also the real part of the NLO contribution to the gluon polarization tensor in the limit of $\mathbf{k} \to 0$ which determines the NLO correction to the gluonic plasma frequency.

The original power-counting arguments of [39] suggested that besides one-loop diagrams with HTL-resummed propagators and vertices, there could be also contributions from two-loop diagrams to relative order g. The explicit (and lengthy) calculation of [117] showed that those contribute only at order $g^2 \ln(1/g)$ rather than g, and the NLO plasma frequency in a pure-glue plasma was obtained as

$$\omega_{\rm pl.} = \omega_{\rm pl.}^{\rm HTL} \left[1 - 0.09\sqrt{N} g\right]. \tag{90}$$

In this particular result, HTL-resummed perturbation theory turns out to give a moderate correction to the leading-order HTL value even for $g \sim 1$; see however below.

While the calculations leading to (89) and (90) contain some interesting physics, in the following we shall go into more detail only for a couple of more tractable cases, which nonetheless will turn out to involve a number of salient points.

[8] For kinematical reasons, there should be no difference between spatially transverse and longitudinal gluonic quasi-particles (cf. Fig. 6), since with $\mathbf{k} \to 0$ one can no longer tell the one from the other. However, the limit $\mathbf{k} \to 0$ involves infrared problems (see further below), and there are even explicit calculations [2,1] that claim to find obstructions to this equality, which are, however, refuted by the recent work of [48].

6.3 NLO Correction to the Non-Abelian Debye Mass

The poles of the (gluon) propagator do not only give the dispersion law of quasi-particles, but also the screening of fields with frequencies below the plasma frequency and in particular of static fields. Below the plasma frequency, there are poles for $\mathbf{k}^2 < 0$, as displayed in Fig. 6, corresponding in configuration space to exponential fall-off with (frequency-dependent) screening mass $\sqrt{|\mathbf{k}^2|}$, i.e. screening length $1/\sqrt{|\mathbf{k}^2|}$.

In the static case, branch A of the gluon propagator describes the screening of (chromo-)magnetostatic fields. While there is a finite screening length as long as $\omega > 0$, the A-branch of the HTL propagator becomes unscreened in the static limit. Whereas in QED, a "magnetic mass" is forbidden by gauge invariance [58,27], some sort of entirely non-perturbative magnetic mass is expected in non-Abelian gauge theories in view of severe infrared problems caused by the self-interactions of magnetostatic gluons [110,96,64].

Branch B, on the other hand, contains the information about screening of (chromo-)electric fields as generated by static charges (Debye screening). The Debye mass given by the leading-order HTL propagator is $\hat{m}_D = \sqrt{3}\omega_{pl.}$. The determination of its NLO correction has a history that is at least as long as the plasmon (damping) puzzle, for it starts already with (ultra-relativistic) QED.

Customarily, the Debye mass (squared) has been *defined* as the infrared limit $\Pi_{00}(\omega = 0, k \to 0)$, which indeed is correct at the HTL level, cf. (68b) and (70).

In QED, this definition has the advantage of being directly related to a derivative of the thermodynamic pressure, so that the higher-order terms known from the latter determine those of $\Pi_{00}^{\rm QED}(\omega = 0, k \to 0)$ through [58,76]

$$\Pi_{00}(0, k \to 0)\Big|_{\mu=0} = e^2 \frac{\partial^2 P}{\partial \mu^2}\Big|_{\mu=0} = \frac{e^2 T^2}{3}\left(1 - \frac{3e^2}{8\pi^2} + \frac{\sqrt{3}e^3}{4\pi^3} + \ldots\right). \tag{91}$$

This result is gauge independent because in QED all of $\Pi_{\mu\nu}$ is.

In the case of QCD, there is no such relation. In fact, one expects $\delta m_D^2/\hat{m}_D^2 \sim g$ rather than g^3 because of gluonic self-interactions and Bose enhancement. The calculation of this quantity should be much easier than the dynamic ones considered above, because in the static limit the HTL effective action collapses to just the local, bilinear HTL Debye mass term,

$$\mathcal{L}^{\rm HTL} \stackrel{\rm static}{\longrightarrow} -\frac{1}{2}\hat{m}_D^2 {\rm tr} A_0^2. \tag{92}$$

This is also gauge invariant, because A_0 behaves like an adjoint scalar under time-independent gauge transformations. Resummed perturbation theory thus boils down to a resummation of the HTL Debye mass in the electrostatic propagator, which is what had been done already since long [63,76].

Using this simple ("ring") resummation in QCD, one finds, however, the gauge dependent result [121]

$$\Pi_{00}(0,0)/m_D^2 = 1 + \alpha \frac{N}{4\pi}\sqrt{\frac{6}{2N+N_f}}g, \tag{93}$$

where α is the gauge parameter of general covariant gauge (which coincides with general Coulomb gauge in the static limit).

This result was interpreted as meaning that the non-Abelian Debye mass could not be obtained in resummed perturbation theory [101] or that one should use a physical gauge instead [73,76]. In particular, temporal axial gauge was put forward, because in this gauge there is, like in QED, a linear relationship between electric field strength correlators and the gauge propagator. However, because static ring resummation clashes with temporal gauge, inconclusive and contradicting results were obtained by different authors [72,61,73], and in fact one cannot do without vertex resummations if one wants to be consistent there [15,106]. But be that as it may be, switching to the chromoelectric field strength correlator is not good enough, for it is gauge variant and its infrared limit is equally gauge dependent [116].

On the other hand, in view of the gauge dependence identities discussed in the previous section, the gauge dependence of (93) is no longer surprising. Gauge independence can only be expected "on-shell", what in this context means $\omega = 0$ but $\mathbf{k}^2 \to -\hat{m}_D^2$.

Indeed, the exponential fall-off of the electrostatic propagator is determined by the position of the singularity of $\Delta_B(0,k)$, and not simply by its infrared limit. This implies in particular that one should use a different definition of the Debye mass already in QED, despite the gauge independence of (91), namely [115]

$$m_D^2 = \Pi_{00}(0,k)\Big|_{\mathbf{k}^2 \to -m_D^2} . \tag{94}$$

For QED (with massless electrons), the Debye mass is thus not given by (91) but rather as

$$\begin{aligned} m_D^2 &= \Pi_{00}(0,k\to 0) + \left[\Pi_{00}(0,k)\big|_{k^2=-m_D^2} - \Pi_{00}(0,k\to 0)\right] \\ &= \frac{e^2T^2}{3}\left(1 - \frac{3e^2}{8\pi^2} + \frac{\sqrt{3}e^3}{4\pi^3} + \ldots - \frac{e^2}{6\pi^2}[\ln\frac{\tilde{\mu}}{\pi T} + \gamma_E - \frac{4}{3}] + \ldots\right), \end{aligned} \tag{95}$$

where $\tilde{\mu}$ is the renormalization scale of the momentum subtraction scheme,[9] i.e. $\Pi_{\mu\nu}(k^2 = -\tilde{\mu}^2)|_{T=0} = 0$. Since $de/d\ln\tilde{\mu} = e^3/(12\pi^2) + O(e^5)$, (95) is a renormalization-group invariant result for the Debye mass in hot QED, which (91) obviously failed to be.

[9] The slightly different numbers in the terms $\propto e^4T^2$ quoted in [27,93] pertain to the minimal subtraction (MS) scheme.

In QCD, where gauge independence is not automatic, the dependence on the gauge fixing parameter α is another indication that (93) is the wrong definition. For (94) we need the full momentum dependence of the correction $\delta\Pi_{00}(k_0 = 0, \mathbf{k})$ to Π_{00}^{HTL}. Since only the electrostatic mode needs to be dressed, this is not difficult to obtain [115]:

$$\delta\Pi_{00}(k_0=0,\mathbf{k}) = \underbrace{g\hat{m}_D N\sqrt{\tfrac{6}{2N+N_f}}}_{g^2T} \int \frac{d^{3-2\varepsilon}p}{(2\pi)^{3-2\varepsilon}} \times \left\{ \frac{1}{\mathbf{p}^2+\hat{m}_D^2} + \frac{1}{\mathbf{p}^2} + \frac{4\hat{m}_D^2 - (\mathbf{k}^2+\hat{m}_D^2)[3+2\mathbf{pk}/\mathbf{p}^2]}{\mathbf{p}^2[(\mathbf{p}+\mathbf{k})^2+\hat{m}_D^2]} + \alpha(\mathbf{k}^2+\hat{m}_D^2)\frac{\mathbf{p}^2+2\mathbf{pk}}{\mathbf{p}^4[(\mathbf{p}+\mathbf{k})^2+\hat{m}_D^2]} \right\}. \tag{96}$$

In accordance with the gauge dependence identities, the last term shows that gauge independence holds algebraically for $\mathbf{k}^2 = -\hat{m}_D^2$. On the other hand, on this "screening mass shell", where the denominator term $[(\mathbf{p}+\mathbf{k})^2 + \hat{m}_D^2] \to [\mathbf{p}^2 + 2\mathbf{pk}]$, we encounter IR-singularities. In the α-dependent term, they are such that they produce a divergent factor $1/[\mathbf{k}^2 + m_D^2]$ so that the gauge dependences no longer disappear even on-shell. This is, however, the very same problem that had to be solved in the above case of the plasmon damping in covariant gauges. Introducing a temporary infrared cut-off (e.g., finite volume), does not modify the factor $[\mathbf{k}^2 + m_D^2]$ in the numerator but defuses the dangerous denominator. Gauge independence thus holds for all values of this cut-off, which can be sent to zero in the end. The gauge dependences are thereby identified as belonging to the (infrared divergent) residue.

The third term in the curly brackets, however, remains logarithmically singular on-shell as the infrared cut-off is to be removed. In contrast to the α-dependent term, closer inspection reveals that these singularities are coming from the massless magnetostatic modes and not from unphysical massless gauge modes.

At HTL level, there is no (chromo-)magnetostatic screening, but, as we have mentioned, one expects some sort of such screening to be generated non-perturbatively in the static sector of hot QCD at the scale $g^2T \sim gm_D$ [110,96,64].

While this singularity prevents evaluating (96) in full, the fact that this singularity is only logarithmic allows one to extract the leading term of (96) under the assumption of an effective cut-off at $p \sim g^2T$ as [115]

$$\frac{\delta m_D^2}{\hat{m}_D^2} = \frac{N}{2\pi}\sqrt{\frac{6}{2N+N_f}}\, g \ln\frac{1}{g} + O(g). \tag{97}$$

The $O(g)$-contribution, however, is sensitive to the physics of the magnetostatic sector at scale g^2T, and is completely non-perturbative in that all loop order ≥ 2 are expected to contribute with equal importance.

Because of the undetermined $O(g)$-term in (97), one-loop resummed perturbation theory only says that for sufficiently small g, where $O(g\ln(1/g)) \gg O(g)$, there is a *positive* correction to the Debye mass of lowest-order perturbation theory following from the pole definition (94), and that it is gauge independent.

On the lattice, the static gluon propagator of pure SU(2) gauge theory at high temperature has been studied in various gauges [68,46] with the result that the electrostatic propagator is exponentially screened with a screening mass that indeed appears to be gauge independent and which is about 60% larger than the leading-order Debye mass for temperatures T/T_c up to about 10^4.

In [116], an estimate of the $O(g)$ contribution to (97) has been made using the crude approximation of a simple massive propagator for the magnetostatic one, which leads to

$$\frac{\delta m_D^2}{\hat{m}_D^2} = \frac{N}{2\pi}\sqrt{\frac{6}{2N+N_f}}\, g \left[\ln\frac{2m_D}{m_m} - \frac{1}{2}\right]. \tag{98}$$

On the lattice one finds strong gauge dependences of the magnetostatic screening function, but the data are consistent with an over-all exponential behaviour corresponding to $m_m \approx 0.5g^2T$ in all gauges [68,45]. Using this number in a self-consistent evaluation of (98) gives an estimate for m_D which is about 20% larger than the leading-order value for $T/T_c = 10\ldots 10^4$.

This shows that there are strong non-perturbative contributions to the Debye screening mass m_D even at very high temperatures. Assuming that these are predominantly of order g^2T, one-loop resummed perturbation theory (which is as far as one can get) is able to account for about 1/3 of this inherently non-perturbative physics already, if one introduces a simple, purely phenomenological magnetic screening mass.

Other Non-Perturbative Definitions of the Debye Mass. A different approach to studying Debye screening non-perturbatively without the complication of gauge fixing is to consider spatial correlation functions of appropriate gauge-invariant operators such as those of the Polyakov loop

$$L(\mathbf{x}) = \frac{1}{N}\mathrm{Tr}\,\mathcal{P}\exp\left\{-ig\int_0^\beta d\tau\, A_0(\tau,\mathbf{x})\right\}. \tag{99}$$

The correlation of two such operators is related to the free energy of a quark-antiquark pair [100]. In lowest order perturbation theory this is given by the square of a Yukawa potential with screening mass $\hat{m}_D$ [101]; at one-loop order one can in fact identify contributions of the form (98) if one assumes magnetic screening [33,116], but there is the problem that through higher loop orders the large-distance behaviour becomes dominated by the magnetostatic modes and their lightest bound states [34].

In [10], Arnold and Yaffe have proposed to use Euclidean time reflection symmetry to distinguish electric and magnetic contributions to screening, and have given a prescription to compute the sublogarithmic contribution of order g^2T to m_D nonperturbatively. This has been carried out in 3-d lattice simulations for SU(2) [74,88] as well as for SU(3) [89]. The Debye mass thus defined shows even larger deviations from the lowest-order perturbative results than that from gauge-fixed lattice propagators. E.g., in SU(2) at $T = 10^4 T_c$ this deviation turns out to be over 100%, while in SU(3) the dominance of g^2T contributions is even more pronounced.

Clearly, (resummed) perturbation theory is of no use here for any temperature of practical interest. However, the magnitude of the contributions from the completely nonperturbative magnetostatic sector depends strongly on the quantity considered. It is significantly smaller in the definition of the Debye mass through the exponential decay of gauge-fixed gluon propagators, which, as we have seen, leads to smaller screening masses on the lattice (and gauge-independent ones, too, apparently). In quantities where the barrier in perturbation theory arising from the magnetostatic sector occurs at higher orders, HTL-resummed perturbation theory should be in much better shape, and we shall find some support for this further below.

6.4 Dynamical Damping and Screening

A logarithmic sensitivity to the nonperturbative physics of the magnetostatic sector has in fact been encountered early on also in the calculation of damping of a heavy fermion [108], and more generally of hard particles [94,95,42,113]. It also turns out to occur for soft quasi-particles as soon as they are propagating [109,56] and not just stationary plasma oscillations.

Because this logarithmic sensitivity arises only if one internal line of (resummed) one-loop diagrams is static, the coefficient of the resulting $g \ln(1/g)$-term is almost as easy to obtain as in the case of the Debye mass, even though the external line is non-static and soft, requiring HTL-resummed vertices (see Fig. 8).

Fig. 8. One-loop diagrams in HTL-resummed perturbation theory. HTL-resummed quantities are marked with a blob

The infrared singularity arises (again) from the dressed one-loop diagram with two propagators, one of which is magnetostatic and thus massless in the HTL approximation, and the other of the same type as the external one, so only the first diagram in Fig. 8 is relevant. The dressed 3-vertices in it are needed only in the limit of one leg being magnetostatic and having zero momentum. Because of the gauge invariance of HTL's, these are determined by the HTL self-energies through a differential Ward identity, e.g.

$$\hat{\Gamma}_{\mu\nu\varrho}(k;-k;0) = -\frac{\partial}{\partial k^\varrho}\hat{\Pi}_{\mu\nu}(k) \tag{100}$$

for the 3-gluon vertex (color indices omitted).

Comparatively simple algebra gives [56]

$$\delta\Pi_I(k) \simeq -g^2 N 4\mathbf{k}^2[1+\partial_{\mathbf{k}^2}\Pi_I(k)]^2 \mathcal{S}_I(k), \qquad I=A,B \tag{101}$$

where

$$\mathcal{S}_I(k) := T\int \frac{d^3p}{(2\pi)^3}\frac{1}{\mathbf{p}^2}\frac{-1}{(k-p)^2-\Pi_I(k-p)}\Big|_{k^2=\Pi_I(k),p^0=0}, \tag{102}$$

and the logarithmic (mass-shell) singularity arises because $(k-p)^2-\Pi_I(k-p)\to -\mathbf{p}^2+2\mathbf{pk}-\Pi_I(k-p)+\Pi_I(k)\sim|\mathbf{p}|$ as $k^2\to\Pi_I(k)$.

The IR-singular part of $\mathcal{S}_I(k)$ is given by

$$\begin{aligned}\mathcal{S}_I(k) &= T\int\frac{d^3p}{(2\pi)^3}\frac{1}{\mathbf{p}^2}\frac{1}{\mathbf{p}^2-2\mathbf{pk}+\Pi_I(k-p)-\Pi_I(k)-\mathrm{i}\varepsilon}\\ &\simeq T[1+\partial_{\mathbf{k}^2}\Pi_I(k)]^{-1}\int\frac{d^3p}{(2\pi)^3}\frac{1}{\mathbf{p}^2}\frac{1}{\mathbf{p}^2-2\mathbf{pk}-\mathrm{i}\varepsilon}\\ &= T[1+\partial_{\mathbf{k}^2}\Pi_I(k)]^{-1}\int_\lambda^\infty\frac{dp}{p}\frac{1}{2|\mathbf{k}|}\ln\frac{p+2|\mathbf{k}|-\mathrm{i}\varepsilon}{p-2|\mathbf{k}|-\mathrm{i}\varepsilon},\end{aligned} \tag{103}$$

where in the last line we have inserted an IR cutoff $\lambda \ll gT$ for the p-integral in order to isolate the singular behaviour.

One finds that (103) has a singular imaginary part for propagating modes,

$$\mathcal{S}_I(k) \simeq \mathrm{i}\frac{T}{8\pi|\mathbf{k}|}[1+\partial_{\mathbf{k}^2}\Pi_I(k)]^{-1}\ln\frac{|\mathbf{k}|}{\lambda}+O(\lambda^0) \tag{104}$$

for $\mathbf{k}^2>0$ ($\mathbf{k}$ real), and a singular real part in screening situations where $|\mathbf{k}|^2\to-\kappa^2$, $\kappa\in\mathbb{R}$ (i.e., $\mathbf{k}$ imaginary):

$$\mathcal{S}_I(k) \simeq +\frac{T}{8\pi\kappa}[1+\partial_{\mathbf{k}^2}\Pi_I(k)]^{-1}\ln\frac{\kappa}{\lambda}+O(\lambda^0). \tag{105}$$

So from one and the same expression we can see that logarithmic IR singularities arise whenever $|\mathbf{k}|\neq 0$, leading to IR singular contributions to damping or (dynamical) screening, depending on whether $\omega > 0$ or < 0 and thus $\mathbf{k}^2 > 0$ or < 0. The case $\mathbf{k} = 0$ is IR-safe, because (101) is proportional to $\mathbf{k}^2$, while

$$\mathcal{S}_I(k) \longrightarrow \frac{T}{4\pi^2\lambda}[1 + \partial_{\mathbf{k}^2} \Pi_I(k)]^{-1} + O\left(\frac{T|\mathbf{k}|}{\lambda^2}\right) \quad \text{for } \mathbf{k} \to 0. \tag{106}$$

This shows that there is a common origin for the infrared sensitivity of screening and damping of HTL quasi-particles. The perturbatively calculable coefficients of the resulting $g \ln(1/g)$-terms are in fact beautifully simple: For the damping of moving quasi-particles one obtains [109,56]

$$\gamma_I(|\mathbf{k}|) \simeq \frac{g^2 NT}{4\pi} \frac{|\mathbf{k}|[1 + \partial_{\mathbf{k}^2} \Pi_I(k)]}{\omega(|\mathbf{k}|)[1 - \partial_{\omega^2} \Pi_I(k)]} \ln \frac{1}{g} \equiv \frac{g^2 NT}{4\pi} v_I(|\mathbf{k}|) \ln \frac{1}{g}, \tag{107}$$

where $v_I(|\mathbf{k}|)$ is the group velocity of mode I (which vanishes at $\mathbf{k} = 0$). The IR-sensitive NLO correction to screening takes its simplest form when formulated as [56]

$$\delta\kappa_I^2(\omega) = \frac{g^2 NT}{2\pi} \kappa_I(\omega) \left(\ln \frac{1}{g} + O(1)\right), \tag{108}$$

where $\kappa_I(\omega)$ is the inverse screening length of mode I at frequency $\omega < \omega_{pl.}$ (which in the static limit approaches the Debye mass and perturbatively vanishing magnetic mass, resp., while approaching zero for both modes as $\omega \to \omega_{pl.}$).

A completely analogous calculation for the fermionic modes (for which there are no screening masses) gives

$$\gamma_\pm(|\mathbf{k}|) = \frac{g^2 C_F T}{4\pi} |v_\pm|(|\mathbf{k}|) \left(\ln \frac{1}{g} + O(1)\right) \tag{109}$$

for $|\mathbf{k}| > 0$. The group velocity $v_\pm$ equals $\pm\frac{1}{3}$ in the limit $(|\mathbf{k}|) \to 0$, and increases monotonically towards $+1$ for larger momenta (with a zero for the $(-)$-branch at $|\mathbf{k}|/\hat{M} \approx 0.41$). For strictly $|\mathbf{k}| = 0$, the IR sensitivity in fact disappears because (109) is no longer valid for $|\mathbf{k}| \ll \lambda$, but one has $\gamma_\pm(|\mathbf{k}|)|_{\text{sing.}} \propto g^2 T|\mathbf{k}|/\lambda$ instead. Thus $\gamma_\pm(0)$ is calculable at order $g^2 T$ in HTL-resummed perturbation theory, and has been calculated in [81,40].

The fermionic result (109) applies in fact equally to QED, for which one just needs to replace $g^2 C_F \to e^2$. This is particularly disturbing as QED does not allow a non-zero magnetic mass as IR cutoff, and it has been conjectured that the damping $\gamma \sim g^2 T$ or $e^2 T$ itself might act as an effective IR cutoff [94,95,109,4], which however led to further difficulties [105]. The solution for QED was finally found by Blaizot and Iancu [23–25] who showed that there the fermionic modes undergo over-exponential damping in the form $\mathrm{e}^{-\gamma t} \to \mathrm{e}^{-\frac{e^2}{4\pi} T t \ln(\omega_{\mathrm{pl.}} t)}$ (for $v \to 1$), so finite time is the actual IR cut-off. The fermion propagator has in fact no simple quasi-particle pole, but nevertheless a sharply peaked spectral density.

In non-Abelian theories, on the other hand, one does expect the static (chromo-)magnetic field to have finite range, and lattice results do confirm this, so the above estimates may be appropriate after all, at least for sufficiently weak coupling.

6.5 NLO Corrections to Real Parts of Dispersion Laws

The above analysis has identified the imaginary parts of the dispersion laws to be sensitive to non-perturbative IR physics except at $\mathbf{k} = 0$ and where the group velocity vanishes (which includes one further point at $|\mathbf{k}| \neq 0$ for the fermionic plasmino branch). On the other hand, the real parts of the dispersion laws of gluonic and fermionic quasi-particles are IR-safe in NLO HTL-resummed perturbation theory. However, such calculations are tremendously difficult, and only some partial results exist so far in QCD [57,54].

In the following, we shall restrict our attention to the case $\mathbf{k}^2/\omega_{\text{pl.}}^2 \gg 1$ and consider the two branches of the gluon/photon propagator in turn. In both cases, interesting physics will be seen to be contained in the NLO corrections.

Longitudinal Plasmons. For momenta $\mathbf{k}^2 \gg \omega_{\text{pl.}}^2$, the longitudinal plasmon branch approaches the light-cone, as can be seen in Fig. 6. From $k^2 = \Pi_B^{\text{HTL}}(k)$ and (68b) one finds

$$\omega_B^2(|\mathbf{k}|) \to \mathbf{k}^2 \left(1 + 4\mathbf{k}^2 \mathrm{e}^{-6\mathbf{k}^2/(e^2T^2)}\right) \tag{110}$$

with $e^2 = g^2(N + N_f/2)$ in QCD, so the light-cone is approached exponentially. If one also calculates the residue, one finds that this goes to zero at the same time, and exponentially so, too.

Instead of QCD, we shall consider the analytically tractable case of massless scalar electrodynamics as a simple toy model with at least some similarities to the vastly more complicated QCD case in that in both theories there are bosonic self-interactions. There are however no HTL vertices in scalar electrodynamics, which makes it possible to do complete momentum-dependent NLO calculations [85].

Comparing HTL values of and NLO corrections to Π_B, one finds that as $k^2 \to 0$ there are collinear singularities in both:

$$\Pi_B^{\text{HTL}}(k)/k^2 \to \frac{3}{2}\omega_{\text{pl.}}^2 \ln \frac{\mathbf{k}^2}{k^2} \tag{111}$$

diverges logarithmically[10], whereas

$$\delta\Pi_B/k^2 \to -e\mu^2_{\text{sc.th.}}\frac{|\mathbf{k}|}{\sqrt{k^2}} \tag{112}$$

(with $\mu_{\text{sc.th.}} \propto eT$ the thermal mass of the scalar). Because (112) diverges stronger than logarithmically, one has $\delta\Pi_B > \Pi_B^{\text{HTL}}$ eventually as $k^2 \to 0$. Clearly, this leads to a breakdown of perturbation theory in the immediate neighbourhood of the light-cone ($k^2/|\mathbf{k}|^2 \lesssim (e/\ln\frac{1}{e})^2$), which this time is not caused by the massless magnetostatic modes, but rather by the massless hard modes contained in the HTL's.

However, a self-consistent gap equation for the scalar thermal mass implies that also the hard scalar modes have a thermal mass $\sim eT$. Including this by extending the resummation of the scalar thermal mass to hard internal lines renders Π_B regular up to and including the light-cone one obtains

$$\lim_{k^2\to 0}\frac{\Pi_B^{\text{resum.}}}{k^2} = \frac{e^2T^2}{3\mathbf{k}^2}\Big[\underbrace{\ln\frac{2T}{\mu_{\text{sc.th.}}}}_{\ln\frac{4}{e}} + \frac{1}{2} - \gamma_E + \frac{\zeta'(2)}{\zeta(2)}\Big] + \ldots \tag{113}$$

The finiteness of (113) makes it possible that there is now a solution to the dispersion law with $k^2 = 0$ at $\mathbf{k}^2/(e^2T^2) = \frac{1}{3}\ln\frac{2.094\ldots}{e} + O(e)$. Because all collinear singularities have disappeared, continuity implies that there are also solutions for space-like momenta $k^2 < 0$, so the longitudinal plasmon branch pierces the light-cone, having group velocity $v < 1$ throughout, though, as shown in Fig. 9. While at HTL level, the strong Landau damping at $k^2 < 0$ switches on discontinuously, it now does so smoothly through an extra factor $\exp[-e\sqrt{|\mathbf{k}|/[8(|\mathbf{k}|-\omega)]}]$, removing the longitudinal plasmons through overdamping for $(|\mathbf{k}|-\omega)/|\mathbf{k}| \gtrsim e^2$.

So the collinear singularities of HTL-resummed perturbation theory on the light-cone were associated with a slight but nevertheless qualitative change of the spectrum of longitudinal plasmons: instead of being time-like throughout and existing for higher momenta, albeit with exponentially small and decreasing residue and effective mass, they become space-like at a particular point $|\mathbf{k}| \sim eT\ln\frac{1}{e}$ and expire through Landau damping soon thereafter.

This phenomenon is in fact known to occur in non-ultrarelativistic ($T < m_e$) QED [122], and has been considered in the case of QCD in a little-known paper by Silin and Ursov [119], who speculated that it may lead to Cherenkov phenomena in the quark-gluon plasma.

In QCD, the situation is in fact much more complicated. Under the assumption that the collinear singularities are removed solely by the resummation of asymptotic gluonic and fermionic thermal masses in hard internal

[10] This is in fact the technical reason why the longitudinal branch approaches the light-cone exponentially when $\mathbf{k}^2 \gg \omega^2_{\text{pl.}}$.

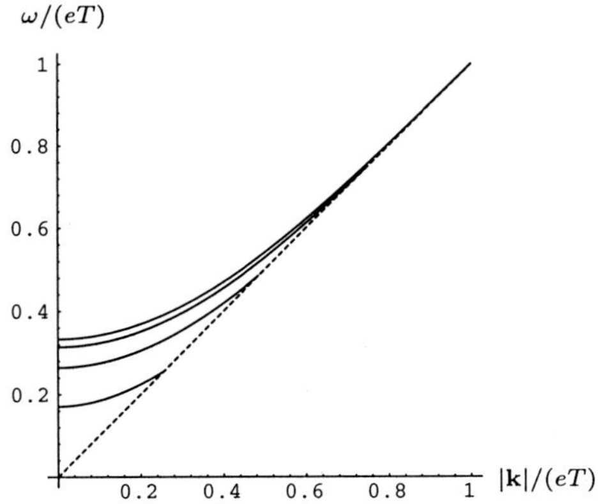

Fig. 9. The longitudinal plasmon branch of scalar electrodynamics including NLO corrections to the HTL result. The upper of the four lines gives the HTL result and the lines below correspond to NLO corrections with $e = 0.3$, 1, and 2, respectively. The latter three lines cross the light-cone such that the phase velocity starts to exceed 1, but with group velocity < 1 throughout. In the space-like region, the plasmon modes are damped by Landau damping, which is strong except in the immediate neighborhood of the light-cone, where it is suppressed by a factor of $\exp\{-e\sqrt{\mathbf{k}/[8(\mathbf{k}-\omega)]}\}$

lines, the value of $|\mathbf{k}|$ where longitudinal plasmons turn space-like has been calculated in [85]. For a pure-glue plasma, it reads

$$\mathbf{k}^2_{\text{crit.}} = g^2T^2[\ln\frac{1.48\ldots}{g} + O(g)]. \tag{114}$$

Such an extended resummation can in fact be related to an improved and still gauge-invariant version of the HTL effective action [55], however it may well be that damping effects are of equal importance here (in contrast to scalar electrodynamics), so that (114) may not be complete. A similar unsolved problem occurs in the calculation of the production rate of real, non-thermalized photons in a quark-gluon plasma from HTL-resummed perturbation theory [18,13,14].

Taken at face value, (114) would imply that propagating longitudinal plasmons do no longer exist for $g \gtrsim 1.48$, and a negative $O(g)$ contribution would even lower this bound.

Energetic Quarks and Transverse Gluons and Their Role in Self-Consistent Thermodynamics At high momenta $\mathbf{k}^2/\omega^2_{\text{pl.}} \gg 1$ the additional collective modes of longitudinal plasmons and "plasminos" disappear. At HTL level, they do so because the residues of the corresponding poles in the gluon and quark propagators vanish exponentially, whereas at NLO, as we have just seen, they cross the light-cone and die from strong Landau damping. The remaining transverse gluonic and normal quark modes on the other hand approach asymptotic mass hyperboloids. For transverse gauge bosons the asymptotic thermal photon/gluon mass of the HTL approximation reads

$$\Pi_A^{\rm HTL} \to m_\infty^2 = \frac{e^2T^2}{6} \tag{115}$$

($e^2 = (N + N_f/2)g^2$ for gluons); whereas in the case of fermions we have

$$2|\mathbf{k}|\Sigma_+^{\rm HTL} \to 2\hat{M}^2 \tag{116}$$

with $\hat{M}$ the HTL fermionic plasma frequency given by (83).

These results remain the correct LO ones even for $\omega, |\mathbf{k}| \sim T$, because the light-cone values of Π_A and Σ_+ are identical to their HTL/HDL values there and do not depend on the HTL approximation that $\omega, |\mathbf{k}| \ll T$ [84,55].

The asymptotic thermal masses play an interesting role in self-consistent (approximations to) thermodynamics [28–30]: The LO ($\propto g^2$) interaction piece of the *entropy density* can be expressed in terms of the light-cone values of the various self-energies and thus the asymptotic thermal masses. E.g., in the pure-glue case, the g^2-contribution to the entropy density reads

$$\begin{aligned} s^{(2)} &= -(N^2-1)\int \frac{d^3k\,d\omega}{(2\pi)^3}\frac{\partial n(\omega)}{\partial T}\mathrm{sgn}(\omega)\delta(\omega^2-k^2)\mathrm{Re}\Pi_T(\omega,k) \\ &= -\frac{(N^2-1)}{6}m_\infty^2 T = -\frac{N(N^2-1)}{36}g^2T^3. \end{aligned} \tag{117}$$

Fermionic contributions give similarly

$$s_f^{(2)} = -\frac{NN_f}{6}M_\infty^2 T, \tag{118}$$

possibly with nonzero chemical potential μ. With nonzero μ, one can also consider the quark density, which likewise is determined by the asymptotic mass:

$$n_f^{(2)} = -\frac{NN_f}{2\pi^2}M_\infty^2\mu. \tag{119}$$

Up to a T- and μ-independent integration constant, entropy and quark densities determine the complete thermodynamical potential, and the above formula give nice, universal formulae for the LO interaction terms.

Remarkably, also the NLO interaction term $\propto g^3$ can be directly related to the properties of HTL/HDL quasiparticles. The so-called plasmon term of the thermodynamic potential $\propto g^3$ is usually understood as arising from the resummation of the static Debye mass, which needs to be kept only in the zero modes of the electrostatic gluon propagator. The resulting coefficient of the order-g^3 contribution to the thermodynamic potential turns out, however, to have an uncomfortably large value,[11] and appears to spoil completely the convergence of perturbation theory for all temperatures smaller than some $10^5 T_c$.

While it is correct that all that is needed for a calculation of the thermodynamic potential through order g^3 is to approximate quarks and gluons by their vacuum spectral densities except for the one massive electrostatic mode [9], this is clearly a cruder approximation than that of HTL-resummed propagators which contain a lot of physics beyond Debye screening.

In [28–30] it has been shown recently that in a self-consistent formulation of the thermodynamic potentials entropy and density one can find a *real-time* description of those using quasi-particles which at soft momenta are described by the HTL effective propagators and at hard momenta by their light-cone limits and NLO corrections thereof. Doing so, it turns out that a larger part (up to $\frac{3}{4}$) of the (soft) plasmon effect $\propto g^3$ comes from the NLO corrections to the *hard* asymptotic masses, reflecting a massive[12] reorganization of usual (Debye-screened) perturbation theory:

$$\begin{aligned}\frac{3}{4}s^{(3)} &= \frac{3}{4}(N^2-1)\frac{\hat{m}_D^3}{3\pi} \\ &= -(N^2-1)\int \frac{d^3k}{(2\pi)^3}\frac{1}{k}\frac{\partial n(k)}{\partial T}\underbrace{\mathrm{Re}\,\delta\Pi_T(\omega=k)}_{\delta m_\infty^2(k)}\end{aligned} \tag{120}$$

(in the case of pure glue).

δm_∞^2 in HTL-resummed perturbation theory is a non-local (momentum-depedent) correction, which is infrared safe and thus calculable. Through the relation (120) one can define the average correction

$$\bar{\delta} m_\infty^2 = -\frac{1}{2\pi}g^2 N T \hat{m}_D, \tag{121}$$

which has a remarkably simple form. Similarly, for fermions one finds

$$\bar{\delta} M_\infty^2 = -\frac{1}{2\pi}g^2 C_f T \hat{m}_D. \tag{122}$$

Now, numerically, this correction is uncomfortably large:

$$\frac{\bar{\delta} m_\infty^2}{m_\infty^2} = 1 - \frac{\sqrt{3N}}{\pi}g \approx 1 - g \tag{123}$$

(pure glue) so that perturation theory seems to become completely useless for $g \gtrsim 1$, i.e., $\alpha_s \gtrsim 0.1$.

---

[11] The same holds true for the order-g^5 contribution which has been calculated for QCD in [11,12,126,35].

[12] Pun intended.

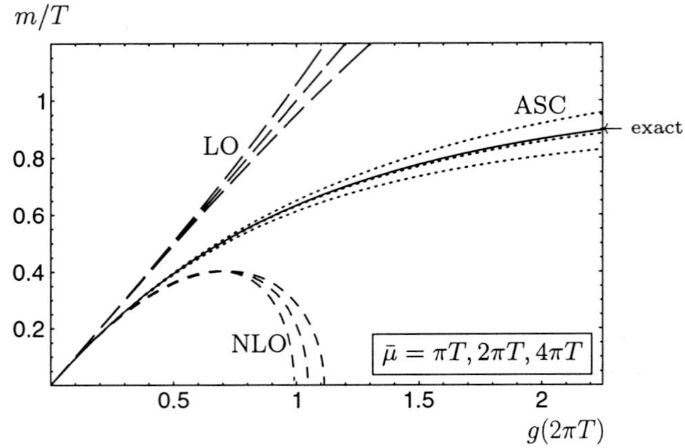

Fig. 10. Various approximations to the thermal mass of a scalar boson in large-N φ^4 theory: leading-order HTL (LO), next-to-leading order (NLO) as given by (125), and the approximately self-consistent (ASC) gap equation (126), which is perturbatively equivalent to NLO. The $\overline{\text{MS}}$ renormalization scale is varied by a factor of 2 about $\bar\mu = 2\pi T$

However, a very similar problem arises already in simple scalar ϕ^4 theory. If one considers the large-N limit of the iso-vector O(N) $g^2\phi^4$ theory, one can write down an exact gap equation of the form [49,50]

$$m^2 = 12g^2 \int \frac{d^3k}{(2\pi)^3} \frac{n(\sqrt{k^2+m^2})}{\sqrt{k^2+m^2}} + \frac{3m^2}{4\pi^2}\left(\ln\frac{m^2}{\bar\mu^2} - 1\right) \tag{124}$$

whose solution has a perturbative expansion beginning as

$$m^2 = g^2T^2(1 - \frac{3}{\pi}g + \ldots), \tag{125}$$

which happens to have the same $O(g)$ coefficent as the QCD result (121), and which likewise gives nonsense such as tachyonic thermal masses for $g \gtrsim 1$.

However, if one instead writes down an approximate gap equation by expanding in powers of m/T and dropping terms of order $(m/T)^2 \sim g^2$:

$$m^2 = g^2T^2 - \frac{3}{\pi}g^2Tm, \tag{126}$$

then one finds that the solution to this simple quadratic equation in m gives a function $m(g)$ that is perturbatively equivalent to (125), but does not go mad for $g \gtrsim 1$. On the contrary, for the standard choice of renormalization scale $\bar\mu = 2\pi T$ in $\overline{\text{MS}}$, it gives a remarkably accurate approximation of the solution to the full gap equation (124), as is shown in Fig. 10.

Implementing analogous "approximately self-consistent" gap equations for the hard modes, a non-perturbative, UV finite and gauge-invariant approximation to entropy and density of hot QCD has been proposed in [28–30].

It is perturbatively equivalent to conventional Debye-resummed perturbation theory but goes beyond the latter in incorporating all of the collective phenomena contained in HTL propagators as well as NLO effects in their asymptotic masses. When compared to available lattice data [32] (see Fig. 11 for the pure-glue case), remarkable agreement is found down to temperatures $\sim 3T_c$. By contrast, conventionally resummed perturbation theory at order g^3 leads to $\mathcal{S}/\mathcal{S}_{\rm SB} > 1$ for all but exceedingly high temperatures.

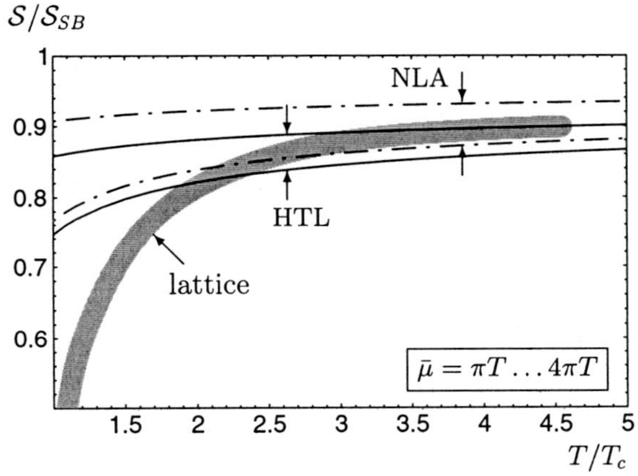

Fig. 11. Comparison of results from approximately self-consistent thermodynamics based on the HTL propagator and a next-to-leading approximation (NLA) using the analogue of (126) for the asymptotic gluon mass correction (121) with lattice data from [32]. The gray band gives roughly the lattice data with their errors. The analytical results are given with two boundaries corresponding to a variation of the $\overline{\rm MS}$ renormalization scale $\bar{\mu}$ from πT to $4\pi T$

An optimistic conclusion one could draw from this is that the transition to gluonic and quark quasi-particles is able to absorb a large part of the strong elementary interaction into the spectral properties of the former, and that, at least in infrared-safe situations, these quasi-particles have comparatively weak residual interactions even in QCD at temperatures a few times above the transition temperature.

Let me recall that even in the infrared-unsafe case of NLO corrections to the Debye mass, the self-consistent NLO result (98) using a phenomenological magnetic mass gives the qualitatively correct result of a substantially increased electric mass, while underestimating the magnitude of the increase by a factor of 3 when compared to lattice simulations of chromoelectrostatic propagators.

7 Conclusions

Let us summarize the findings that are of specific interest to a perturbative formulation of non-Abelian gauge theories at finite temperature and/or density:

The leading-order results for self-energies and vertices in a high-temperature/density expansion, the so-called hard thermal (dense) loops, form a gauge-invariant and gauge-independent effective action, which is the basis for a systematic perturbative expansion in powers of g (rather than g^2), as long as one does not run into the perturbative barrier formed by the completely non-perturbative self-interacting chromomagnetostatic modes.

Beyond leading order, gauge dependences appear in all Green functions of the fundamental fields. In particular, the propagators which are expected to carry information on gluonic or fermionic quasi-particles depend on gauge-fixing parameters. The gauge dependence identities that we have discussed above imply, however, that under certain conditions the location of the singularities which define the dispersion laws of these quasi-particles are gauge-independent, though not, e.g., residues or even the type of the singularities, which need not be simple poles.

Already at NLO, screening lengths and damping constants are logarithmically infrared sensitive to the nonperturbative magnetostatic sector, with the exception of zero 3-momentum. Infrared-safe quantities are also the real corrections to the dispersion laws, which in the case of longitudinal plasmons (and also of the plasmino branch of fermions) lead to a finite 3-momentum range, and, at its upper end, to space-like phase velocities. In the case of transverse gluonic quasiparticles and the normal branch of fermionic ones, the NLO corrections play an important role in self-consistent formulations of thermodynamics (the equation of state).

Particularly in QCD, one faces the problem that corrections to LO results are rather large for almost all values of the coupling of interest. However, we have seen indications that this poor convergence of thermal perturbation theory may be overcome in approximately self-consistent reformulations.[13] Where those can be implemented, the picture of weakly interacting quasi-particles even in strong interactions seems to have some support from comparison with lattice data (where the latter are available), and may remain valid down to a few times the deconfinement phase transition temperature.

[13] There are alternative methods to reorganize thermal perturbation theory which aim to improve its convergence. A particularly interesting one is "screened" or "optimized" perturbation theory [77,43,8] which employs a single mass parameter in a variational ansatz. In [5–7] an extension of this method to gauge theories has been proposed which uses the HTL effective action uniformly at soft as well as hard momenta with the thermal-mass prefactors $\omega_{\rm pl.}^2$ and $\hat{M}^2$ turned into variational parameters. In contrast to the entropy-based approach, this requires explicit dressed 2-loop contributions (involving HTL vertices) in order to contain the correct LO interaction coefficients in the thermodynamic pressure.

This picture, being primarily set up in real Minkowski space, is complementary to lattice or dimensional reduction formulations, and allows (analytical) calculations from first principles also where lattice gauge theory calculations are not (yet) feasible. Its potentials, in particular when combined with results from other nonperturbative approaches, are, in my opinion, not yet fully explored.

Acknowledgements

I would like to thank all the organizers of the 40th Schladming Winter School on "Dense Matter" for their efforts and for having once again set up a marvellous meeting. Furthermore, I would like to thank Jean-Paul Blaizot, Ian Drummond, Frithjof Flechsig, Ron Horgan, Edmond Iancu, Randy Kobes, Ulli Kraemmer, Gabor Kunstatter, Peter Landshoff, and Hermann Schulz for past and present collaborations that provided a large part of the material I had the privilege to present at this meeting.

References

1. A. Abada, O. Azi: Phys. Lett. **B463**, 117 (1999)
2. A. Abada, O. Azi, K. Benchallal: Phys. Lett. **B425**, 158 (1998)
3. L. F. Abbott: Nucl. Phys. **B185**, 189 (1981)
4. T. Altherr, E. Petitgirard, T. del Rio Gaztelurrutia: Phys. Rev. **D47**, 703 (1993)
5. J. O. Andersen, E. Braaten, M. Strickland: Phys. Rev. Lett. **83**, 2139 (1999)
6. J. O. Andersen, E. Braaten, M. Strickland: Phys. Rev. **D61**, 014017 (2000)
7. J. O. Andersen, E. Braaten, M. Strickland: Phys. Rev. **D61**, 074016 (2000)
8. J. O. Andersen, E. Braaten, M. Strickland: Phys. Rev. **D63**, 105008 (2001)
9. P. Arnold, O. Espinosa: Phys. Rev. **D47**, 3546 (1993)
10. P. Arnold, L. G. Yaffe: Phys. Rev. **D52**, 7208 (1995)
11. P. Arnold, C.-X. Zhai: Phys. Rev. **D50**, 7603 (1994)
12. P. Arnold, C.-X. Zhai: Phys. Rev. **D51**, 1906 (1995)
13. P. Aurenche, F. Gelis, R. Kobes, E. Petitgirard: Phys. Rev. **D54**, 5274 (1996)
14. P. Aurenche, F. Gelis, R. Kobes, E. Petitgirard: Z. Phys. **C75**, 315 (1997)
15. R. Baier, O. K. Kalashnikov: Phys. Lett. **B328**, 450 (1994)
16. R. Baier, G. Kunstatter, D. Schiff: Nucl. Phys. **B388**, 287 (1992)
17. R. Baier, G. Kunstatter, D. Schiff: Phys. Rev. **D45**, 4381 (1992)
18. R. Baier, S. Peigné, D. Schiff: Z. Phys. **C62**, 337 (1994)
19. P. M. Bakshi, K. T. Mahanthappa: J. Math. Phys. **4**, 1 (1963)
20. C. W. Bernard: Phys. Rev. **D9**, 3312 (1974)
21. J. P. Blaizot, E. Iancu: Phys. Rev. Lett. **70**, 3376 (1993)
22. J. P. Blaizot, E. Iancu: Nucl. Phys. **B417**, 608 (1994)
23. J.-P. Blaizot, E. Iancu: Phys. Rev. Lett. **76**, 3080 (1996)
24. J.-P. Blaizot, E. Iancu: Phys. Rev. **D55**, 973 (1997)
25. J.-P. Blaizot, E. Iancu: Phys. Rev. **D56**, 7877 (1997)
26. J.-P. Blaizot, E. Iancu: *The quark-gluon plasma: Collective dynamics and hard thermal loops*, hep-ph/0101103 (2001)

27. J.-P. Blaizot, E. Iancu, R. R. Parwani: Phys. Rev. **D52**, 2543 (1995)
28. J. P. Blaizot, E. Iancu, A. Rebhan: Phys. Rev. Lett. **83**, 2906 (1999)
29. J. P. Blaizot, E. Iancu, A. Rebhan: Phys. Lett. **B470**, 181 (1999)
30. J. P. Blaizot, E. Iancu, A. Rebhan: Phys. Rev. **D63**, 065003 (2001)
31. N. N. Bogoliubov, D. V. Shirkov: *Introduction to the Theory of Quantized Fields* (Interscience Publishers, New York, 1959)
32. G. Boyd, et al.: Nucl. Phys. **B469**, 419 (1996)
33. E. Braaten, A. Nieto: Phys. Rev. Lett. **73**, 2402 (1994)
34. E. Braaten, A. Nieto: Phys. Rev. Lett. **74**, 3530 (1995)
35. E. Braaten, A. Nieto: Phys. Rev. **D53**, 3421 (1996)
36. E. Braaten, R. D. Pisarski: Phys. Rev. **D42**, 2156 (1990)
37. E. Braaten, R. D. Pisarski: Nucl. Phys. **B339**, 310 (1990)
38. E. Braaten, R. D. Pisarski: Phys. Rev. Lett. **64**, 1338 (1990)
39. E. Braaten, R. D. Pisarski: Nucl. Phys. **B337**, 569 (1990)
40. E. Braaten, R. D. Pisarski: Phys. Rev. **D46**, 1829 (1992)
41. E. Braaten, R. D. Pisarski: Phys. Rev. **D45**, 1827 (1992)
42. C. P. Burgess, A. L. Marini: Phys. Rev. **D45**, 17 (1992)
43. S. Chiku, T. Hatsuda: Phys. Rev. **D58**, 076001 (1998)
44. J. M. Cornwall, W.-S. Hou, J. E. King: Phys. Lett. **B153**, 173 (1985)
45. A. Cucchieri, F. Karsch, P. Petreczky: Phys. Lett. **B497**, 80 (2001)
46. A. Cucchieri, F. Karsch, P. Petreczky: *Propagators and dimensional reduction of hot SU(2) gauge theory*, hep-lat/0103009 (2001)
47. B. S. DeWitt: Phys. Rev. **162**, 1195 (1967)
48. M. Dirks, A. Niegawa, K. Okano: Phys. Lett. **B461**, 131 (1999)
49. L. Dolan, R. Jackiw: Phys. Rev. **D9**, 3320 (1974)
50. I. T. Drummond, R. R. Horgan, P. V. Landshoff, A. Rebhan: Nucl. Phys. **B524**, 579 (1998)
51. I. T. Drummond, R. R. Horgan, P. V. Landshoff, A. Rebhan: Phys. Lett. **B460**, 197 (1999)
52. R. Efraty, V. P. Nair: Phys. Rev. Lett. **68**, 2891 (1992)
53. R. Efraty, V. P. Nair: Phys. Rev. **D47**, 5601 (1993)
54. F. Flechsig: Nucl. Phys. **B547**, 239 (1999)
55. F. Flechsig, A. K. Rebhan: Nucl. Phys. **B464**, 279 (1996)
56. F. Flechsig, A. K. Rebhan, H. Schulz: Phys. Rev. **D52**, 2994 (1995)
57. F. Flechsig, H. Schulz: Phys. Lett. **B349**, 504 (1995)
58. E. S. Fradkin: Proc. Lebedev Inst. **29**, 1 (1965)
59. J. Frenkel, J. C. Taylor: Nucl. Phys. **B334**, 199 (1990)
60. J. Frenkel, J. C. Taylor: Nucl. Phys. **B374**, 156 (1992)
61. T. Furusawa, K. Kikkawa: Phys. Lett. **B128**, 218 (1983)
62. F. Gelis: Phys. Lett. **B455**, 205 (1999)
63. M. Gell-Mann, K. A. Brueckner: Phys.Rev. **106**, 364 (1957)
64. D. J. Gross, R. D. Pisarski, L. G. Yaffe: Rev. Mod. Phys. **53**, 43 (1981)
65. T. H. Hansson, I. Zahed: Nucl. Phys. **B292**, 725 (1987)
66. H. Hata, T. Kugo: Phys. Rev. **D21**, 3333 (1980)
67. U. Heinz, K. Kajantie, T. Toimela: Ann. Phys. **176**, 218 (1987)
68. U. M. Heller, F. Karsch, J. Rank: Phys. Rev. **D57**, 1438 (1998)
69. W. Israel: Physica **106A**, 204 (1981)
70. K. A. James: Z. Phys. **C48**, 169 (1990)
71. K. A. James, P. V. Landshoff: Phys. Lett. **B251**, 167 (1990)

72. K. Kajantie, J. Kapusta: Phys. Lett. **B110**, 299 (1982)
73. K. Kajantie, J. Kapusta: Ann. Phys. **160**, 477 (1985)
74. K. Kajantie, et al.: Phys. Rev. Lett. **79**, 3130 (1997)
75. O. K. Kalashnikov, V. V. Klimov: Sov. J. Nucl. Phys. **31**, 699 (1980)
76. J. I. Kapusta: *Finite-Temperature Field Theory* (Cambridge University Press, Cambridge, UK, 1989)
77. F. Karsch, A. Patkós, P. Petreczky: Phys. Lett. **B401**, 69 (1997)
78. L. V. Keldysh: Zh. Eksp. Teor. Fiz. **47**, 1515 (1964)
79. P. F. Kelly, Q. Liu, C. Lucchesi, C. Manuel: Phys. Rev. **D50**, 4209 (1994)
80. V. V. Klimov: Sov. J. Nucl. Phys. **33**, 934 (1981)
81. R. Kobes, G. Kunstatter, K. Mak: Phys. Rev. **D45**, 4632 (1992)
82. R. Kobes, G. Kunstatter, A. Rebhan: Phys. Rev. Lett. **64**, 2992 (1990)
83. R. Kobes, G. Kunstatter, A. Rebhan: Nucl. Phys. **B355**, 1 (1991)
84. U. Kraemmer, M. Kreuzer, A. Rebhan: Ann. Phys. **201**, 223 (1990)
85. U. Kraemmer, A. K. Rebhan, H. Schulz: Ann. Phys. **238**, 286 (1995)
86. T. Kugo, I. Ojima: Phys. Lett. **B73**, 459 (1978)
87. T. Kugo, I. Ojima: Prog. Theor. Phys. **61**, 644 (1979)
88. M. Laine, O. Philipsen: Nucl. Phys. **B523**, 267 (1998)
89. M. Laine, O. Philipsen: Phys. Lett. **B459**, 259 (1999)
90. P. V. Landshoff, A. Rebhan: Nucl. Phys. **B383**, 607 (1992)
91. P. V. Landshoff, A. Rebhan: Nucl. Phys. **B410**, 23 (1993)
92. N. P. Landsman, C. G. van Weert: Phys. Rept. **145**, 141 (1987)
93. M. Le Bellac: *Thermal Field Theory* (Cambridge University Press, Cambridge, UK, 1996)
94. V. V. Lebedev, A. V. Smilga: Ann. Phys. **202**, 229 (1990)
95. V. V. Lebedev, A. V. Smilga: Phys. Lett. **B253**, 231 (1991)
96. A. D. Linde: Phys. Lett. **B96**, 289 (1980)
97. T. Matsubara: Prog. Theor. Phys. **14**, 351 (1955)
98. H. Matsumoto, Y. Nakano, H. Umezawa: Phys. Rev. **D28**, 1931 (1983)
99. H. Matsumoto, Y. Nakano, H. Umezawa: J. Math. Phys. **25**, 3076 (1984)
100. L. D. McLerran, B. Svetitsky: Phys. Rev. **D24**, 450 (1981)
101. S. Nadkarni: Phys. Rev. **D33**, 3738 (1986)
102. S. Nadkarni: Phys. Rev. Lett. **61**, 396 (1988)
103. V. P. Nair: Phys. Rev. **D48**, 3432 (1993)
104. A. Niegawa: Phys. Rev. **D40**, 1199 (1989)
105. S. Peigné, E. Pilon, D. Schiff: Z. Phys. **C60**, 455 (1993)
106. S. Peigné, S. M. H. Wong: Phys. Lett. **B346**, 322 (1995)
107. R. D. Pisarski: Nucl. Phys. **A498**, 423C (1989)
108. R. D. Pisarski: Phys. Rev. Lett. **63**, 1129 (1989)
109. R. D. Pisarski: Phys. Rev. **D47**, 5589 (1993)
110. A. M. Polyakov: Phys. Lett. **B72**, 477 (1978)
111. K. Rajagopal, E. Shuster: Phys. Rev. **D62**, 085007 (2000)
112. A. Rebhan: Nucl. Phys. **B288**, 832 (1987)
113. A. Rebhan: Phys. Rev. **D46**, 482 (1992)
114. A. Rebhan: Phys. Rev. **D46**, 4779 (1992)
115. A. K. Rebhan: Phys. Rev. **D48**, 3967 (1993)
116. A. K. Rebhan: Nucl. Phys. **B430**, 319 (1994)
117. H. Schulz: Nucl. Phys. **B413**, 353 (1994)
118. J. Schwinger: J. Math. Phys. **2**, 407 (1961)

119. V. P. Silin, V. N. Ursov: Sov. Phys. - Lebedev Inst. Rep. **5**, 43 (1988)
120. J. C. Taylor, S. M. H. Wong: Nucl. Phys. **B346**, 115 (1990)
121. T. Toimela: Z. Phys. **C27**, 289 (1985)
122. V. N. Tsytovich: Sov. Phys. (JETP) **13**, 1249 (1961)
123. G. A. Vilkovisky: Nucl. Phys. **B234**, 125 (1984)
124. H. A. Weldon: Phys. Rev. **D26**, 1394 (1982)
125. H. A. Weldon: Phys. Rev. **D40**, 2410 (1989)
126. C.-X. Zhai, B. Kastening: Phys. Rev. **D52**, 7232 (1995)

Lattice QCD at High Temperature and Density

Frithjof Karsch

Fakultät für Physik, Universität Bielefeld
D-33615 Bielefeld, Germany

Summary. After a brief introduction into basic aspects of the formulation of lattice regularized QCD at finite temperature and density we discuss our current understanding of the QCD phase diagram at finite temperature. We present results from lattice calculations that emphasize the deconfining as well as chiral symmetry restoring features of the QCD transition, and discuss the thermodynamics of the high temperature phase.

1 Introduction

Almost immediately after the ground-breaking demonstration that the numerical analysis of lattice regularized quantum field theories [1] can also provide quantitative results on fundamental non-perturbative properties of QCD [2] it has been realized that this approach will also allow to study the QCD phase transition [3,4] and the equation of state of the quark-gluon plasma [5]. During the last 20 years we have learned a lot from lattice calculations about the phase structure of QCD at finite temperature. In fact, we do understand quite well the thermodynamics in the heavy quark mass limit of QCD, the pure $SU(3)$ gauge theory, and even have calculated the critical temperature and the equation of state in this limit with an accuracy of a few percent. However, it is only now that we start to reach a level of accuracy in numerical calculations of QCD thermodynamics that allows to seriously consider quantitative studies of QCD with a realistic light quark mass spectrum. An important ingredient in the preparation of such calculations is the development of new regularization schemes in the fermion sector of the QCD Lagrangian, which allows to reduce discretization errors and also improves the flavour symmetry of the lattice actions. The currently performed investigations of QCD thermodynamics provide first results with such improved actions and prepare the ground for calculations with a realistic light quark mass spectrum.

The interest in analyzing the properties of QCD under extreme conditions is twofold. On the one hand it is the goal to reach a quantitative description of the behaviour of matter at high temperature and density. This does provide important input for a quantitative description of experimental signatures for the occurrence of a phase transition in heavy ion collisions and should also help to understand better the phase transitions that occurred during the early times of the evolution of the universe. Eventually it also may allow to answer the question whether a quark-gluon plasma can exist in the interior of dense neutron stars or did exist in early stages of supernova explosions. For this reason one would like to reach a quantitative understanding of the QCD equation of state, determine critical parameters such as the critical temperature and the critical energy density and predict the modification of basic hadron properties (masses, decay widths) with temperature. On the other hand the analysis of a complicated quantum field theory like QCD at non-zero temperature can also help to improve our understanding of its non-perturbative properties at zero temperature. The introduction of external control parameters (temperature, chemical potential) allows to observe the response of different observables to this and may provide a better understanding of their interdependence [6]. As such one would, for instance, like to clarify the role of confinement and chiral symmetry breaking for the QCD phase transition. In which respect is the QCD phase transition deconfining and/or chiral symmetry restoring? In how far can the critical behaviour be described by intuitive pictures based on percolation, bag or resonance gas models which have been developed for the QCD transition? We will discuss these qualitative aspects of the QCD thermodynamics and also present results on basic questions concerning the equation of state and the critical temperature of the transition which ask for quantitative answers.

In the next section we give a short introduction into the lattice formulation of QCD thermodynamics. In Section 3 we discuss the basic structure of the QCD phase diagram at finite temperature as it is known from lattice calculations. Section 4 is devoted to a discussion of basic thermodynamic observables which characterize the QCD transition to the plasma phase and we will identify general properties which show the deconfining and chiral symmetry restoring features of this transition. In Section 5 we comment on different length scales characterizing the QCD plasma and try to establish the temperature regime where lattice calculations may make contact with perturbative approaches. A description of recent results on the QCD equation of state and the critical temperature of the QCD transition which emphasizes the quark mass and flavour dependence of these quantities is given in Sections 6 and 7, respectively. A brief discussion of the problems arising in lattice formulations of QCD at non-zero baryon number density or chemical potential is given in Section 8. Finally we give our conclusions in Section 9 and describe a specific set of improved gauge and fermion lattice actions in an Appendix.

2 The Lattice Formulation of QCD Thermodynamics

2.1 The Basic Steps from the Continuum to the Lattice ...

Starting point for the discussion of the equilibrium thermodynamics of QCD on the lattice is the QCD partition function, which explicitly depends on the volume (V), the temperature (T) and the quark number chemical potential (μ). It is represented in terms of a Euclidean path integral over gauge (A_ν) and fermion ($\bar{\psi}$, ψ) fields,

$$Z(V,T,\mu) = \int \mathcal{D}A_\nu \mathcal{D}\bar{\psi}\mathcal{D}\psi \; \mathrm{e}^{-S_E(V,T,\mu)} \quad , \tag{1}$$

where A_ν and $\bar{\psi}$, ψ obey periodic and anti-periodic boundary conditions in Euclidean time, respectively. The Euclidean action $S_E \equiv S_G + S_F$ contains a purely gluonic contribution (S_G) expressed in terms of the field strength tensor, $F_{\mu\nu} = \partial_\mu A_\nu - \partial_\nu A_\mu - \mathrm{i}g[A_\mu, A_\nu]$, and a fermion part ($S_F$), which couples the gauge and fermions field through the standard minimal substitution,

$$S_E(V,T,\mu) \equiv S_G(V,T) \; + \; S_F(V,T,\mu) \tag{2}$$

with

$$S_G(V,T) = \int_0^{1/T} \Delta x_0 \int_V \Delta^3\mathbf{x} \; \frac{1}{2}\mathrm{Tr}\; F_{\mu\nu}F_{\mu\nu} \tag{3}$$

$$S_F(V,T,\mu) = \int_0^{1/T} \Delta x_0 \int_V \Delta^3\mathbf{x} \sum_{f=1}^{n_f} \bar{\psi}_f \left(\gamma_\mu[\partial_\mu - \mathrm{i}gA_\mu] + m_f - \mu\gamma_0\right)\psi_f . \tag{4}$$

Here m_f are the different quark masses for the n_f different quark flavours and g denotes the QCD coupling constant.

The path integral appearing in Eq. (1) is regularized by introducing a four-dimensional space-time lattice of size $N_\sigma^3 \times N_\tau$ with a lattice spacing a. Volume and temperature are then related to the number of points in space and time directions, respectively,

$$V = (N_\sigma \; a)^3 \quad , \quad T^{-1} = N_\tau \; a \quad . \tag{5}$$

While the discretization of the fermion sector, at least on the naive level, is straightforwardly achieved by replacing derivatives by finite differences, the gauge sector is a bit more involved. Here we introduce *link variables* $U_\mu(x)$ which are associated with the link between two neighbouring sites of the lattice and describe the parallel transport of the field $\mathcal{A}$ from site x to $x + \hat{\mu}a$,

$$U_{x,\mu} = \mathrm{P}\exp\left(ig\int_x^{x+\hat{\mu}a} \Delta x^\mu \, A_\mu(x)\right) \quad , \tag{6}$$

where P denotes the path ordering. The link variables $U_\mu(x)$ are elements of the $SU(3)$ colour group. A product of these link variables around an elementary plaquette may be used to define an approximation to the gauge action,

$$W^{(1,1)}_{n,\mu\nu} = 1 - \frac{1}{3}\mathrm{Re}\,\square_{n,\mu\nu} \equiv \mathrm{Re\ Tr}\, U_{n,\mu}U_{n+\hat{\mu},\nu}U^\dagger_{n+\hat{\nu},\mu}U^\dagger_{n,\nu}$$
$$= \frac{g^2a^4}{2}F^a_{\mu\nu}F^a_{\mu\nu} + \mathcal{O}(a^6) \quad . \tag{7}$$

A discretized version of the Euclidean gauge action, which reproduces the continuum version up to cut-off errors of order a^2, thus is given by the *Wilson action* [1],

$$\beta S_G = \beta \sum_{\substack{n \\ 0\le\mu<\nu\le 3}} W^{(1,1)}_{n,\mu\nu} \quad \Longrightarrow \quad \int \Delta^4 x\, \mathcal{L}_E + \mathcal{O}(a^2) \quad , \tag{8}$$

where we have introduced the gauge coupling $\beta = 6/g^2$.

As is well known the naive discretization of the fermionic part of the action, which is obtained by introducing the simple finite difference scheme to discretize the derivative appearing in the fermion Lagrangian, i.e. $\partial_\mu\psi_f(x) = (\psi_{n+\hat{\mu}} - \psi_{n-\hat{\mu}})/2a$, does in the continuum limit not reproduce the particle content one started with. The massless lattice fermion propagator has poles not only at zero momentum but also at all other corners of the Brillouin zone and thus generates 16 rather than a single fermion species in the continuum limit. One thus faces a severe species doubling problem. The way out has been to either introduce an explicit chiral symmetry breaking term, which is proportional to $a\partial^2_\mu\psi_f(x)$ and thus vanishes in the continuum limit (Wilson fermions [1]), or to distribute the components of the fermion Dirac spinors over several lattice sites (staggered fermions) [7]. The staggered fermion formulation does not eliminate the species doubling problem completely. One still gets four degenerate fermion species. However, it has the advantage that it preserves a continuous subgroup of the original global chiral symmetry. In the massless limit the chiral condensate thus still is an order parameter for the occurrence of a phase transition at finite temperature.

Progress has been made in formulating lattice QCD also with chiral fermion actions, which do avoid the species doubling and at the same time preserve the chiral symmetry of the QCD Lagrangian. This can, for instance, be achieved by introducing an extra fifth dimension [8]. At present, however, very little has been done to study QCD thermodynamics on the lattice with these actions [9]. Much more is known on the QCD thermodynamics from calculations with Wilson and staggered fermions. We will in the following present results from both approaches. However, to be specific we will restrict ourselves here to a discussion of the staggered fermion formulation introduced by Kogut and Susskind [7]. The fermion action can be written as

$$S_F^{KS} = \sum_{nm} \bar{\chi}_n Q^{KS}_{nm}\chi_m \quad , \tag{9}$$

where the staggered fermion matrix Q^{KS} is given by

$$Q^{KS}_{nm}(m_q,\tilde{\mu}) = \frac{1}{2}\sum_{\mu=1}^{3}(-1)^{n_0+\ldots+n_{\mu-1}}(\delta_{n+\hat{\mu},m}U_{n,\mu} - \delta_{n,m+\hat{\mu}}U^\dagger_{m,\mu})$$
$$+\frac{1}{2}(\delta_{n+\hat{0},m}U_{n,0}\,\mathrm{e}^{\tilde{\mu}} - \delta_{n,m+\hat{0}}U^\dagger_{m,0})\,\mathrm{e}^{-\tilde{\mu}} + \delta_{nm}m_q \quad . \quad (10)$$

Here we have introduced the chemical potential $\tilde{\mu}$ on the temporal links [10]. As the fermion action is quadratic in the Grassmann valued quark fields $\bar{\chi}$ and χ we can integrate them out in the partition function and finally arrive at a representation of $Z(V,T,\mu)$ on a 4-dimensional lattice of size $N_\sigma^3 \times N_\tau$,

$$Z(N_\sigma, N_\tau, \beta, m_q, \tilde{\mu}) = \int \prod_{n\nu} \Delta U_{n,\nu} (\det Q^{KS}(m_q,\tilde{\mu}))^{n_f/4} \mathrm{e}^{-\beta S_G} \quad . \qquad (11)$$

We have made explicit the fact that the staggered fermion action does lead to four degenerate fermion flavours in the continuum limit, i.e. taking the continuum limit with the action given in Eqs. 9 and 10 corresponds to $n_f = 4$ in Eq. (11). As the number of fermion species does appear only as an appropriate power of the fermion determinant, which is true also in the continuum limit, one also may choose $n_f \neq 4$ in Eq. (11). This is the approach used to perform simulations with different number of flavours in the staggered fermion formulation.

For $\tilde{\mu} = 0$ the fermion determinant appearing in Eq. (11) is real and positive. Standard numerical techniques, which rely on a probability interpretation of the integrand in Eq. (11), thus can be applied. For $\tilde{\mu} \neq 0$ the determinant, however, becomes complex. Although the contribution of the imaginary part can easily be shown to be zero, as it should to give a real partition function, the real part is no longer strictly positive. This *sign problem* so far still constitutes a major problem in the application of numerical techniques to studies of QCD at non-zero baryon number density or non-zero chemical potential. We therefore will restrict our discussion of QCD thermodynamics mainly to the case $\tilde{\mu} \equiv 0$ and will come back to the problems one faces for $\tilde{\mu} \neq 0$ in Section 8.

2.2 ... and Back from the Lattice to the Continuum

The lattice discretized QCD action discussed above reproduces the continuum action up to discretization errors of $\mathcal{O}(a^2)$. In order to perform the continuum limit at constant temperature, we will have to take the limit ($a \to 0$, $N_\tau \to \infty$) with $T = 1/N_\tau a$ fixed. In particular, for bulk thermodynamic observables like the pressure and energy density, which have dimension $[T^4]$ this limit is rather cumbersome. All lattice observables are dimensionless and are thus calculated in appropriate units of the lattice spacing a. As a consequence a calculation of, e.g., the pressure will provide pa^4 and thus yields a numerical result which decreases in magnitude like N_τ^{-4}. Numerical calculations, however, are always based on the analysis of a finite set of suitably generated gauge field configurations and thus produce results which have a statistical error. It therefore rapidly becomes difficult to calculate bulk thermodynamic quantities on lattices with large temporal extent N_τ. For this reason it is of particular importance for finite temperature calculations to be able to use actions which have small discretization errors and thus allow to perform calculations on lattices with moderate temporal extent. Such actions have been developed and successfully applied in thermodynamic calculations for the pure $SU(3)$ gauge theory. In the fermion sector appropriate actions, which reduce cut-off effects in the high temperature ideal gas limit, so far have only been constructed for staggered fermions. As an example we describe a specific choice of improved gauge and staggered fermion actions in more detail in an Appendix.

As mentioned above we have to perform the continuum limit in order to eliminate lattice discretization errors and to arrive finally at quantitative predictions for the QCD thermodynamics. Eventually we thus have to analyze our observables on different size lattices and extrapolate our results to $N_\tau \to \infty$ at fixed temperature. Unless we perform calculations at a well defined temperature, e.g., the critical temperature, we will have to determine the temperature scale from an additional (zero-temperature) calculation of an observable for which we know its physical value (in MeV). This requires a calculation at the same value of the cut-off (same values of the bare couplings). Of course, we know such a quantity only for the physical case realized in nature, i.e. QCD with two light up and down quark flavours and a heavier strange quark. Nonetheless, we have good reason to believe that certain observables are quite insensitive to changes in the quark masses, e.g., quenched hadron masses[1] ($\tilde{m}_H$) or the string tension ($\tilde{\sigma}$) are believed to be suitable observables to set a physical scale even in the limit of infinite quark masses (pure $SU(3)$ gauge theory). We thus may use calculations of these quantities to define a temperature scale,

$$T/\sqrt{\sigma} = 1/\sqrt{\tilde{\sigma}} N_\tau \quad \text{or} \quad T/m_H = 1/\tilde{m}_H N_\tau \quad . \tag{12}$$

In the pure $SU(3)$ gauge theory as well as in the massless limit the lattice spacing is controlled through β, the only bare coupling appearing in the Euclidean action. Asymptotically a and β are then related through the leading order renormalization group equation,

$$a\Lambda_L \simeq (6b_0/\beta)^{-b_1/2b_0^2} \mathrm{e}^{-\beta/12b_0} \quad , \tag{13}$$

[1] A physical observable O is calculated on the lattice as dimensionless quantity, which we denote here by $\tilde{O}$. Quite often, however, we will also adopt the customary lattice notation, which explicitly specifies the cut-off dependence in the continuum limit, e.g., $\tilde{m}_H \equiv m_H a$ or $\tilde{\sigma} \equiv \sigma a^2$.

where the two universal coefficients are given by

$$b_0 = \frac{1}{16\pi^2}\left(11 - \frac{2}{3}n_f\right) \quad , \quad b_1 = \left(\frac{1}{16\pi^2}\right)^2 \left[102 - \left(10 + \frac{8}{3}\right) n_f\right] \tag{14}$$

and Λ_L is a scale parameter, which unambiguously can be related to the scale parameter in other regularization schemes, e.g., to $\Lambda_{\overline{MS}}$. The continuum limit thus is reached with increasing β.

In the case of non-zero quark masses one has in addition to insure that the continuum limit is taken along a *line of constant physics*. This can be achieved by keeping a ratio of hadron masses, for instance the ratio of pseudo-scalar and vector meson masses, m_{PS}/m_V, constant while varying the couplings (β, m_q). In the limit $\beta \to \infty$ this requires a tuning of the bare quark masses such that $m_q \to 0$. We also note that for small quark masses the vector meson mass m_V approaches a constant, $m_V = m_\rho + \mathcal{O}(m_q)$, while the pseudo-scalar is the Goldstone-particle corresponding to the broken chiral symmetry of QCD (pion). Its mass is proportional to the square root of m_q. In the following we will quite often quote results as a function of m_{PS}/m_V which is just another way for quoting results for different values of the quark mass.

3 The QCD Phase Diagram at Finite Temperature

At vanishing baryon number density (or zero chemical potential) the properties of the QCD phase transition depend on the number of quark flavours and their masses. While it is a detailed quantitative question at which temperature the transition to the high temperature plasma phase occurs, we do expect that the nature of the transition, e.g., its order and details of the critical behaviour, are controlled by global symmetries of the QCD Lagrangian. Such symmetries only exist in the limits of either infinite or vanishing quark masses. For any non-zero, finite value of quark masses the global symmetries are explicitly broken. In fact, in the case of QCD the explicit symmetry breaking induced by the finite quark masses is very much similar to that induced by an external ferromagnetic field in spin models. We thus expect that a continuous phase transition, which may exist in the zero or infinite quark mass limit, will turn into a non-singular crossover behaviour for any finite value of the quark mass. First order transitions, on the other hand, may persist for some time before they end in a continuous transition. Whether a true phase transition exists in QCD with the physically realized spectrum of quark masses or whether in this case the transition is just a (rapid) crossover, again becomes a quantitative question which we have to answer through direct numerical calculations.

Our current understanding of the qualitative aspects of the QCD phase diagram is based on universality arguments for the symmetry breaking patterns in the heavy [11] as well as the light quark mass regime [12,13]. In the limit of infinitely heavy quarks, the pure $SU(3)$ gauge theory, the large distance behaviour of the heavy quark free energy, $F_{\bar{q}q}$, provides a unique distinction between confinement below T_c and deconfinement for $T > T_c$. On a lattice of size $N_\sigma^3 \times N_\tau$ the heavy quark free energy[2] can be calculated from the expectation value of the Polyakov loop correlation function

$$\exp\left(-\frac{F_{\bar{q}q}(r,T)}{T}\right) = \langle \mathrm{Tr} L_{\mathbf{x}} \mathrm{Tr} L_{\mathbf{y}}^\dagger \rangle \quad , \quad rT = |\mathbf{x}-\mathbf{y}| N_\tau \ , \tag{15}$$

where $L_{\mathbf{x}}$ and $L_{\mathbf{y}}^\dagger$ represent static quark and anti-quark sources located at the spatial points $\mathbf{x}$ and $\mathbf{y}$, respectively,

$$L_{\mathbf{x}} = \prod_{x_0=1}^{N_\tau} U_{n,0} \quad , \quad n \equiv (x_0, \mathbf{x}) \quad . \tag{16}$$

For large separations ($r \to \infty$) the correlation function approaches $|\langle L \rangle|^2$, where $\langle L \rangle = N_\sigma^{-3} \langle \sum_{\mathbf{x}} \mathrm{Tr} L_{\mathbf{x}} \rangle$ denotes the Polyakov loop expectation value, which therefore characterizes the behaviour of the heavy quark free energy at large distances and is an order parameter for deconfinement in the $SU(3)$ gauge theory,

$$\langle L \rangle \begin{cases} = 0 \Leftrightarrow \text{confined phase}, & T < T_c \\ > 0 \Leftrightarrow \text{deconfined phase}, & T > T_c \end{cases} \quad . \tag{17}$$

The effective theory for the order parameter is a 3-dimensional spin model with global $Z(3)$ symmetry. Universality arguments then suggest that the phase transition is first order in the infinite quark mass limit [11].

In the limit of vanishing quark masses the classical QCD Lagrangian is invariant under chiral symmetry transformations; for n_f massless quark flavours the symmetry is

$$U_A(1) \times SU_L(n_f) \times SU_R(n_f).$$

However, only the $SU(n_f)$ flavour part of this symmetry is spontaneously broken in the vacuum, which gives rise to $(n_f^2 - 1)$ massless Goldstone particles, the pions. The axial $U_A(1)$ only is a symmetry of the classical Lagrangian. It is explicitly broken due to quantum corrections in the QCD partition function, the axial anomaly, and therefore gets replaced by a discrete $Z(n_f)$ symmetry at low temperature. The basic observable which reflects the chiral properties of QCD is the chiral condensate,

$$\langle \bar{\chi}\chi \rangle = \frac{1}{N_\sigma^3 N_\tau} \frac{\partial}{\partial m_q} \ln Z \quad . \tag{18}$$

[2] In the $T \to 0$ limit this is just the heavy quark potential; at non-zero temperature $F_{\bar{q}q}$ does, however, also include a contribution resulting from the overall change of entropy that arises from the presence of external quark and anti-quark sources.

In the limit of vanishing quark masses the chiral condensate stays non zero as long as chiral symmetry is spontaneously broken. The chiral condensate thus is an obvious order parameter in the chiral limit,

$$\langle\bar{\chi}\chi\rangle \begin{cases} > 0 \Leftrightarrow \text{symmetry broken phase}, & T < T_c \\ = 0 \Leftrightarrow \text{symmetric phase}, & T > T_c \end{cases} . \tag{19}$$

For light quarks the global chiral symmetry is expected to control the critical behaviour of the QCD phase transition. In particular, the order of the transition is expected to depend on the number of light or massless flavours. The basic aspects of the n_f-dependence of the phase diagram have been derived by Pisarski and Wilczek [12] from an effective, 3-dimensional Lagrangian for the order parameter[3],

$$\begin{aligned} \mathcal{L}_{eff} = & -\frac{1}{2}\mathrm{Tr}(\partial_\mu \Phi^\dagger \partial^\mu \Phi) - \frac{1}{2}m^2\mathrm{Tr}(\Phi^\dagger \Phi) + \frac{\pi^2}{3}g_1 \left(\mathrm{Tr}(\Phi^\dagger \Phi)\right)^2 \\ & + \frac{\pi^2}{3}g_2 \mathrm{Tr}\left((\Phi^\dagger \Phi)^2\right) + c\left(\det \Phi + \det \Phi^\dagger\right) \quad , \end{aligned} \tag{20}$$

with $\Phi \equiv (\Phi_{ij})$, $i,j = 1,...,n_f$. $\mathcal{L}_{eff}$ has the same global symmetry as the QCD Lagrangian. A renormalization group analysis of this Lagrangian suggests that the transition is first order for $n_f \geq 3$ and second order for $n_f = 2$. The latter, however, is expected to hold only if the axial $U_A(1)$ symmetry breaking, related to the $\det \Phi$ terms in Eq. (20), does not become too weak at T_c so that the occurrence of a fluctuation induced first order transition would also become possible.

This basic pattern has indeed been observed in lattice calculations. So far no indication for a discontinuous transition has been observed for $n_f = 2$. The transition is found to be first order for $n_f \geq 3$. Moreover, the transition temperature is decreasing with increasing n_f and there are indications that chiral symmetry is already restored in the vacuum above a critical number of flavours [15].

The anticipated phase diagram of 3-flavour QCD at vanishing baryon number density is shown in Fig. 1. An interesting aspect of the phase diagram is the occurrence of a second order transition line in the light quark mass regime, the boundary of the region of first order phase transitions. On this line the transition is controlled by an effective 3-dimensional theory with global $Z(2)$ symmetry [13], which is not a symmetry of the QCD Lagrangian. As this boundary lies in the light quark mass regime it may well be that this second order transition, for which neither the chiral condensate nor the Polyakov loop will be the order parameter, is equally important for the critical or crossover behaviour of QCD with a realistic quark mass spectrum as the nearby critical

[3] It should be noted that this ansatz assumes that chiral symmetry is broken at low temperatures. Instanton model calculations suggest that the vacuum, in fact, is chirally symmetric already for $n_f \geq 5$ [14].

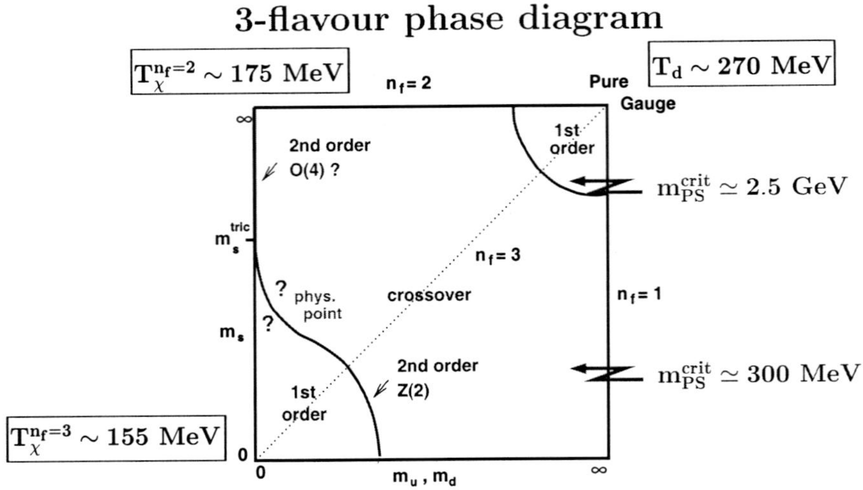

Fig. 1. The QCD phase diagram of 3-flavour QCD with degenerate (u,d)-quark masses and a strange quark mass m_s

point in the chiral limit. In particular, we note that the critical exponent α is positive for the 3-d, $Z(2)$ symmetric models whereas it is negative for the $O(4)$ model. A nearby $Z(2)$ symmetric critical point in the QCD phase diagram will thus induce larger density fluctuations than would be expected in the vicinity of the chiral critical point. It therefore will be important to determine in detail the location of the physical point in the QCD phase diagram.

4 Deconfinement versus Chiral Symmetry Restoration

As outlined in the previous section the two properties of QCD, which explain the basic features of the observed spectrum of hadrons, are also of central importance for the structure of the QCD phase diagram at finite temperature – confinement and chiral symmetry breaking. While the former explains why we observe only colourless states in the spectrum the latter describes the presence of light Goldstone particles, the pions. The confining property of QCD manifests itself in the long range behaviour of the heavy quark potential. At zero temperature the potential rises linearly at large distances[4], $V_{\bar{q}q}(r) \sim \sigma r$, where $\sigma \simeq (425\ \mathrm{MeV})^2$ denotes the string tension, and forces the quarks and gluons to be confined to a *hadronic bag*. Chiral symmetry breaking leads to a non-vanishing quark anti-quark condensate, $\langle\bar{q}q\rangle \simeq (250\ \mathrm{MeV})^3$ in the

[4] Here large distances actually refer to $r \simeq 1$ fm. For larger distances the spontaneous creation of quark anti-quark pairs from the vacuum leads to a breaking of the string, i.e. the potential tends to a constant value for $r \to \infty$ (see Fig. 4).

vacuum. Inside the hadron bag, however, the condensate vanishes. At high temperatures the individual hadronic bags are expected to merge to a single large bag, in which quarks and gluons can move freely. This bag picture is closely related to percolation models for the QCD phase transition [16]. It provides an intuitive argument for the occurrence of deconfinement and chiral symmetry restoration. A-priori it is, however, not evident that both non-perturbative properties have to get lost at the same temperature. It has been speculated that two distinct phase transitions leading to deconfinement at T_d and chiral symmetry restoration at T_χ could occur in QCD [17]. General arguments about the scales involved[5] suggest that $T_d \leq T_\chi$. Two distinct phase transitions indeed have been found in QCD related models like the $SU(3)$ gauge theory with adjoint fermions [18]. In QCD, however, there seems to be only one transition from the low temperature hadronic regime to the high temperature plasma phase. In fact, as can be seen from Fig. 1 there is a wide range of parameters (quark masses) for which the transition is not related to any singular behaviour in thermodynamic observables; instead of a phase transition one observes just a rapid crossover behaviour. It thus is legitimate to ask which thermodynamic properties change when one moves from the low to the high temperature regime and to what extent these changes are related to deconfinement and/or chiral symmetry restoration.

In the previous section we have introduced order parameters for deconfinement in the infinite quark mass limit, $\langle L \rangle$, and chiral symmetry restoration in the limit of vanishing quark masses, $\langle \bar{\psi}\psi \rangle$. Related observables, which also signal a sudden change in the long distance behaviour of the heavy quark potential or the chiral condensate as function of temperature, are the corresponding susceptibilities, the Polyakov loop susceptibility (χ_L) and the chiral susceptibility (χ_m),

$$\chi_L = N_\sigma^3 \left(\langle L^2 \rangle - \langle L \rangle^2 \right) \quad , \quad \chi_m = \frac{\partial}{\partial m_q} \langle \bar{\psi}\psi \rangle \quad . \tag{21}$$

The behaviour of these observables is shown in Fig. 2 for the case of two flavour QCD with light quarks. This clearly shows that the gauge coupling at which the different susceptibilities attain their maxima, or correspondingly the points of most rapid change in $\langle L \rangle$ and $\langle \bar{\psi}\psi \rangle$ coincide. Calculations of these observables for QCD with three degenerate quark flavours[6] have been performed for a wide range of quark masses [19]. They confirm that the location of maxima in both susceptibilities are indeed strongly correlated. Within statistical accuracy they occur at the same temperature, although the height of these maxima is strongly quark mass dependent. This is shown in Fig. 3. For large ($m_{PS}/m_V \gtrsim 0.9$) and small ($m_{PS}/m_V \lesssim 0.3$) quark masses

[5] The hadronic bag is larger than the constituent quark bag of a current quark surrounded by its gluon cloud.

[6] This corresponds to calculations along the dotted, diagonal line in the phase diagram shown in Fig. 1.

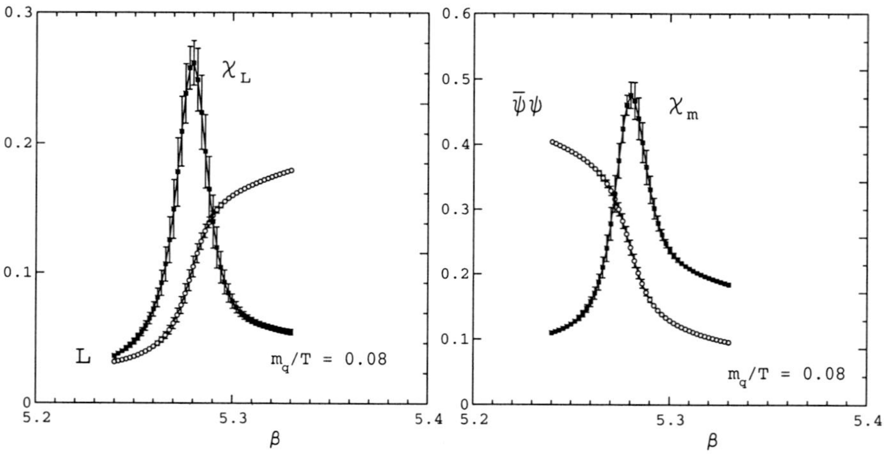

Fig. 2. Deconfinement and chiral symmetry restoration in 2-flavour QCD: Shown is $\langle L \rangle$ (left), which is the order parameter for deconfinement in the pure gauge limit ($m_q \to \infty$), and $\langle \bar{\psi}\psi \rangle$ (right), which is the order parameter for chiral symmetry breaking in the chiral limit ($m_q \to 0$). Also shown are the corresponding susceptibilities as a function of the coupling $\beta = 6/g^2$

the Polyakov loop and chiral susceptibility, respectively, show a strong volume dependence, which is indicative for the presence of first order phase transitions in these corners of the phase diagram. Using zero temperature string tension calculations the pseudo-scalar meson masses have been estimated at which the first order transitions end in a second order transition. For better orientation in the phase diagram these estimates, which at present are not well established and certainly are still subject to lattice artifacts (discretization errors, flavour symmetry breaking), are shown in Fig. 1. As can be seen there is a broad range of quark (or meson) masses for which the QCD transition to the high temperature phase is a non-singular crossover.

As expected the first order phase transition in the large quark mass regime is most clearly visible in the behaviour of the Polyakov loop susceptibility, i.e. the fluctuation of the order parameter for the confinement-deconfinement transition in the pure gauge ($m_q \to \infty$) limit. Similarly, the transition in the chiral limit is most pronounced in the behaviour of the chiral condensate and its susceptibility. This emphasizes the chiral aspects of the QCD transition. One thus may wonder in what respect this transition in the light quark mass regime is a deconfining transition.

4.1 Deconfinement

When talking about deconfinement in QCD we have in mind that a large number of new degrees of freedom gets liberated at a (phase) transition temperature; quarks and gluons, which at low temperature are confined in colourless

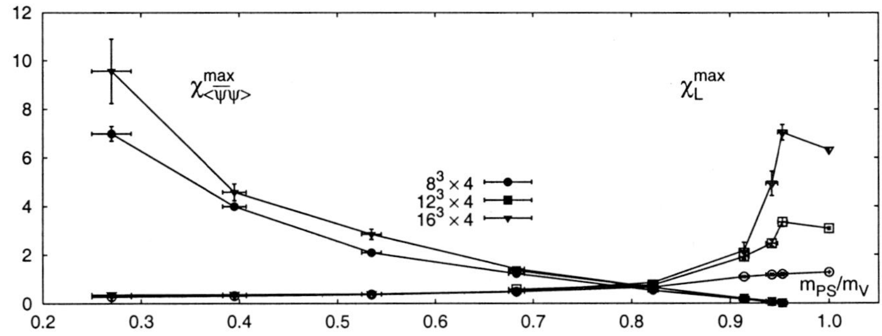

Fig. 3. Quark mass dependence of the Polyakov loop and chiral susceptibilities versus m_{PS}/m_V for 3-flavour QCD. Shown are results from calculations with the improved gauge and staggered fermion action discussed in the Appendix

hadrons and thus do not contribute to the thermodynamics, suddenly become liberated and start contributing to bulk thermodynamic observables like the energy density or pressure. In the heavy quark mass limit ($m_q \equiv \infty$) a more rigorous statement is based on the analysis of the long distance behaviour of the heavy quark free energy, which approaches a constant above T_c but diverges for $T < T_c$. For finite quark masses, however, the long distance behaviour of $F_{\bar{q}q}$ can no longer serve as an order parameter, the heavy quark free energy stays finite for all temperatures. This is shown in Fig. 4 where we compare results from calculations in the pure gauge theory [20] with results from a calculation in three flavour QCD [19].

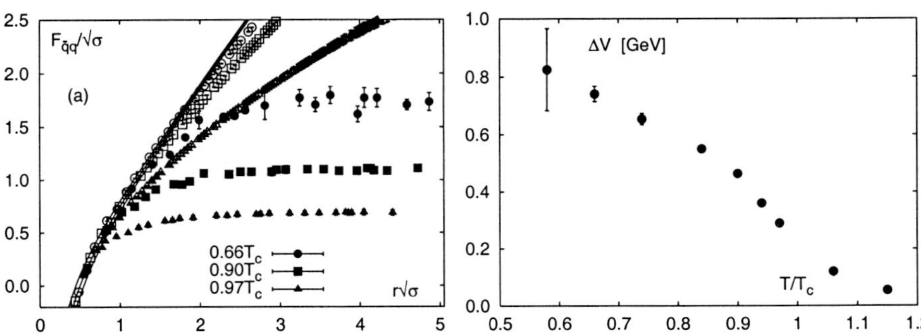

Fig. 4. The left hand figure shows the heavy quark free energy in units of the square root of the string tension for the SU(3) gauge theory (open symbols) and three flavour QCD with light quarks (full symbols). The right hand figure gives the limiting value of the free energy normalized to the value at distance $r \simeq 0.23$ fm [19] as a function of temperature for the case of three flavour QCD. The quark mass used in the $n_f = 3$ calculations corresponds to a ratio of pseudo-scalar and vector meson masses of $m_{PS}/m_V \simeq 0.7$

It is apparent from this figure that there is no drastic qualitative change in the structure of the heavy quark free energy as one crosses the transition temperature. Although the most rapid change in $\Delta V \equiv F_{\bar{q}q}(\infty) - F_{\bar{q}q}(r = 0.5/\sqrt{\sigma})$ occurs for $T \simeq T_c$ and thus resembles the behaviour of the Polyakov loop expectation value and its susceptibility shown in Fig. 2, the heavy quark free energy does not seem to be a good indicator for deconfinement in the presence of light quarks. It looses its role as a rigorous order parameter and also does not reflect changes in the number of partonic degrees of freedom contributing to the thermodynamics.

On the other hand, we expect that bulk thermodynamic quantities like the pressure do reflect the relevant number of degrees of freedom contributing to the thermodynamics in the high temperature limit. Due to asymptotic freedom the QCD pressure will approach the ideal gas value at infinite temperature. In this limit the number of degrees of freedom (quarks+gluons) is much larger than the three light pions which dominate the thermodynamics at low temperature,

$$\frac{p}{T^4} = \begin{cases} 3\frac{\pi^2}{90} & , \ T \to 0 \\ (16 + \frac{21}{2}n_f)\frac{\pi^2}{90} & , \ T \to \infty \end{cases} \quad . \tag{22}$$

This change of active degrees of freedom is clearly visible in calculations of, e.g., the pressure in the pure gauge sector and for QCD with different numbers of flavours. As can bee seen in Fig. 5 the pressure strongly reacts to changes in the number of degrees of freedom. It is this drastic change in the behaviour of the pressure or the energy density which indicates that the QCD (phase) transition to the plasma phase indeed is deconfining. However, it also is worthwhile to note that the transition does, in fact, take place at rather small values of the pressure (and energy density). Only for temperatures $T \gtrsim 2T_c$ does the pressure come close to the ideal gas limit so that one can, with some justification, identify the corresponding light degrees of freedom. This is the case for QCD with light quarks as well as in the quenched limit. At least for temperatures up to a few times T_c the dynamical degrees of freedom are certainly not just weakly interacting partons.

4.2 Chiral Symmetry Restoration

As chiral symmetry restoration does not lead to a significant change of light degrees of freedom, it also is not expected to have an appreciable effect on bulk thermodynamic observables – apart from controlling details of the transition very close to T_c. In particular, we expect that in the case of a continuous transition for $n_f = 2$, the chiral order parameter and its derivative, the chiral susceptibility, show critical behaviour which is characteristic for $O(4)$ spin models in three dimensions [12]. The expected critical behaviour follows from standard scaling arguments derived from the singular part of the free

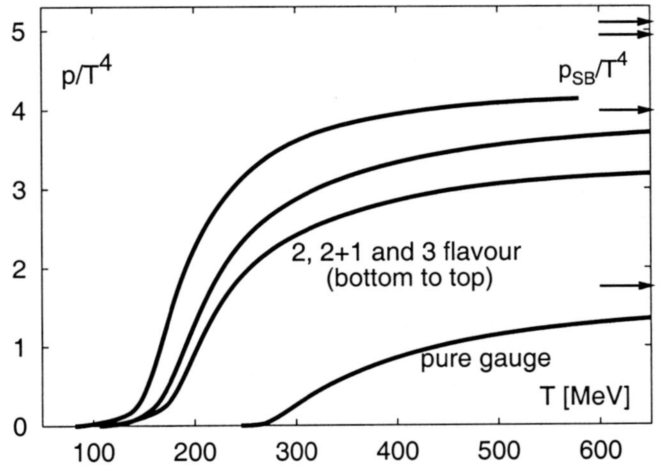

Fig. 5. The pressure in QCD with different number of degrees of freedom as a function of temperature. The curve labeled (2+1)-flavour corresponds to a calculation with two light and a four times heavier strange quark mass [21]

energy density,

$$f_s(t,h) \equiv -\frac{T}{V}\ln Z_s = b^{-d} f_s(b^{y_t}t, b^{y_h}h) \quad , \tag{23}$$

where $t = (T - T_c)/T_c \sim (\beta - \beta_c)$ is the reduced temperature, $h = m/T \sim m_q N_\tau$ the scaled quark mass and b is an arbitrary scale factor. For the chiral order parameter, $\langle\bar{\psi}\psi\rangle$, and the chiral susceptibility, χ_m, one finds from Eq. (23),

$$\langle\bar{\psi}\psi\rangle = h^{1/\delta} F(z) \tag{24}$$

$$\chi_m(t,h) = \frac{1}{\delta} h^{1/\delta - 1}\left[F(z) - \frac{z}{\beta}F'(z)\right] \quad , \tag{25}$$

with scaling functions F and F' that only depend on a specific combination of the reduced temperature and scaled quark mass, $z = th^{-1/\beta\delta}$. The critical exponents β and δ are given in terms of y_t and y_h as $\beta = (1 - y_h)/y_t$ and $\delta = y_h/(1 - y_h)$. As the t-dependence enters in $\chi_m(t,h)$ only through z one also deduces that the line of pseudo-critical couplings defined through the location of the maximum of $\chi_m(t,h)$ at fixed h is described by a universal scaling function,

$$t_c(h) \equiv z_c\, h^{1/\beta\delta} \quad . \tag{26}$$

Although there is ample evidence that the phase transition in 2-flavour QCD is continuous in the chiral limit, the evidence for the expected $O(4)$ scaling is, at present, ambiguous. The behaviour of the pseudo-critical couplings is, in general, consistent with the expected scaling behaviour. The information on the magnetic equation of state, Eq. (24), however, seems to depend on the fermion discretization scheme used to analyze the critical behaviour. While calculations with Wilson fermions yield almost perfect agreement with the universal form of the $O(4)$ magnetic equation of state [22], significant deviations have been found in the case of staggered fermions [23]. The failure of the scaling analysis in the case of staggered fermions is surprising as this staggered fermion action has a global $O(2)$ symmetry even for finite values of the lattice cut-off and as the $O(4)$ and $O(2)$ magnetic equations of state are quite similar [24]. This suggests that finite size effects still play an important role, which is supported by a recent finite size scaling analysis [25]. In Fig. 6 we show the finite size scaling behaviour of the chiral condensate, which has been reanalyzed in [25]. It is consistent with $O(4)$ (or $O(2)$) scaling behaviour. This aspect, however, clearly needs further studies.

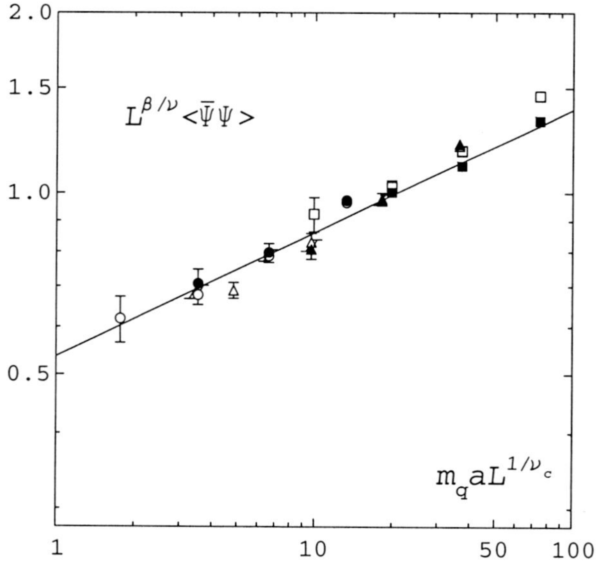

Fig. 6. Finite size scaling of the chiral condensate in 2-flavour QCD [25]. Shown are data from calculations with standard staggered fermions on lattices of size $8^3 \times 4$ (circles), $12^3 \times 4$ (triangles) and $16^3 \times 4$ (squares). The calculations have been performed with different values of the quark mass at fixed value of the scaling variable z_c corresponding to the pseudo-critical line, see Eq. (26)

The changes of the chiral condensate below T_c and chiral symmetry restoration at T_c will have a strong influence on the light hadron spectrum. At T_c the pseudo-scalar mesons (pions) will no longer be Goldstone particles, they turn into massive modes (quasi-particle excitations?) above T_c. Long distance correlations of the chiral condensate decay exponentially with a characteristic length scale proportional to the inverse scalar meson mass. A diverging chiral susceptibility at T_c thus indicates that the scalar meson mass vanishes at T_c. The mass splitting between parity partners thus will decrease when the symmetry breaking reduces and finally will become degenerate at T_c.

As indicated above the modifications of the hadron spectrum are reflected by the temperature dependence of appropriately chosen susceptibilities, which are the space-time integral over hadronic correlation functions in a given quantum number channel,

$$\chi_H = \int_0^{1/T} \mathrm{d}\tau \int \mathrm{d}^3r \; G_H(\tau, \mathbf{r}) \quad , \tag{27}$$

where the hadronic correlation function $G_H(\tau, \mathbf{r})$ for mesons is given by,

$$G_H(\tau, \mathbf{r}) = \langle \bar{\chi}(0) \Gamma_H \chi(0) \bar{\chi}(\tau, \mathbf{r}) \Gamma_H \chi(\tau, \mathbf{r}) \rangle \quad , \tag{28}$$

and Γ_H is an appropriate combination of γ-matrices that projects onto a chosen quantum number channel. In particular, we note that the chiral susceptibility, χ_m, defined in Eq. (21) is the susceptibility of the scalar correlation function. These susceptibilities define generalized masses, $m_H^{-2} \equiv \chi_H$, which are shown in Fig. 7. They, indeed, show the expected behaviour; scalar (f_0) and pseudo-scalar (π) partners become degenerate at T_c whereas the vector meson (δ) which is related to the scalar meson through a $U_A(1)$ rotation only gradually approaches the other masses. The axial $U_A(1)$ symmetry thus remains broken at T_c.

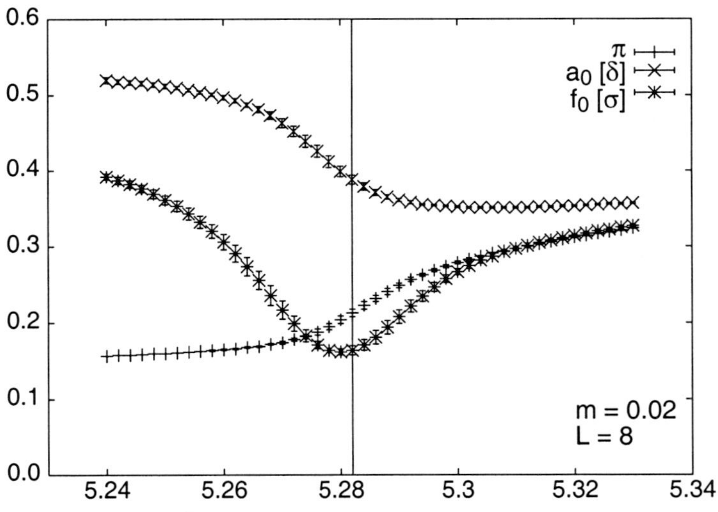

Fig. 7. Temperature dependence of generalized hadron masses extracted from hadronic susceptibilities. Shown are results from calculations in 2-flavour QCD performed on lattices of size $8^3 \times 4$ with staggered fermions of mass $m_q = 0.02$

5 Screening at High Temperature – Short- versus Long-Distance Physics

Our picture of the thermodynamics in the high temperature phase of QCD is largely influenced by perturbative concepts – asymptotic freedom and the screening of electric and magnetic components of the gluon fields. Asymptotic freedom suggests that the temperature dependent running coupling, $g(T)$, becomes small at high temperatures and eventually vanishes in the limit $T \to \infty$. This in turn will lead to a separation of the thermal length scale, $1/T$, from the electric, $1/g(T)T$, and magnetic, $1/g^2(T)T$, screening length scales. The experience gained from lattice calculations in the pure $SU(3)$ gauge theory, however, suggests that this separation of scales, unfortunately, will set in only at asymptotically large temperatures. For all interesting temperatures reachable in heavy ion experiments or even covering the temperature interval between the strong and electroweak phase transitions that occurred in the early universe, the coupling $g(T)$ is of $\mathcal{O}(1)$ and, moreover, the Debye screening mass is significantly larger than the leading order perturbative value,

$$m_{\rm D} = \sqrt{1 + \frac{n_{\rm f}}{6}}\, g(T)\, T \quad . \tag{29}$$

In fact, for $T_c \lesssim T \lesssim 100\, T_c$ one finds that m_D is still about three times larger than this leading order value [26]. A consequence of this large value of the screening mass is that also short distance properties of the plasma are strongly influenced by non-perturbative screening effects; the Debye screening length, $r_D \equiv 1/m_D$, becomes compatible with the characteristic length scale r_{SB} in a free gas where the main contribution to the Stefan-Boltzmann law originates from particles with momenta $p \sim 3T$, i.e. $r_{SB} \sim 1/3T$, and is of the same order as the mean separation between partons in a quark-gluon plasma. We thus must expect that non-perturbative screening effects also have an influence on bulk thermodynamics properties (pressure, energy density) above T_c and even up to quite large temperatures. In fact, the calculations of the pressure and energy density in the $SU(3)$ gauge theory and in QCD with light quarks, which we are going to discuss in the next section, show that at temperatures a few times T_c deviations from the ideal gas limit are still too large to be understood in terms of conventional high temperature perturbation theory, which converges badly at these temperatures just because of the large contribution arising from Debye screening [30].

The screening of static quark and anti-quark sources is commonly analyzed in terms of Polyakov-loop correlation functions, which define the heavy quark free energy introduced in Eq. (15). The leading perturbative contribution to $F_{\bar{q}q}(r,T)$ results from the exchange of two gluons,

$$\frac{V_{\bar{q}q}(r,T)}{T} \equiv \frac{F_{\bar{q}q}(r,T) - F_{\bar{q}q}^{\infty}}{T} = -\ln\left(\frac{\langle {\rm Tr} L_{\mathbf{x}} {\rm Tr} L_{\mathbf{y}}^{\dagger}\rangle}{|\langle L\rangle|^2}\right)$$

$$= -\frac{1}{16}\left(\frac{g^2(T)}{3\pi}\frac{1}{rT}\right)^2 + \mathcal{O}(g^5) \quad . \tag{30}$$

Higher order contributions will lead to screening of this powerlike large distance behaviour, $1/rT \rightarrow \exp(-m_D r)/rT$, and also result in an exponentiation of the leading order contribution. We then may split $F_{\bar{q}q}$ in contributions arising from quark anti-quark pairs in singlet (F_1) and octet (F_8) configurations [3],

$$\mathrm{e}^{-F_{\bar{q}q}(r,T)/T} = \frac{1}{9}\,\mathrm{e}^{-F_1(r,T)/T} + \frac{8}{9}\,\mathrm{e}^{-F_8(r,T)/T} \quad . \tag{31}$$

In accordance with zero temperature perturbation theory the singlet free energy is attractive whereas the octet free energy is repulsive. Their relative strength is such that it leads to a cancellation of the leading $\mathcal{O}(g^2)$ contributions to the colour averaged heavy quark free energy $F_{\bar{q}q}$. From Eq. (31) it is, however, apparent that the cancellation of singlet and octet contributions only occurs at large distances. At short distances the contribution from the attractive singlet channel will dominate the heavy quark free energy,

$$\begin{aligned} \frac{F_{\bar{q}q}(r,T)}{T} &= \frac{F_1(r,T)}{T} + \text{const.} \\ &= -\frac{g^2(T)}{3\pi}\frac{1}{rT} + \text{const.} \qquad \text{for} \quad rT << 1 \quad . \end{aligned} \tag{32}$$

In order to eliminate the subleading power-like behaviour at large distances we show in Fig. 8 $(rT)^2\, V_{\bar{q}q}(r,T)/T$ calculated for the $SU(3)$ gauge theory. As can be seen, the change from the Coulomb-like behaviour at short distances to the exponential screening at large distances can be well localized. For $T_c \le T \lesssim 2T_c$ it occurs already for $rT \simeq 0.2$ or $r \simeq 0.15\ (T_c/T)$ fm and shifts slightly to smaller rT with increasing temperature. A consequence of this efficient screening at short distances is that even heavy quark bound states get destroyed close to T_c in the plasma phase (J/ψ-suppression [28]).

The perturbative analysis of the heavy quark free energy also suggests that for fixed rT the only temperature dependence of $V(r,T)/T$ arises from the running of the coupling $g(T)$. The rapid change of $V(r,T)/T$ at fixed rT which is apparent in Fig. 8 thus also suggests that for temperatures $T \lesssim 3T_c$ the coupling $g(T)$ varies much more rapidly than the asymptotically expected logarithmic running with T.

We thus conclude that non-perturbative screening effects are important for the thermodynamics in the plasma phase also for short distance observables which are sensitive to the physics at distances $r \gtrsim 1/5T$. The strong temperature dependence observed for $T \lesssim 3T_c$, moreover, suggests that the system cannot be described at a weakly coupled, asymptotically free plasma at these temperatures. These general features will carry over to the temperature dependence of the QCD equation of state which we are going to discuss in the next section.

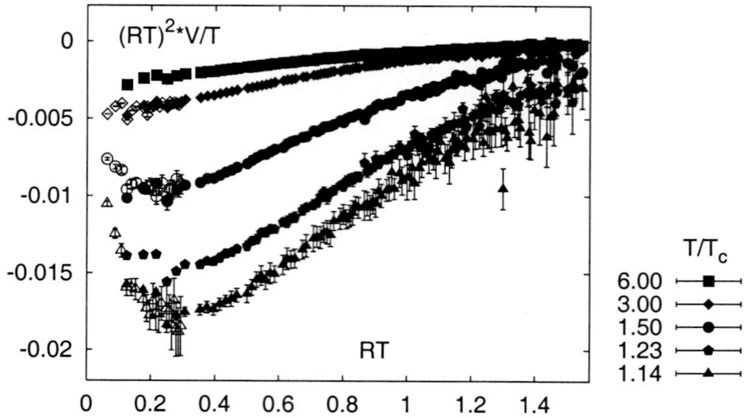

Fig. 8. The heavy quark free energy at various temperatures in the deconfined phase of the $SU(3)$ gauge theory. Calculations have been performed on lattices of size $32^3 \times 8$ (filled symbols) and $32^3 \times 16$ (open symbols) [27]

6 The QCD Equation of State

The most fundamental quantity in equilibrium thermodynamics is, of course, the partition function itself, or the free energy density,

$$f = -\frac{T}{V} \ln Z(T,V) \quad . \tag{33}$$

All basic bulk thermodynamic observables can be derived from the free energy density. In the thermodynamic limit we obtain directly the pressure, $p = -f$ and subsequently also other quantities like the energy (ϵ) and entropy (s) densities or the velocity of sound (c_s),

$$\frac{\epsilon - 3p}{T^4} = T\frac{\mathrm{d}}{\mathrm{d}T}\left(\frac{p}{T^4}\right) \quad , \quad \frac{s}{T^3} = \frac{\epsilon + p}{T^4} \quad , \quad c_s^2 = \frac{\mathrm{d}p}{\mathrm{d}\epsilon} \quad . \tag{34}$$

In the limit of infinite temperature asymptotic freedom suggests that these observables approach the ideal gas limit for a gas of free quarks and gluons, $\epsilon = 3p = -3f$ with p/T^4 given by Eq. (22). Deviations from this ideal gas values have been studied in high temperature perturbation theory. However, it was well-known that this expansion is no longer calculable perturbatively at $\mathcal{O}(g^6)$ [29]. By now all calculable orders up to $\mathcal{O}(g^5 \ln g)$ have been calculated [30]. Unfortunately it turned out that the information gained from this expansion is rather limited. The expansion shows bad convergence behaviour and suggests that it is of use only at temperatures several orders of magnitude larger than the QCD transition temperature. In analytic approaches one thus has to go beyond perturbation theory which currently is being attempted by either using hard thermal loop resummation techniques [31,32] or perturbative dimensional reduction combined with numerical simulations of the resulting effective 3-dimensional theory [33].

In order to make use of the basic thermodynamic relations, Eqs. (33) and (34), in numerical calculations on the lattice we have to go through an additional intermediate step. The free energy density itself is not directly accessible in Monte Carlo calculations; e.g., only expectation values can be calculated easily. One thus proceeds by calculating differences of the free energy density at two different temperatures. These are obtained by taking a suitable derivative of $\ln Z$ followed by an integration, e.g.,

$$\frac{f}{T^4}\bigg|_{T_o}^{T} = -\frac{1}{V}\int_{T_o}^{T} \mathrm{d}x\, \frac{\partial x^{-3}\ln Z(x,V)}{\partial x} \quad . \tag{35}$$

This ansatz readily translates to the lattice. Taking derivatives with respect to the gauge coupling, $\beta = 6/g^2$, rather than the temperature as was done in Eq. (35), we obtain expectation values of the Euclidean action which can be integrated again to give the free energy density,

$$\frac{f}{T^4}\bigg|_{\beta_o}^{\beta} = N_\tau^4 \int_{\beta_o}^{\beta} \mathrm{d}\beta' \left(\langle\tilde{S}\rangle - \langle\tilde{S}\rangle_{T=0}\right) \quad . \tag{36}$$

Here,

$$\langle\tilde{S}\rangle = -\frac{1}{N_\sigma^3 N_\tau}\frac{\partial \ln Z}{\partial\beta} \tag{37}$$

is calculated on a lattice of size $N_\sigma^3 \times N_\tau$ and $\langle ... \rangle_{T=0}$ denotes expectation values calculated on zero temperature lattices, which usually are approximated by symmetric lattices with $N_\tau \equiv N_\sigma$. The lower integration limit is chosen at low temperatures so that f/T_o^4 is small and may be ignored[7].

A little bit more involved is the calculation of the energy density as we have to take derivatives with respect to the temperature, $T = 1/N_\tau a$. On lattices with fixed temporal extent N_τ we rewrite this in terms of a derivative with respect to the lattice spacing a which in turn is controlled through the bare couplings of the QCD Lagrangian, $a \equiv a(\beta, m_q)$. We thus find for the case of n_f degenerate quark flavours of mass m_q

$$\frac{(\epsilon - 3p)}{T^4} = N_\tau^4 \left[\left(\frac{\mathrm{d}\beta(a)}{\mathrm{d}\ln a}\right)\left(\langle\tilde{S}\rangle - \langle\tilde{S}\rangle_{T=0}\right)\right.$$

[7] In the gluonic sector the relevant degrees of freedom at low temperature are glueballs. Even the lightest ones calculated on the lattice have large masses, $m_G \simeq 1.5$ GeV. The free energy density thus is exponentially suppressed already close to T_c. In QCD with light quarks the dominant contribution to the free energy density comes from pions. As long as we are dealing with massive quarks also this contribution gets suppressed exponentially. However, in the massless limit clearly some care has to be taken with the normalization of the free energy density.

$$- \left(\frac{\mathrm{d} m_q(a)}{\mathrm{d} \ln a} \right) \left(\langle \bar{\chi}\chi \rangle - \langle \bar{\chi}\chi \rangle_{T=0} \right) \Big] \quad . \tag{38}$$

An evaluation of the energy density thus, e.g., requires the knowledge of two β-functions. These may be determined by calculating two physical observables in lattice units for given values of β and m_q; for instance, the string tension, σa^2 and a ratio of hadron masses, $m_{PS}/m_V \equiv m_\pi/m_\rho$. These quantities will have to be calculated at zero temperature which then also allows to determine a temperature scale in physical units as given in Eq. (12). This forms the basis for a calculation of the pressure as shown already in Fig. 5.

The numerical calculation of thermodynamic quantities is done on finite lattices with spatial extent N_σ and temporal extent N_τ. In order to perform calculations close to the thermodynamic limit we want to use a large spatial extent of the lattice. In general it has been found that lattices with $N_\sigma \gtrsim 4N_\tau$ provide a good approximation to the infinite volume limit. In addition, we want to get close to the continuum limit in order to eliminate discretization errors. Taking the continuum limit at fixed temperature requires to perform the limit $N_\tau \to \infty$. In order to perform this limit in a controlled way we have to analyze in how far lattice calculations of bulk thermodynamic observables are influenced by the introduction of a finite lattice cut-off, i.e. we have to understand the systematic cut-off effects introduced through the non-zero lattice spacing. These cut-off effects are largest in the high (infinite) temperature limit which can be analyzed analytically in weak coupling lattice perturbation theory. We thus will discuss this limiting case first.

6.1 High-Temperature Limit of the QCD Equation of State

In the high temperature limit bulk thermodynamic observables are expected to approach their free gas values (Stefan-Boltzmann constants). In this limit cut-off effects in the pressure and in turn also in the energy density ($\epsilon_{\rm SB} = 3\, p_{\rm SB}$) become most significant. Momenta of the order of the temperature, i.e. short distance properties, dominate the ideal gas behaviour.

As discussed in Section 2 the most straightforward lattice representation of the QCD partition function in terms of the standard Wilson gauge and fermion actions as well as the staggered fermion action leads to a systematic $\mathcal{O}(a^2)$ cut-off dependence of physical observables. At finite temperature the temperature itself sets the scale for these $\mathcal{O}(a^2)$ effects, which thus give rise to $\mathcal{O}((aT)^2 \equiv 1/N_\tau^2)$ deviations of, e.g., the pressure from the continuum Stefan-Boltzmann value,

$$\left. \frac{p}{T^4} \right|_{N_\tau} = \left. \frac{p}{T^4} \right|_{\infty} + \frac{c}{N_\tau^2} + \mathcal{O}(N_\tau^{-4}) \quad . \tag{39}$$

One can eliminate these leading order cut-off effects by using improved actions which greatly reduces the cut-off dependence in the ideal gas limit. In the Appendix we discuss a specific set of improved gauge and fermion actions. In the gauge sector one may in addition to the standard Wilson plaquette term ((1 × 1)-action) also include planar 6-link terms in the action. This is done in the $O(a^2)$ tree level improved (1 × 2)-action, which eliminates the leading order cut-off dependence completely. On lattices with temporal extent N_τ one finds for the deviation of the gluonic part of the pressure [34],

$$\frac{p_G(N_\tau)}{p_{G,SB}} = \begin{cases} 1 + \frac{8}{21}\left(\frac{\pi}{N_\tau}\right)^2 + \frac{5}{21}\left(\frac{\pi}{N_\tau}\right)^4 + \mathcal{O}(N_\tau^{-6}), & (1\times 1)\text{-action} \\ 1 + \mathcal{O}(N_\tau^{-4}), & (1\times 2)\text{-action} \end{cases} \tag{40}$$

A similar reduction of cut-off effects can be achieved in the fermion sector through the use of improved actions. So far, however, improved fermion actions, which reduce or eliminate the leading order cut-off effects have only been constructed in the staggered fermion formulation. The Naik action [35], which in addition to the ordinary one-link term in the staggered action also includes straight three-link terms, completely eliminates the $\mathcal{O}(N_\tau^{-2})$ errors on the tree level (ideal gas limit). The p4-action discussed in the Appendix does not eliminate this correction completely. It, however, reduces its contribution drastically over those present in the standard staggered action[8],

$$\frac{p_F(N_\tau)}{p_{F,SB}} = \begin{cases} 1 + 1.57\left(\frac{\pi}{N_\tau}\right)^2 + 8.47\left(\frac{\pi}{N_\tau}\right)^4 + & \text{1-link standard} \\ \mathcal{O}(N_\tau^{-6}), & \text{staggered action,} \\ & \\ 1 + 0.007\left(\frac{\pi}{N_\tau}\right)^2 + 1.07\left(\frac{\pi}{N_\tau}\right)^4 + & \\ \mathcal{O}(N_\tau^{-6}) & \text{p4-action,} \end{cases} \tag{41}$$

as can be seen in Fig. 9. Moreover, the p4-action has the advantage that it improves the rotational symmetry of the fermion propagator which in turn also reduces violations of rotational symmetry in the heavy quark potential.

Using the tree level improved gauge action in combination with the improved staggered fermion action in numerical simulations at finite temperature it is possible to perform calculations with small systematic cut-off errors already on lattices with small temporal extent, e.g., $N_\tau = 4$ or 6. In actual calculations performed with various actions in the pure gauge sector one finds that for temperatures $T \lesssim 5T_c$ the cut-off dependence of thermodynamic shows the pattern predicted by the infinite temperature perturbative calculation. The absolute magnitude of the cut-off effects, however, is smaller by about a factor of two. This, of course, is reassuring for the numerical calculations performed with light quarks, where such a detailed systematic study of the cut-off dependence at present does not exist.

6.2 Thermodynamics of the SU(3) Gauge Theory

Before entering a discussion of bulk thermodynamics in two and three flavour QCD it is worthwhile to discuss some results on the equation of state in the

[8] We quote here only an approximation to the N_τ-dependence obtained from a fit in the interval $10 \leq N_\tau \leq 16$.

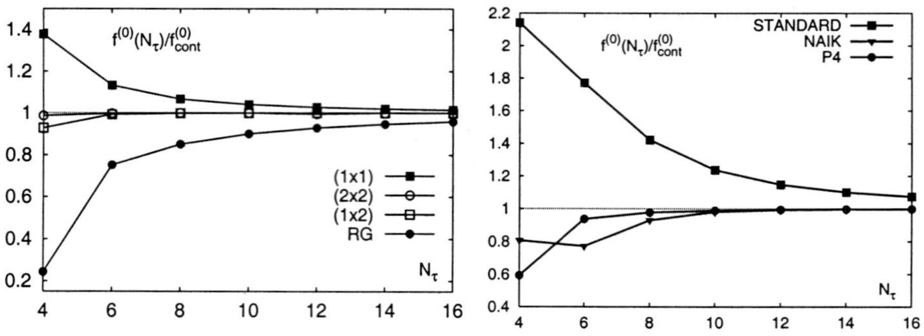

Fig. 9. Cut-off dependence of the ideal gas pressure for the $SU(3)$ gauge theory (left) and several staggered fermion actions (right). These actions are defined in the Appendix. Cut-off effects for the Wilson fermion action are compatible with those of the standard staggered fermion action

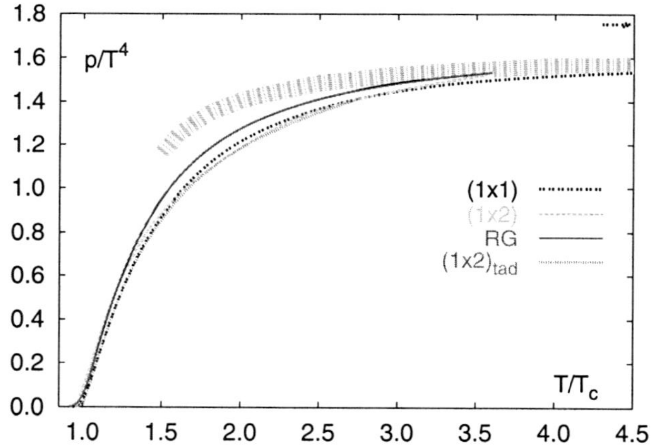

Fig. 10. Pressure of the SU(3) gauge theory calculated on lattices with different temporal extent and extrapolated to the continuum limit. Shown are results from calculations with the standard Wilson (1 × 1)-action [36] and several improved actions [38,39], which are defined in the Appendix. The broad band shows the approximately self-consistent HTL calculation of [41]

heavy quark mass limit of QCD – the SU(3) gauge theory. In this case the temperature dependence of the pressure and energy density has been studied in great detail, calculations with the standard action [36] and various improved actions [37–39] have been performed, the cut-off dependence has explicitly been analyzed through calculations on lattices with varying temporal extent N_τ and results have been extrapolated to the continuum limit. In Fig. 10 we show some results for the pressure obtained from such detailed analyzes with different actions [36,38,39].

This figure shows the basic features of the temperature dependence of bulk thermodynamic quantities in QCD, which also carry over to the case of QCD with light quarks. The pressure stays small for almost all temperatures below T_c; this is expected, as the only degrees of freedom in the low temperature phase are glueballs which are rather heavy and thus lead to an exponential suppression of pressure and energy density at low temperature. Above T_c the pressure rises rapidly and reaches about 70% of the asymptotic ideal gas value at $T = 2\,T_c$. For even larger temperatures the approach to this limiting value proceeds rather slowly. In fact, even at $T \simeq 4\,T_c$ deviations from the ideal gas value are larger than 10%. This is too much to be understood in terms of weakly interacting gluons as they are described by ordinary high temperature perturbation theory [30]. Even at these high temperatures non-perturbative effects have to be taken into account which may be described in terms of interactions among quasi-particles [31,40]. In Fig. 10 we show the result of a self-consistent HTL resummation [41], which leads to good agreement with the lattice calculations for $T \gtrsim 3T_c$. Other approaches [33,40] reach a similarly good agreement in the high temperature regime.

Compared to the pressure the energy density rises much more rapidly in the vicinity of T_c. In fact, as the transition is first order in the $SU(3)$ gauge theory the energy density is discontinuous at T_c with a latent heat of about $1.5T_c^4$ [42]. In Fig. 11 we show results for the energy density, entropy density and the pressure obtained from calculations with the Wilson action which have been extrapolated to the continuum limit [36].

The delayed rise of the pressure compared to that of the energy density has consequences for the velocity of sound in the QCD plasma in the vicinity of T_c. It is substantially smaller than in the high temperature ideal gas limit.

6.3 Flavour Dependence of the QCD Equation of State

As shown in Eq. (36) the pressure in QCD with light quarks can be calculated along the same line as in the pure gauge sector. Unlike in the pure gauge case it, however, will be difficult to perform calculations on lattices with large temporal extent. In fact, at present all calculations of the equation of state are restricted to lattices with $N_\tau = 4$ and 6 [21,43,44]. The use of an improved fermion action thus seems to be even more important in this case. Of course, an additional problem arises from insufficient chiral properties of staggered and Wilson fermion actions. This will mainly be of importance in the low temperature phase and in the vicinity of the transition temperature. The continuum extrapolation thus will be more involved in the case of QCD with light quarks than in the pure gauge theory and we will have to perform calculations closer to the continuum limit. Nonetheless, in particular for small number of flavours, we may expect that the flavour symmetry breaking only has a small effect on the overall magnitude of bulk thermodynamic observables. After all, for $n_f = 2$, the pressure of an ideal massless pion gas contributes less than 10% of that of an ideal quark-gluon gas in the

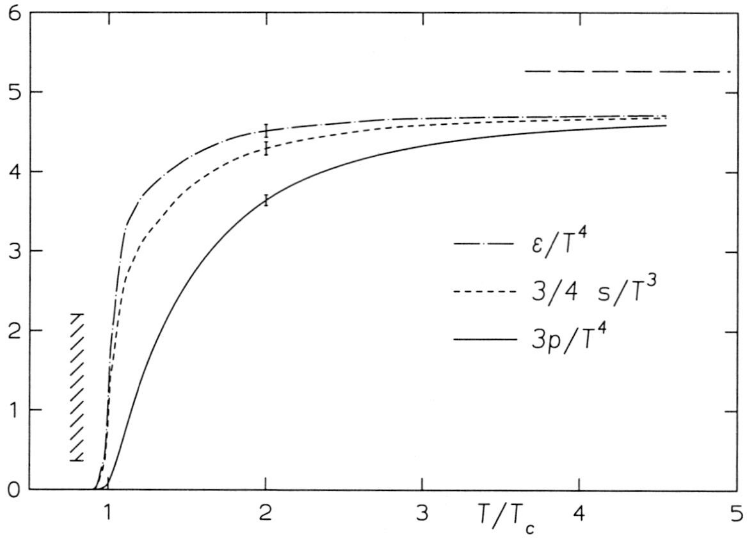

Fig. 11. Energy density, entropy density and pressure of the SU(3) gauge theory calculated on lattices with different temporal extent and extrapolated to the continuum limit. The dashed band indicates the size of the latent heat gap in energy and entropy density

high temperature limit. For our discussion of bulk thermodynamic observables the main source for lattice artifacts thus still seems to arise from the short distance cut-off effects, which we have to control. Additional confidence in the numerical results can be gained by comparing simulations performed with different fermion actions.

The importance of an improved lattice action, which leads to small cut-off errors at least in the high temperature ideal gas limit is apparent from Fig. 12, where we compare the results of a calculation of the pressure in 2-flavour QCD performed with unimproved gauge and staggered fermion actions [43] and the RG-improved gauge and clover improved Wilson action [44] with results obtained with the p4-action discussed in the Appendix. At temperatures above $T \simeq 2\ T_c$ these actions qualitatively reproduce the cut-off effects calculated analytically in the infinite temperature limit (see Section 6.1). In particular, it is evident that also the Clover improved Wilson action leads to an overshooting of the continuum ideal gas limit. This is expected as the Clover term in the Wilson action does eliminate $\mathcal{O}(ag^2)$ cut-off effects but does not improve the high temperature ideal gas limit, which is $\mathcal{O}(g^0)$. The clover improved Wilson action thus leads to the same large $\mathcal{O}(a^2)$ cut-off effects as the unimproved Wilson action. The influence of cut-off effects in bulk thermodynamic observables thus is similar in calculations with light quarks and in the $SU(3)$ gauge theory. This observation may also help to estimate the cut-off effects still present in current calculations with light quarks. In

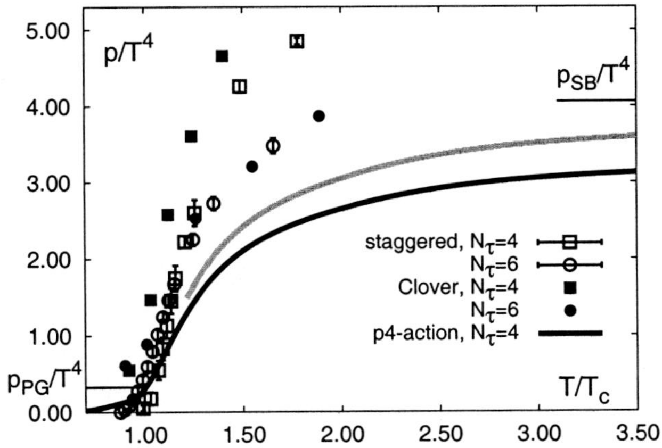

Fig. 12. The pressure in two flavour QCD calculated with unimproved gauge and staggered fermion actions (open symbols) [43], RG-improved gauge and clover improved Wilson action (full symbols) [44] and the p4-action (improved gauge and improved staggered fermions, see Appendix) (full line) [21]. The grey line estimates the results in the continuum limit as described in the text. The horizontal lines to the right and left show the Stefan-Boltzmann values for an ideal pion gas and a free quark-gluon gas, respectively

particular, we know from the analysis performed in the pure gauge sector that in the interesting temperature regime of a few times T_c the cut-off dependence seems to be about a factor two smaller than calculated analytically in the infinite temperature limit; we may expect that this carries over to the case of QCD with light quarks. This is the basis for the estimated continuum extrapolation of the $n_f = 2$ results shown as a dashed band in Fig. 12.

In Fig. 5 we have already shown results for the pressure calculated in QCD with different number of flavours. This figure clearly shows that the transition region shifts to smaller temperatures as the number of degrees of freedom is increased. Such a conclusion, of course, requires the determination of a temperature scale that is common to all *QCD-like* theories which have a particle content different from that realized in nature. We have determined this temperature scale by assuming that the string tension is flavour and quark mass independent. This assumption is supported by the observation that already in the heavy quark mass limit the string tension calculated in units of quenched hadron masses, e.g., $m_\rho/\sqrt{\sigma} = 1.81\ (4)$ [45], is in good agreement with values required in QCD phenomenology, $\sqrt{\sigma} \simeq 425$ MeV.

At high temperature the magnitude of p/T^4 clearly reflects the change in the number of light degrees of freedom present in the ideal gas limit. When we rescale the pressure by the corresponding ideal gas values it becomes, however, apparent that the overall pattern of the temperature dependence of p/T^4 is quite similar in all cases. This is shown in Fig. 13. In particular, when one takes into account that a proper continuum extrapolation in QCD with light quarks is still missing this agreement achieved with improved staggered fermions is quite remarkable.

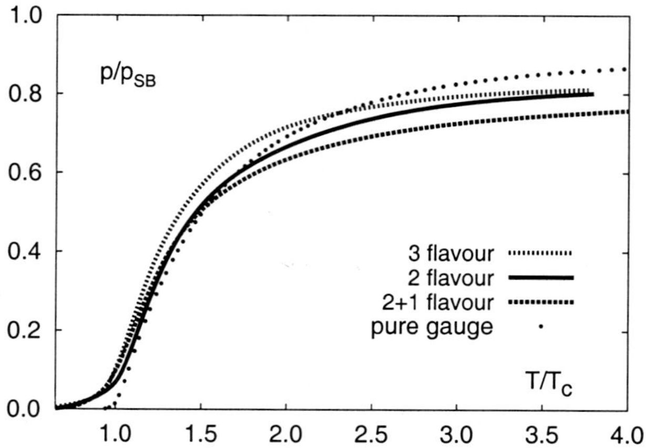

Fig. 13. The pressure in units of the ideal gas pressure for the $SU(3)$ gauge theory and QCD with various number of flavours. The latter calculations have been performed on lattices with temporal extent $N_\tau = 4$ using the p4-action defined in the Appendix. Results are not yet extrapolated to the continuum limit

We also note that the pressure at low temperature is enhanced in QCD with light quarks compared to the pure gauge case. This is an indication for the contribution of hadronic states, which are significantly lighter than the heavy glueballs of the $SU(3)$ gauge theory.

This behaviour is even more clearly visible in the behaviour of the energy density which is shown in Fig. 14, where we show results obtained with improved staggered[9] and Wilson [44] fermions. We note that these calculations yield consistent estimates for the energy density at T_c,

$$\epsilon_c \simeq (6 \pm 2) T_c^4 \quad . \tag{42}$$

This estimate for ϵ_c/T_c^4, which also is consistent with results obtained from calculations with a standard staggered fermion action [43], is an order of magnitude larger than the critical value on the hadronic side of the transition in the pure gauge theory (see Fig. 11). It is, however, interesting to note that when we convert this result for ϵ_c in physical units, [MeV/fm^3], this difference gets to a large extent compensated by the shift in T_c to smaller

[9] The figure for staggered fermions is based on data from Ref. [21]. Here a contribution to ϵ/T^4 which is proportional to the bare quark mass and vanishes in the chiral limit is not taken into account.

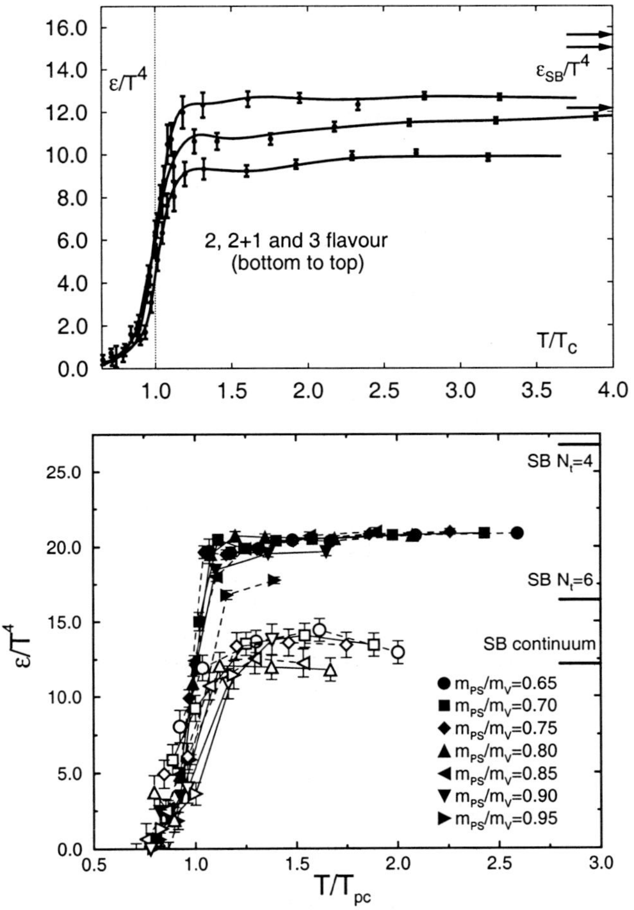

Fig. 14. The energy density in QCD. The upper (lower) figure shows results from a calculation with improved staggered [21] (Wilson [44]) fermions on lattices with temporal extent $N_\tau = 4$ ($N_\tau = 4,\ 6$). The staggered fermion calculations have been performed for a pseudo-scalar to vector meson mass ratio of $m_{PS}/m_V = 0.7$

values. When going from the infinite quark mass limit to the light quark mass regime the QCD transition thus seems to take place at compatible values of the energy density, $\epsilon_c \simeq (0.5 - 1)\mathrm{GeV/fm}^3$. The largest uncertainty on this number at present arises from uncertainties on the value of T_c (see next section). However, also the magnitude of ϵ_c/T_c^4 still has to be determined more accurately. Here two competing effects will be relevant. On the one hand we expect ϵ_c/T_c^4 to increase with decreasing quark masses, i.e. closer to the chiral limit. On the other hand, it is likely that finite volume effects are similar to those in the pure gauge sector, which suggests that ϵ_c/T_c^4 will still decrease closer to the thermodynamic limit, i.e. for $N_\sigma \to \infty$.

In the 2-flavour calculations performed with improved Wilson fermions [44] the pressure and energy density have been calculated for several values of the quark mass, which corresponds to different ratios of the pseudo-scalar to vector meson mass m_{PS}/m_V. The results show no significant quark mass dependence up to $m_{PS}/m_V \simeq 0.9$. This meson mass ratio corresponds to pseudo-scalar meson masses of about 1.5 GeV, which is somewhat larger than the Φ-meson mass. This suggests that the corresponding quark mass is compatible with that of the strange quark. The approximate quark mass independence of the equation of state observed in the high temperature phase thus is consistent with our expectation that quark mass effects should become significant only when the quark masses get larger than the temperature.

7 The Critical Temperature of the QCD Transition

As discussed in Section 3 the transition to the high temperature phase is continuous and non-singular for a large range of quark masses. Nonetheless, for all quark masses this transition proceeds rather rapidly in a small temperature interval. A definite transition point thus can be identified, for instance through the location of peaks in the susceptibilities of the Polyakov loop or the chiral condensate defined in Eq. (21). For a given value of the quark mass one thus determines pseudo-critical couplings, $\beta_{pc}(m_q)$, on a lattice with temporal extent N_τ. An additional calculation of an experimentally or phenomenologically known observable at zero temperature, e.g., a hadron mass or the string tension, is still needed to determine the transition temperature from Eq. (12). In the pure gauge theory the transition temperature again has been analyzed in great detail and the influence of cut-off effects has been examined through calculations on different size lattices and with different actions. From this one finds for the critical temperature of the first order phase transition in the pure $SU(3)$ gauge theory,

$$\begin{aligned} \underline{\text{SU(3) gauge theory}:}\quad & T_c/\sqrt{\sigma} = 0.637 \pm 0.005 \\ & T_c = (271 \pm 2)\ \text{MeV} \end{aligned} \tag{43}$$

Already the early calculations for the transition temperature with light quarks [46,47] indicated that the inclusion of light quarks leads to a significant decrease of the transition temperature. However, these early calculations, which have been performed with standard Wilson [46] and staggered [47] fermion actions, also led to significant discrepancies in the results for T_c as well as the order of the transition. These differences strongly diminished in the newer calculations which are based on improved Wilson fermions (Clover action) [47–49], domain wall fermions [50] as well as improved staggered fermions (p4-action) [19]. A compilation of these newer results is shown in Fig. 15 for various values of the quark masses. In order to compare calculations performed with different actions the results are presented in terms of a *physical observable*, the meson mass ratio m_{PS}/m_V. In Fig. 15a we show T_c/m_V obtained for 2-flavour QCD while Fig. 15b gives a comparison of results obtained with improved staggered fermions [19] for 2 and 3-flavour QCD. Also shown there is a result for the case of (2+1)-flavour QCD, i.e. for two light and one heavier quark flavour degree of freedom. Unfortunately the quark masses in this latter case are still too large to be compared directly with the situation realized in nature. We note however, that the results obtained so far suggest that the transition temperature in (2+1)-flavour QCD is close to that of 2-flavour QCD. The 3-flavour theory, on the other hand, leads to consistently smaller values of the critical temperature, $T_c(n_f = 2) - T_c(n_f = 3) \simeq 20$ MeV. The extrapolation of the transition temperatures to the chiral limit gave

$$\underline{2-\text{flavour QCD}:}\ T_c = \begin{cases} (171 \pm 4)\ \text{MeV}, & \text{clover-improved Wilson fermions [48]} \\ (173 \pm 8)\ \text{MeV}, & \text{improved staggered fermions [19]} \end{cases}$$

$$\underline{3-\text{flavour QCD}:}\ T_c = (154 \pm 8)\ \text{MeV}, \quad \text{improved staggered fermions [19]}$$

Here m_ρ has been used to set the scale for T_c. Although the agreement between results obtained with Wilson and staggered fermions is striking, one should bear in mind that all these results have been obtained on lattice with temporal extent $N_\tau = 4$, i.e. at rather large lattice spacing, $a \simeq 0.3$ fm. Moreover, there are uncertainties involved in the ansatz used to extrapolate to the chiral limit. We thus estimate that the systematic error on the value of T_c/m_ρ still is of similar magnitude as the purely statistical error quoted above.

We note from Fig. 15 that T_c/m_V drops with increasing ratio m_{PS}/m_V, i.e. with increasing quark mass. This may not be too surprising as m_V, of course, does not take on the physical ρ-meson mass value as long as m_{PS}/m_V did not reach is physical value (vertical line in Fig. 15a). In fact, we know that T_c/m_V will approach zero for $m_{PS}/m_V = 1$ as T_c will stay finite and take on the value calculated in the pure $SU(3)$ gauge theory whereas m_V will diverge in the heavy quark mass limit. Fig. 15 thus does not yet allow to quantify how T_c depends on the quark mass. A simple percolation picture for the QCD transition would suggest that $T_c(m_q)$ or better $T_c(m_{PS})$ will increase with increasing m_q; with increasing m_q also the hadron masses increase and it becomes more difficult to excite the low lying hadronic states. It thus becomes more difficult to create a sufficiently high particle/energy density in the hadronic phase that can trigger a phase (percolation) transition. Such a picture also follows from chiral model calculations [51].

As argued previously we should express T_c in units of an observable, which itself is not dependent on m_q; the string tension (or also a quenched hadron mass) seems to be suitable for this purpose. In fact, this is what tacitly has

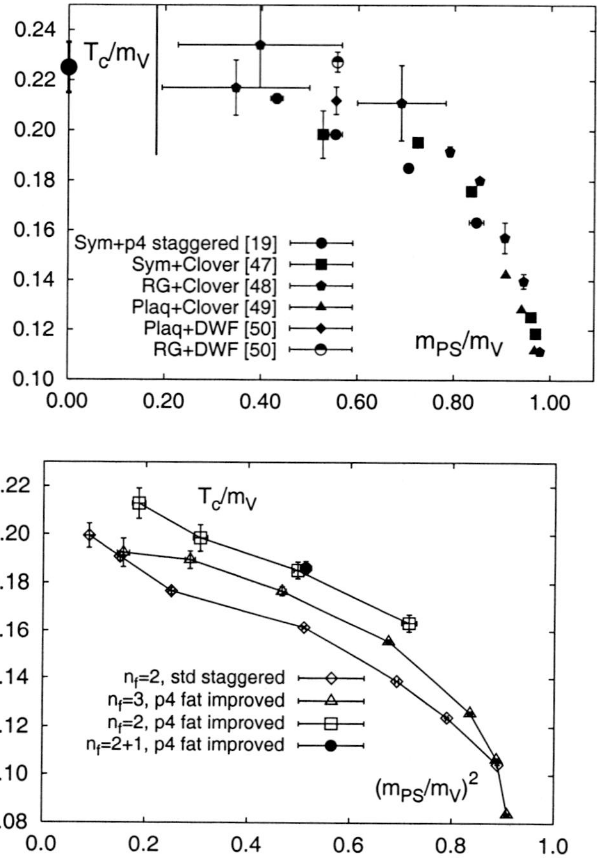

Fig. 15. Transition temperatures in units of m_V. The upper figure shows a collection of results obtained for 2-flavour QCD with various fermion actions while in the lower figure we compare results obtained in 2 and 3-flavour QCD with the p4-action described in the Appendix. All results are from simulations on lattices with temporal extent $N_\tau = 4$. The large dot drawn for $m_{PS}/m_V = 0$ indicates the result of chiral extrapolations based on calculations with improved Wilson [48] as well as improved staggered [19] fermions. The vertical line in the upper figure shows the location of the physical limit, $m_{PS} \equiv m_\pi = 140$ MeV

been assumed when one converts the critical temperature of the SU(3) gauge theory $T_c/\sqrt{\sigma} \simeq 0.63$ into physical units as has also been done in Eq. (43).

To quantify the quark mass dependence of the transition temperature one may express T_c in units of $\sqrt{\sigma}$. This ratio is shown in Fig. 16 as a function of $m_{PS}/\sqrt{\sigma}$. As can be seen the transition temperature starts deviating from the quenched values for $m_{PS} \lesssim (6-7)\sqrt{\sigma} \simeq 2.5$ GeV. We also note that the dependence of T_c on $m_{PS}/\sqrt{\sigma}$ is almost linear in the entire mass interval. Such a behaviour might, in fact, be expected for light quarks in the vicinity

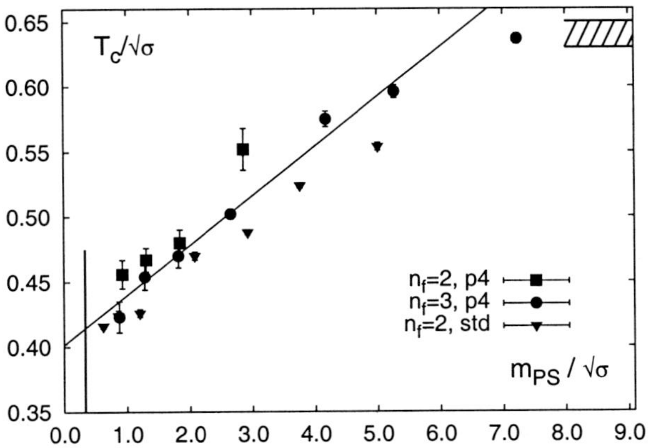

Fig. 16. The transition temperature in 2 (filled squares) and 3 (circles) flavour QCD versus $m_{PS}/\sqrt{\sigma}$ using an improved staggered fermion action (p4-action). Also shown are results for 2-flavour QCD obtained with the standard staggered fermion action (open squares). The dashed band indicates the uncertainty on $T_c/\sqrt{\sigma}$ in the quenched limit. The straight line is the fit given in Eq. (45)

of a 2nd order chiral transition where the dependence of the pseudo-critical temperature on the mass of the Goldstone-particle follows from the scaling relation, Eq. (26),

$$T_c(m_\pi) - T_c(0) \sim m_\pi^{2/\beta\delta} \ . \tag{44}$$

For 2-flavour QCD the critical indices are expected to belong to the universality class of 3-d, $O(4)$ symmetric spin models and one thus would indeed expect $1/\beta\delta = 0.55$. However, this clearly cannot be the origin of the quasi linear behaviour which is observed for rather large hadron masses and seems to be independent of n_f. Moreover, unlike in chiral models [51] the dependence of T_c on m_{PS} turns out to be rather weak. The line shown in Fig. 16 is a fit to the 3-flavour data, which gave

$$\left(\frac{T_c}{\sqrt{\sigma}}\right)_{m_{PS}/\sqrt{\sigma}} = \left(\frac{T_c}{\sqrt{\sigma}}\right)_0 + 0.04(1)\left(\frac{m_{PS}}{\sqrt{\sigma}}\right) \ . \tag{45}$$

It thus seems that the transition temperature does not react strongly on changes of the lightest hadron masses. This favours the interpretation that the contributions of heavy resonance masses are equally important for the occurrence of the transition. In fact, this also can explain why the transition still sets in at quite low temperatures even when all hadron masses, including the pseudo-scalars, attain masses of the order of 1 GeV or more. Such an interpretation also is consistent with the weak quark mass dependence of the critical energy density we found from the analysis of the QCD equation of state in the previous section.

For the quark masses currently used in lattice calculations a resonance gas model combined with a percolation criterion thus provides an appropriate to describe the thermodynamics close to T_c. It remains to be seen whether the role of the light meson sector becomes more dominant when we get closer to the chiral limit.

8 Finite Density QCD

Finite density calculations in QCD are affected by the well known sign problem, i.e. the fermion determinant appearing in the QCD partition function, Eq. (11), becomes complex for non-zero values of the chemical potential μ and thus prohibits the use of conventional numerical algorithms. The most detailed studies of this problem have so far been performed using the Glasgow algorithm [52], which is based on a fugacity expansion of the grand canonical partition function at non-zero μ,

$$Z_{GC}(\mu/T,T,V) = \sum_{B=-\alpha V}^{\alpha V} z^B Z_B(T,V) \ , \tag{46}$$

where $z = \exp(\mu/T)$ is the fugacity and Z_B are the canonical partition functions for fixed quark number B; $\alpha = 3,6$ for one species of staggered or Wilson fermions, respectively. After introducing a complex chemical potential in Z_{GC} the canonical partition functions can be obtained via a Fourier transformation[10],

$$Z_B(T,V) = \frac{1}{2\pi}\int_0^{2\pi} \mathrm{d}\phi \ \mathrm{e}^{i\phi B} \ Z_{GC}(i\phi,T,V) \tag{47}$$

$$\equiv \int \prod_{n\nu} \Delta U_{n,\nu} a_B \mathrm{e}^{-\beta S_G} \quad , \tag{48}$$

with

$$a_B = \frac{1}{2\pi}\int_0^{2\pi} \mathrm{d}\phi \ \mathrm{e}^{i\phi B} \ (\det Q^F(m_q,i\phi))^{n_f/4} \quad . \tag{49}$$

One thus may evaluate the canonical partition functions as expectation values with respect to a trial partition function that can be handled numerically, for instance the partition function of the pure $SU(3)$ gauge theory,

$$Z_{GC}(\mu/T,T,V) = Z_{\mathrm{SU(3)}} \sum_{B=-\alpha V}^{\alpha V} z^B \langle a_B \rangle_{\mathrm{SU(3)}} \quad . \tag{50}$$

[10] The use of this ansatz for the calculation of canonical partition functions as expansion coefficients for Z_{GC} has been discussed in [55,56]. A new approach has been suggested recently, which combines simulations with imaginary chemical potential with an analytic continuation based on the Ferrenberg-Swendsen multi-histogram method [57].

However, this approach so far did not overcome the severe numerical difficulties. Like other approaches it suffers from the problem that expectation values have to be calculated with respect to another ensemble so that the importance sampling which is at the heart of every numerical approach samples the wrong region of phase space and thus may become quite inefficient.

It thus may be helpful to approach the finite density problems from another perspective. A reformulation of the original ansatz may lead to a representation of the partition function which, in the ideal case, would require the averaging over configurations with strictly positive weights only, or at least would lead to a strong reduction of configurations with negative weights.

An alternative formulation of finite density QCD is given in terms of canonical rather than grand canonical partition functions [53], i.e. rather than introducing a non-zero chemical potential through which the number density is controlled one introduces directly a non-zero baryon number (or quark number B) through Eq. (47) from which the baryon number density on lattices of size $N_\sigma^3 \times N_\tau$ is obtained as $n_B/T^3 = \frac{B}{3}(N_\tau/N_\sigma)^3$. Also this formulation is by no means easy to use in general, i.e. for QCD with light quarks. In particular, it also still suffers from a sign problem. It, however, leads to a quite natural and useful formulation of the quenched limit of QCD at non-zero density [54] which may be a good starting point for generalizing this approach to finite values of the quark mass. In the following we briefly outline the basic ideas of this approach.

8.1 Quenched Limit of Finite Density QCD

It had been noticed early that the straightforward replacement of the fermion determinant by a constant does not lead to a meaningful static limit of QCD [58]. In fact, this simple replacement corresponds to the static limit fermion flavours carrying baryon number B and $-B$, respectively [59]. This should not be too surprising. When one starts with QCD at a non-zero baryon number and takes the limit of infinitely heavy quarks something should be left over from the determinant that represents the objects that carry the baryon number. In the canonical formulation this becomes obvious. For $m_q \to \infty$ one ends up with a partition function, which for baryon number $B/3$ still includes the sum over products of B Polyakov loops, i.e. the static quark propagators which carry the baryon number [54]. This limit also has some analogy in the grand canonical formulation where the coupled limit $m_q, \mu \to \infty$ with $\exp(\mu)/2m_q$ kept fixed has been performed [60,61][11].

As the baryon number is carried by the rather heavy nucleons in the confined phase of QCD we may expect that it is quite reasonable to approximate

[11] This is a well known limit in statistical physics. When deriving the non-relativistic gas limit from a relativistic gas of particles with mass $\bar{m}$, the rest mass is splitted off from the chemical potential, $\mu \equiv \mu_{nr} + \bar{m}$, in order to cancel the corresponding rest mass term in the particle energies. On the lattice $\bar{m} = \ln(2m_q)$ for large bare quark masses.

them by static objects. This may already provide valuable insight into the thermodynamics of QCD at non-zero baryon number density already from quenched QCD.

In the canonical approach simulations at non-zero B can be performed on relatively large lattices and the use of baryon number densities up to a few times nuclear matter density is possible [54,62]. The simulations performed so far in the static limit show the basic features expected at non-zero density. As can be seen from the behaviour of the Polyakov loop expectation value shown in Fig. 17 the transition region gets shifted to smaller temperatures (smaller coupling β). The broadening of the transition region suggests a smooth crossover behaviour at non-zero density. However, in a canonical simulation it also may indicate the presence of a region of coexisting phases and thus would signal the existence of a 1st order phase transition. This deserves further analysis.

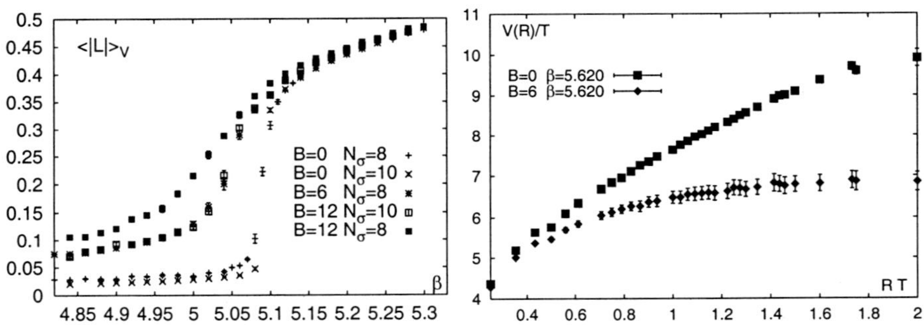

Fig. 17. Polyakov loop expectation value (left) calculated on $N_\sigma^3 \times 2$ and the heavy quark potential (right) calculated on $16^3 \times 4$ lattices in quenched QCD at zero and non-zero baryon number, $B/3$

Even more interesting is the behaviour of the heavy quark free energy in the low temperature phase. As shown in the right frame of Fig. 17 the free energy does get screened at non-vanishing number density. The influence of static quark sources on the heavy quark free energy is similar in magnitude to the screening (string breaking) seen in QCD simulations at finite temperature in the low temperature hadronic phase (see Fig. 4). At non-zero baryon number density we thus may expect similarly strong medium effects as at finite temperature.

9 Conclusions

We have given a brief introduction into the lattice formulation of QCD thermodynamics and presented a few of the basic results on the QCD equation of state, the critical parameters for the transition to the QCD plasma phase and properties of this new phase of matter.

The thermodynamics of the heavy quark mass limit is quite well under control; we know the equation of state and the transition temperature with an accuracy of a few percent. We now also have reached a first quantitative understanding of QCD with light quarks, which at present still corresponds to a world in which the pion would have a mass of about (300-500) MeV. This still is too heavy to become sensitive to details of the physics of chiral symmetry breaking. Nonetheless, lattice calculations performed with different lattice fermion formulations start to produce a consistent picture for the quark mass dependence of the equation of state as well as the influence of the number of light flavours on the phase transition and they yield compatible results for the transition temperature. In these calculations we learn to control the systematic errors inherent to lattice calculations performed with a finite lattice cut-off and start getting control over the effects resulting from the explicit breaking of continuum symmetries in the fermion sector. With these experience at hand we soon will be able to study the thermodynamics of QCD with a realistic spectrum of light up and down quarks and a heavier strange quark.

Acknowledgement

I would like to thank the organizers of the 40th Schladming Winter School on "Dense Matter" for arranging a very stimulating topical meeting. Furthermore, I would like to thank all my colleagues in Bielefeld, in particular Jürgen Engels, Edwin Laermann, Bengt Petersson and Helmut Satz, for the extremely fruitful collaboration I had with them for many years on most of the topics discussed in these lectures.

Appendix: Improved Gauge and Fermion Actions

Improved Gauge Actions

When formulating a discretized version of QCD one has a great deal of freedom in choosing a lattice action. Different formulations may differ by subleading powers of the lattice cut-off, which vanish in the continuum limit. This has, for instance, been used by Symanzik to systematically improve scalar field theories [63] and has then been applied to lattice regularized $SU(N)$ gauge theories [64,65]. In addition to the elementary plaquette term appearing in the standard Wilson formulation of lattice QCD larger loops can be added to the action in such a way that the leading $\mathcal{O}(a^2g^0)$ deviations from the continuum formulation are eliminated and corrections only start in $\mathcal{O}(a^4g^0, a^2g^2)$. A simple class of improved actions is, for instance, obtained by adding planar loops of size (k,l) to the standard Wilson action (one-plaquette action) [34]. The simplest extension of the Wilson one-plaquette action thus is to include an additional contribution from a planar six-link Wilson loop,

$$W^{(1,2)}_{n,\mu\nu} = 1 - \frac{1}{6}\,\mathrm{Re}\left(\boxed{} + \boxed{\begin{matrix}\\ \end{matrix}} \right)_{n,\mu\nu} \quad . \tag{51}$$

Combining this six-link contribution with the four-link plaquette term in a suitable way one can eliminate the leading $\mathcal{O}(a^2)$ corrections and arrives at a formulation that reproduces the continuum action up to $\mathcal{O}(a^4)$ corrections at least on the classical level at $\mathcal{O}(g^0)$,

$$\beta S_G = \beta \sum_{\substack{n \\ 0\le\mu<\nu\le 3}} c_{1,1} W^{(1,1)}_{n,\mu\nu} + c_{1,2} W^{(1,2)}_{n,\mu\nu} \quad , \tag{52}$$

with $c_{1,1} = 5/3$ and $c_{1,2} = -1/6$. We call this action the tree-level improved (1×2)-action. It may be further improved perturbatively by eliminating the leading lattice cut-off effects also at $\mathcal{O}(g^2)$, i.e. $c_{i,j} \Rightarrow c^{(0)}_{i,j} + g^2 c^{(1)}_{i,j}$, or by introducing non-perturbative modifications. A well-studied gluon action with non-perturbative corrections is the RG-improved action introduced by Y. Iwasaki [66]. This RG-action also has the structure of Eq. (52) but with coefficients $c^{RG}_{1,1} = 3.648$ and $c^{RG}_{1,2} = -0.662$. Of course, this action will still lead to $\mathcal{O}(a^2)$ corrections in the ideal gas limit. The N_τ-dependence of cut-off effects resulting from these actions is shown in Fig. 9. In Section 6 we also show some results from calculations with a tadpole improved action [67]. This non-perturbative improvement amounts to $c^{\mathrm{tad}}_{1,1} \equiv c_{1,1}$ and a replacement of $c_{1,2}$ by $c^{\mathrm{tad}}_{1,2} = 1/6u_0^2(\beta)$ where,

$$u_0^4 = \frac{1}{6N_\sigma^3 N_\tau} \langle \sum_{x,\nu>\mu} (1 - W^{1,1}_{\mu,\nu}(x)) \rangle \quad . \tag{53}$$

In the ideal gas limit this action still has the same cut-off dependence as the tree-level improved (1×2) action.

Improved Staggered Fermion Actions

When discussing the improvement of fermion actions there are at least two aspects one has to take into account. On the one hand one faces problems with cut-off effects similar to the pure gauge sector; on the tree level the standard Wilson and Kogut-Susskind discretization schemes introduce $\mathcal{O}(a^2)$ which will influence the short distance properties of physical observables. On the other hand also the global symmetries of the continuum Lagrangian are explicitly broken at non-zero lattice spacing. This influences the long distance properties of these actions, e.g., the light particle sector (Goldstone modes) of the lattice regularized theory. Both aspects are of importance for thermodynamic calculations. The latter problem certainly is of importance in the vicinity of the QCD phase transition while the former will show up when analyzing the high temperature limit of the equation of state.

In the case of staggered fermions both problems have been addressed and schemes have been developed that lead to a reduction of cut-off effects at short distances, i.e. high temperature, and also allow to reduce the explicit flavour symmetry breaking of the staggered discretization scheme.

A particular form of improved action used in recent calculations of the Bielefeld group is a staggered fermion action, which in addition to the standard one-link term includes a set of bended three-link terms,

$$S_F(m_q) = c_1^F S_{1-link,fat}(\omega) + c_3^F S_{3-link} + m_q \sum_x \bar{\chi}_x^f \chi_x^f$$

$$\equiv \sum_x \bar{\chi}_x^f \sum_\mu \eta_\mu(x) \left(\frac{3}{8} \left[\quad + \ \omega \sum_{\nu\neq\mu} \quad \right] \right.$$

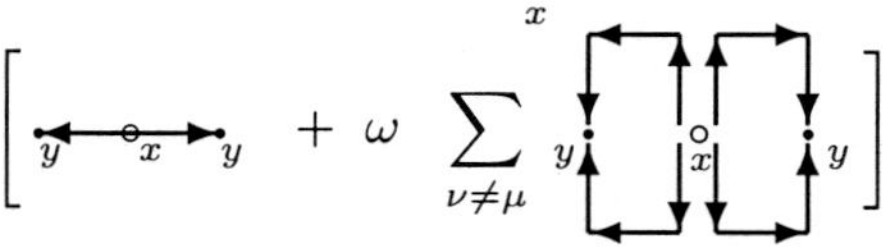

$$+\frac{1}{96} \sum_{\nu\neq\mu} \left[\quad + \quad + \quad + \quad \right] \Bigg) \chi_y^f$$

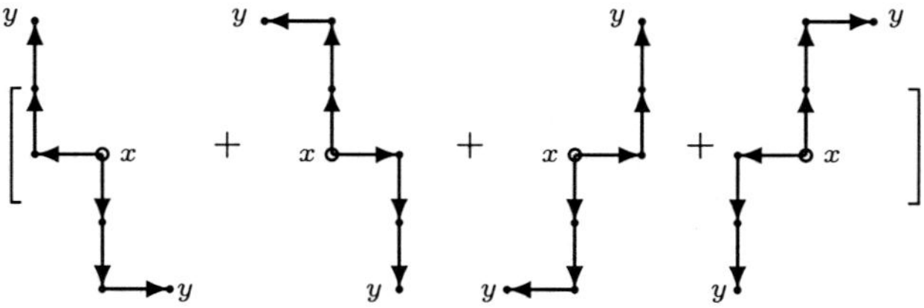

$$+m_q \sum_x \bar{\chi}_x^f \chi_x^f \quad . \tag{54}$$

Here $\eta_\mu(x) \equiv (-1)^{x_0+..+x_{\mu-1}}$ denotes the staggered fermion phase factors. Furthermore, we have made explicit the dependence of the fermion action on different quark flavours q, and the corresponding bare quark masses m_q, and give an intuitive graphical representation of the action. The tree level coefficients c_1^F and c_3^F appearing in S_F have been fixed by demanding rotational invariance of the free quark propagator at $\mathcal{O}(p^4)$ ("p4-action") [69]. In addition the 1-link term of the fermion action has been modified by introducing "fat" links [68] with a weight $\omega = 0.2$. The use of fat links does lead to a reduction of the flavour symmetry breaking close to T_c and at the same time does not modify the good features of the p4-action at high temperature, i.e. it does not modify the cut-off effects at tree level and has little influence on the cut-off dependence of bulk thermodynamic observables at $\mathcal{O}(g^2)$ in the high temperature phase [69]. Further details on the definition of the action are given in [69].

We refer to this action with a fat 1-link term combined with the tree level improved gauge action as *the p4-action.* The N_τ-dependence of cut-off effects resulting from this action is shown in Fig. 9.

References

1. K.G.Wilson: Phys. Rev. **D10**, 2445 (1974);
   K.G.Wilson: in *New Phenomena in Subnuclear Physics, Part A*, Proceedings of the International School of Subnuclear Physics, Erice, 1975, ed. A. Zichichi: Plenum Press, New York 1977
2. M. Creutz: Phys. Rev. **D21**, 2308 (1980)
3. L.D. McLerran, B. Svetitsky: Phys. Lett. **B98**, 195 (1981) and Phys. Rev. **D24**, 450 (1981)
4. J. Kuti, J. Polonyi, K. Szlachanyi: Phys. Lett. **B98**, 199 (1981)
5. J. Engels, F. Karsch, I. Montvay, H. Satz: Phys. Lett. **B101**, 89 (1981)
6. F. Wilczek: Physics Today, April 1998, p.11
7. J. Kogut, L. Susskind: Phys. Rev. **D11**, 395 (1975)
8. D.B. Kaplan: Phys. Lett. **B288**, 342 (1992)
9. P. Chen et al.: *The finite temperature QCD phase transition with domain wall fermions*, hep-lat/0006010
10. F. Karsch, P. Hasenfratz: Phys. Lett. **125B**, 308 (1983)
11. B. Svetitsky, L.G. Yaffe: Nucl. Phys. **B210**, 423 (1982)
12. R.D. Pisarski, F. Wilczek: Phys. Rev. **D29**, 338 (1984)
13. S. Gavin, A. Gocksch, R.D. Pisarski: Phys. Rev. **D49**, 3079 (1994)
14. E. Shuryak, T. Schäfer: Phys. Rev. Lett **75**, 1707 (1995)
15. Y. Iwassaki, K. Kanaya, S. Sakai, T. Yoshie: Nucl. Phys. (Proc. Suppl.) **42**, 261 (1995)
16. H. Satz: Nucl. Phys. **A642**, 130 (1998)
17. E. Shuryak: Phys. Lett. **B107**, 103 (1981)
18. F. Karsch, L. Lütgemeier: Nucl. Phys. **B550** 449 (1999)
19. F. Karsch, E. Laermann, A. Peikert: *Quark Mass and Flavour Dependence of the QCD Phase Transition*, hep-lat/0012023, to appear in Nucl. Phys. B
20. O. Kaczmarek, F. Karsch, E. Laermann, M. Lütgemeier: Phys. Rev. **D62**, 034021 (2000)
21. F. Karsch, E. Laermann, A. Peikert: Phys. Lett. **B478**, 447 (2000)
22. Y. Iwasaki, K. Kanaya, S. Kaya, T. Yoshié: Phys. Rev. Lett. **78**, 179 (1997)
23. C. Bernard et al. (MILC Collaboration): Phys. Rev. **D61**, 054503 (2000)
24. J. Engels, S. Holtmann, T. Mendes, T. Schulze: Phys. Lett. **B492**, 219 (2000)
25. J. Engels, S. Holtmann, T. Mendes, T. Schulze: *Finite-size-scaling functions for 3d $O(4)$ and $O(2)$ spin models and QCD*, hep-lat/0105028
26. K. Kajantie et al.: Phys. Rev. Lett. **79**, 3130 (1997)
27. O. Kaczmarek, F. Karsch, P. Petreczky, F. Zantow: in preparation
28. T. Matsui, H. Satz: Phys. Lett. **B178**, 416 (1986)
29. A.D. Linde: Phys. Lett. **B96**, 289 (1980)
30. P. Arnold, C.-X. Zhai: Phys. Rev. **D50**, 7603 (1994);
    B. Kastening, C. Zhai: Phys. Rev. **D52**, 7232 (1995)
31. J.-P. Blaizot, E. Iancu, A. Rebhan: Phys. Rev. Lett. **83**, 2906 (1999)
32. For a review see J.-P. Blaizot, E. Iancu, hep-ph/0101103; for further details see also the contributions of J.-P. Blaizot and A. Rebhan to this volume
33. K. Kajantie, M. Laine, K. Rummukainen, Y. Schröder: Phys. Rev. Lett. **86**, 10 (2001)
34. B. Beinlich, F. Karsch, E. Laermann: Nucl. Phys. **B462**, 415 (1996)
35. S. Naik: Nucl. Phys. **B316**, 238 (1989)

36. G. Boyd et al.: Phys. Rev. Lett. **75**, 4169 (1995) and Nucl. Phys. **B469**, 419 (1996)
37. A. Papa: Nucl. Phys. **B478**, 335 (1996)
38. M. Okamoto et al. (CP-PACS): Phys. Rev. **D60**, 094510 (1999)
39. B. Beinlich et al.: Eur. Phys. J. **C6**, 133 (1999)
40. A. Peshier, B. Kämpfer, O.P. Pavlenko, G. Soff: Phys. Rev. **D54**, 2399 (1996); P. Lévai, U. Heinz: Phys. Rev. **C57**, 1879 (1998)
41. J.-P. Blaizot, E. Iancu, A. Rebhan: Phys. Lett. **B470**, 181 (1999)
42. B. Beinlich, F. Karsch, A. Peikert: Phys. Lett. **B390**, 268 (1997)
43. C. Bernard et al.: Phys. Rev. **D55**, 6861 (1997)
44. A. Ali Khan et al. (CP-PACS): *Equation of state in finite temperature QCD with two flavors of improved Wilson quarks*, hep-lat/0103028
45. H. Wittig: Int. J. Mod. Phys. **A12**, 4477 (1997)
46. K.M. Bitar et al.: Phys. Rev. **D43**, 2396 (1991)
47. C. Bernard et al.: Phys. Rev. **D56**, 5584 (1997) and references therein
48. A. Ali Khan et al. (CP-PACS): Phys. Rev. **D63**, 034502 (2001)
49. R.G. Edwards, U.M. Heller: Phys. Lett. **B462**, 132 (1999)
50. N. Christ: private communication
51. H. Meyer-Ortmanns, B.-J. Schaefer: Phys. Rev. **D53**, 6586 (1996); J. Berges, D.U. Jungnickel, C. Wetterich: Phys. Rev. **D59**, 034010 (1999)
52. I.M. Barbour et al.: Nucl. Phys. B (Proc. Suppl.) **60A**, 220 (1998)
53. D.E. Miller, K. Redlich: Phys. Rev. **D35**, 2524 (1987)
54. J. Engels et al.: hep-lat/9903030
55. A. Hasenfratz, D. Toussaint: Nucl. Phys. **B371**, 539 (1992) and references therein
56. M. Alford et al.: Phys. Rev. **D59**, 054502 (1999)
57. Z. Fodor, S.D. Katz: *A new method to study QCD at finite temperature and chemical potential*, hep-lat/0104001
58. I. Barbour et al.: Nucl. Phys. **B275**, 296 (1986)
59. M.A. Stephanov: Phys. Rev. Lett. **76**, 4472 (1996)
60. I. Bender: Nucl. Phys. B (Proc. Suppl.) **26**, 323 (1992)
61. T. Blum, J.E. Hetrick, D. Toussaint: Phys. Rev. Lett. **76**, 1019 (1996)
62. J. Engels, O. Kaczmarek, F. Karsch, E. Laermann: Nucl. Phys. **B558**, 307 (1999) and ERRATUM Nucl. Phys. **B576**, 657 (2000)
63. K. Symanzik: Nucl. Phys. **B226**, 187 (1983) and Nucl. Phys. **B226**, 205 (1983)
64. P. Weisz: Nucl. Phys. **B212**, 1 (1983); P. Weisz, R. Wohlert: Nucl. Phys. **B236**, 397 (1984) and Nucl. Phys. **B247**, 544 (1984)
65. M. Lüscher, P. Weisz: Comm. Math. Phys. **97**, 59 (1985)
66. Y. Iwasaki: Nucl. Phys. **B258**, 141 (1985) and Univ. of Tsukuba Report UTHEP-118 (1983)
67. G.P. Lepage, P.B Mackenzie: Phys. Rev. **D48**, 2250 (1993)
68. T. Blum, C. De Tar, S. Gottlieb, K. Rummukainen, U.H. Heller, J.E. Hetrick, D. Toussaint, R.L. Sugar, M. Wingate: Phys. Rev. **D55**, 1133 (1997)
69. U.M. Heller, F. Karsch, B. Sturm: Phys. Rev. **D60**, 114502 (1999)

Nonperturbative Phenomena and Phases of QCD

Edward V. Shuryak

Department of Physics, SUNY
Stony Brook, NY 11794-3800, USA

1 Introduction

1.1 An Outline

The article provides a brief overview of what we have learned about the QCD vacuum, hadrons and hadronic matter during the last two decades. A systematic description of the topic would need a large book, not an article. Some material is available in reviews written in a more technical way: instantons and chiral symmetry breaking are treated in [1,2], correlators, OPE etc. in [3]. A lot of other material has been published recently: one can only consult the original papers.

In this article there are not so many formulae. I tried to clarify the main physics point, the main questions debated today, and to show a few recent examples. Admittedly, the title of this article is very general: but it covers a lot of different phenomena. We will start with the QCD *vacuum structure*, move to *hadronic structure*, discuss *phases* of hot/dense QCD and eventually consider *high energy collisions* of hadrons and heavy ions.

The main line in all the discussion will be a systematic use of semiclassical methods, specifically of instantons. The reasons for that are: (i) they are the only truly non-perturbative effects understood by now; (ii) they lead to large and probably even dominant effects in many cases; (iii) due to the progress during the last decade, we have a nearly quantitative theory of instanton effects, solved numerically to *all orders* in the so called 't Hooft interaction.

Although we still do not understand confinement, its companion problem - chiral symmetry breaking in the QCD vacuum - is now understood to a significant degree. Not only have we a simple qualitative understanding of where those quasi-zero modes of the Dirac operator come from, but we can calculate their density, space-time shape and eventually *QCD correlation functions* with surprising accuracy. So, in a way, the problem of hadronic structure is nearly solved for light-quark hadrons[1].

As we will see below, the QCD phase diagram can also be well understood in the instanton framework. The boundaries of the three basic phases of QCD, (i) the hadronic phase, (ii) the Quark-Gluon Plasma (QGP), and (iii)

[1] Medium-heavy-quark hadrons, such as $\bar{c}c$ or $\bar{b}b$, do care about the confining potential, while very-heavy quarkonia need only the Coulomb forces.

the Color Superconductor (CS) phases, appear as a balance between three fundamental pairing channels, namely (i) the attraction in the scalar colorless $\bar{q}q$ channel, (ii) the instanton-anti-instanton pairing induced by light quark exchanges, and (iii) the attraction in scalar but colored qq channels.

The last part deals with heavy ion collisions: those are related to the other articles of this book since this is how we try to access hot QCD experimentally. We will discuss first results coming from RHIC, and show that the matter produced seems to behave macroscopically (namely, hydrodynamically) according to a proper Equation of State. We will also try to connect the rapid onset of QGP equilibration with existing perturbative and non-perturbative estimates.

1.2 Scales of QCD

Let me start with an introductory discussion of various "scales" of non-perturbative QCD. The major reason I do this is the following: some naive simplistic ideas we had in the early days of QCD, in the 70's, are still alive today. I would strongly argue against the picture of non-perturbative objects as some structure-less fields with typical momenta of the order of $p \sim \Lambda_{QCD} \sim (1\,\mathrm{fm})^{-1}$. In the mid-70's people considered hadrons to be structure-less "bags" filled with near-massless perturbative quarks, with mild non-perturbative effects appearing at its boundaries and confining them at the scale of 1 fm.

One logical consequence of this picture would be the applicability of the derivative expansion of the non-perturbative fields or Operator Product Expansion (OPE), the basis of QCD sum rules. However, after the first successful applications of the method [4] rather serious problems [6] have surfaced. All spin-zero channels (as we will see, those are the ones directly coupled to instantons) related to quark or gluon-based operators alike, indicate unexpectedly large non-perturbative effects and deviate from the OPE predictions at very small distances.

One learned a very important lesson: *the non-perturbative fields form structures with sizes significantly smaller than 1 fm* and a local field strength much larger than Λ^2. Instantons are one of them: in order to describe many of these phenomena in a consistent way one needs instantons of a small size $\rho \sim 1/3$ fm [5]. We have direct confirmation of them from the lattice, but we do not really understand why there are no large-size instantons.

Furthermore, the instanton is not the only such small-scale gluonic object. We also learned from the lattice-based works that QCD flux tubes (or confining strings) also have a small radius, of only about $r_{string} \approx 1/5$ fm. So, all hadrons (and clearly the QCD vacuum itself) have a *substructure*, with "constituent quarks" generated by instantons connected by such flux tubes.

Clearly this substructure should play an important role in hadronic physics. We would like to know why the usual quark model has been so successful in spectroscopy, and why so little of exotic states have been seen. Also, high energy hadronic collisions must tell us a lot about a substructure, since the famous Pomeron also belongs to the list of those surprisingly small non-perturbative objects.

At the opposite end of the spectrum, people have found that QCD seems to have also a surprisingly *small energy/momentum scale*, several times lower than Λ. It was found that the behavior of the so called "quenched" and the true QCD is very different, but only if the quark mass is below some scale of the order of 20-50 MeV. As we will see below, this surprisingly low scale has been explained by properties of the instanton ensemble.

2 Chiral Symmetry Breaking and Instantons

2.1 Brief History

Let me start around 1961, when the ideas about chiral symmetry and what it may take to break it spontaneously have appeared. The NJL model [12] was the first microscopic model which attempted to derive dynamically the properties of chiral symmetry breaking and pions, starting from some *hypothetical 4-fermion interaction*

$$L_{NJL} = G(\pi^2 + \sigma^2), \tag{1}$$

where π and σ denote the corresponding scalar-isovector and scalar-isoscalar currents, respectively.

Let me make a few comments about this.

(i) It was the first bridge between the BCS theory of superconductivity and quantum field theory, leading the way to the Standard Model. It first showed that the vacuum can be truly nontrivial, a superconductor of a kind, with the mass gap $\Delta = 330 - 400$ MeV, known as "constituent quark mass".

(ii) The NJL model has 2 parameters: the strength of its 4-fermion interaction G and the cutoff $\Lambda \sim 0.8 - 1$ GeV. The latter regulates the loops (the model is non-renormalizable, which is all right for an effective theory) and is directly the "chiral scale" we are discussing. We will relate Λ to the typical instanton size ρ, and G to a combination $n\rho^2$ of the size and density of instantons.

(iii) One non-trivial prediction of the NJL model was that the mass of the scalar is $m_\sigma \approx 2m_{const.quark}$. Because this state is the P-wave in non-relativistic language, it means that there is a strong attraction which is able to compensate exactly for the rotational kinetic energy. For decades, simpler hadronic models failed to get this effect, and even now spectroscopists still argue that this (40-year-old!) result is incorrect. However, lattice results in fact show that it is exactly right and theoretically understood through instantons. Moreover, the phenomenological σ meson is being revived now, so possibly it will even come back to its proper place in the Particle Data Table, after decades of absence.

Let me now jump to instantons. We will show below that they generate quite a specific 4-fermion 't Hooft interaction [11] (for a 2-flavor theory: for pedagogical reasons we ignore strange quarks altogether now). Furthermore, its Lagrangian includes the NJL one, but it also has 2 new terms:

$$L_{tHooft} = G(\pi^2 + \sigma^2 - \eta^2 - \delta^2) \tag{2}$$

with the isoscalar-pseudoscalar η and the isovector-scalar δ. 't Hooft's minus sign is crucial here: it shows that the axial U(1) symmetry (e.g. the rotation of sigma into eta) is *not* a symmetry. That is why η (actually η' if strangeness is included) is *not* a massless Goldstone particle like the pion.

The most important next development happened in 1980's: it has been shown in [5,13] that an instanton-induced interaction does break *spontaneously* the $SU(N_f)$ chiral symmetry. Unlike the NJL model, the instanton-induced interaction has a natural cut-off parameter ρ, and the coupling constants are not free parameters, but determined by a physical quantity, the instanton density. That eventually allowed to solve the 't Hooft interaction in all orders, and get quantitative results, see [1].

2.2 General Things About Instantons

I will omit from this paper general things about instantons, well covered elsewhere. Let me just briefly mention that the topologically-nontrivial 4d solution was found by Polyakov and collaborators [7], and soon it was interpreted as a semi-classical tunneling between topologically non-equivalent vacua. The name itself was suggested by 't Hooft, meaning "existing for an instant". Formally, instantons appear in the context of the semi-classical approximation to the (Euclidean) QCD partition function

$$Z = \int DA_\mu \, \exp(-S) \prod_f^{N_f} \det\left(\not{D} + m_f\right), \tag{3}$$

$$S = \frac{1}{4g^2} \int d^4x \, G^a_{\mu\nu} G^a_{\mu\nu}. \tag{4}$$

Here, S is the gauge field action, and the determinant of the Dirac operator $\not{D} = \gamma_\mu(\partial_\mu - iA_\mu)$ accounts for the contribution of fermions. In the semi-classical approximation, we look for saddle points of the functional integral (3), i.e. configurations that minimize the classical action S. This means that saddle point configurations are solutions of the classical equations of motion.

These solutions can be found using the identity

$$S = \frac{1}{4g^2} \int d^4x \left[\pm G^a_{\mu\nu} \tilde{G}^a_{\mu\nu} + \frac{1}{2} \left(G^a_{\mu\nu} \mp \tilde{G}^a_{\mu\nu} \right)^2 \right], \tag{5}$$

where $\tilde{G}_{\mu\nu} = 1/2 \epsilon_{\mu\nu\rho\sigma} G_{\rho\sigma}$ is the dual field strength tensor (the field strength tensor in which the roles of electric and magnetic fields are reversed). Since the first term is a topological invariant (see below) and the last term is always positive, it is clear that the action is minimal if the field is (anti) self-dual

$$G^a_{\mu\nu} = \pm\tilde{G}^a_{\mu\nu}. \tag{6}$$

The action of a self-dual field configuration is determined by its topological charge

$$Q = \frac{1}{32\pi^2}\int d^4x\, G^a_{\mu\nu}\tilde{G}^a_{\mu\nu}. \tag{7}$$

From (5), we have $S = (8\pi^2|Q|)/g^2$. For finite action configurations, Q has to be an integer. The instanton is a solution with $Q = 1$ [7]

$$A^a_\mu(x) = \frac{2\eta_{a\mu\nu}x_\nu}{x^2+\rho^2}, \tag{8}$$

where the 't Hooft symbol $\eta_{a\mu\nu}$ is defined by

$$\eta_{a\mu\nu} = \begin{cases} \epsilon_{a\mu\nu} & \mu,\, \nu = 1,\, 2,\, 3 \\ \delta_{a\mu} & \nu = 4 \\ -\delta_{a\nu} & \mu = 4 \end{cases} \tag{9}$$

and ρ is an arbitrary parameter characterizing the size of the instanton. This original instanton has its non-trivial topology at large distances, but if we are to consider an instanton ensemble, it is another form, the so called *singular gauge*,

$$A^a_\mu(x) = \frac{2\overline{\eta}_{a\mu\nu}x_\nu\rho^2}{(x^2+\rho^2)x^2}, \tag{10}$$

which is needed because in this case the non-trivial topology is at the point singularity.

The classical instanton solution has a number of degrees of freedom, known as collective coordinates. In addition to the size, the solution is characterized by the instanton position z_μ and the color orientation matrix R^{ab} (corresponding to color rotations $A^a_\mu \to R^{ab}A^b_\mu$). A solution with topological charge $Q = -1$ can be constructed by replacing $\eta_{a\mu\nu} \to \overline{\eta}_{a\mu\nu}$, where $\overline{\eta}_{a\mu\nu}$ is defined by changing the sign of the last two equations in (9).

The physical meaning of the instanton solution becomes clear if we consider the classical Yang-Mills Hamiltonian (in the temporal gauge, $A_0 = 0$)

$$H = \frac{1}{2g^2}\int d^3x\,(E_i^2 + B_i^2), \tag{11}$$

where E_i^2 is the kinetic and B_i^2 the potential energy term. The classical vacua correspond to configurations with zero field strength. For non-abelian gauge fields this limits the gauge fields to be "pure gauge" $A_i = iU(\mathbf{x})\partial_i U(\mathbf{x})^\dagger$. Such configurations are characterized by a topological winding number n_W, which distinguishes between gauge transformations U that are not continuously connected.

This means that there is an infinite set of classical vacua enumerated by an integer n. Instantons are tunneling solutions that connect the different vacua. They have potential energy $B^2 > 0$ and kinetic energy $E^2 < 0$, their sum being zero at any moment in time. Since the instanton action is finite, the barrier between the topological vacua can be penetrated, and the true vacuum is a linear combination $|\theta\rangle = \sum_n e^{in\theta}|n\rangle$, called the theta vacuum. In QCD, the value of θ is an external parameter. If $\theta \neq 0$ the QCD vacuum breaks CP invariance. Experimental limits on CP violation require[2] $\theta < 10^{-9}$.

The rate of tunneling between different topological vacua is determined by the semi-classical (WKB) method. From the single instanton action one expects

$$P_{tunneling} \sim \exp(-8\pi^2/g^2). \tag{12}$$

The factor in front of the exponent can be determined by taking into account fluctuations $A_\mu = A_\mu^{cl} + \delta A_\mu$ around the classical instanton solution. This calculation was performed in a classic paper by 't Hooft [11]. The result is

$$dn_I = \frac{0.47\exp(-1.68N_c)}{(N_c-1)!(N_c-2)!}\left(\frac{8\pi^2}{g^2}\right)^{2N_c}\exp\left(-\frac{8\pi^2}{g^2(\rho)}\right)\frac{d^4z d\rho}{\rho^5}, \tag{13}$$

where $g^2(\rho)$ is the running coupling constant at the scale of the instanton size. Taking into account quantum fluctuations, the effective action depends on the instanton size. This is a sign of the conformal (scale) anomaly in QCD. Using the one-loop beta function the result can be written as $dn_I/(d^4z) \sim d\rho\rho^{-5}(\rho\Lambda)^b$, where $b = (11N_c/3) = 11$ is the first coefficient of the beta function. Since b is a large number, small size instantons are strongly suppressed. On the other hand, there appears to be a divergence at large ρ. In this regime, however, the perturbative analysis based on the one loop beta function is not applicable.

2.3 Zero Modes and the $U(1)_A$ Anomaly

In the last section we showed that instantons interpolate between different topological vacua in QCD. It is then natural to ask if the different vacua can be physically distinguished. This question is answered most easily in the presence of light fermions, because the different vacua have different axial charges. This observation is the key element in understanding the mechanism of chiral anomalies.

[2] The question why θ happens to be so small is known as the "strong CP problem". Most likely, the resolution of the strong CP problem requires physics outside QCD and we will not discuss it any further.

Anomalies first appeared in the context of perturbation theory [8,9]. From the triangle diagram involving an external axial vector current one finds that the flavor singlet current, which is conserved on the classical level, develops an anomalous divergence on the quantum level

$$\partial_\mu j^5_\mu = \frac{N_f}{16\pi^2} G^a_{\mu\nu} \tilde{G}^a_{\mu\nu}. \tag{14}$$

This anomaly plays an important role in QCD, because it explains the absence of a ninth Goldstone boson, what is the so called $U(1)_A$ puzzle.

The mechanism of the anomaly is intimately connected with instantons. First, we recognize the integral of the RHS of (14) as $2N_f Q$, where Q is the topological charge. This means that in the background field of an instanton we expect axial charge conservation to be violated by $2N_f$ units. The crucial property of instantons, originally discovered by 't Hooft, is that the Dirac operator has a zero mode $iD\!\!\!/\psi_0(x) = 0$ in the instanton field. For an instanton in the singular gauge, the zero mode wave function is

$$\psi_0(x) = \frac{\rho}{\pi} \frac{1}{(x^2+\rho^2)^{3/2}} \frac{\gamma\cdot x}{\sqrt{x^2}} \frac{1+\gamma_5}{2}\phi, \tag{15}$$

where $\phi^{\alpha m} = \epsilon^{\alpha m}/\sqrt{2}$ is a constant spinor, which couples the color index α to the spin index $m = 1, 2$. Note that the solution is left handed, $\gamma_5\psi_0 = -\psi_0$. Analogously, in the field of an anti-instanton there is a right handed zero mode.

We can now see how the axial charge is violated during tunneling. For this purpose, let us consider the Dirac Hamiltonian $i\alpha\cdot\mathbf{D}$ in the field of the instanton. The presence of a 4-dimensional normalizable zero mode implies that there is one left-handed state that crosses from positive to negative energy during the tunneling event. This can be seen as follows: In the adiabatic approximation, solutions of the Dirac equation are given by

$$\psi_i(\mathbf{x},t) = \psi_i(\mathbf{x},t=-\infty)\exp\left(-\int_{-\infty}^{t} dt'\,\epsilon(t')\right). \tag{16}$$

The only way we can have a 4-dimensional normalizable wave function is if ϵ_i is positive for $t \to \infty$ and negative for $t \to -\infty$. This explains how the axial charge can be violated during tunneling. No fermion ever changes its chirality, all states simply move one level up or down. The axial charge comes, so to say, from the "bottom of the Dirac sea".

2.4 The Effective Interaction Between Quarks

Proceeding from pure glue theory to QCD with light quarks, one has to deal with the much more complicated problem of quark-induced interactions. Indeed, on the level of a single instanton we cannot even understand the presence of instantons in full QCD. The reason is again related to the existence of zero modes. In the presence of light quarks, the tunneling rate is proportional to the fermion determinant, which is given by the product of the eigenvalues of the Dirac operator. This means that (as $m \to 0$) the tunneling amplitude vanishes and individual instantons cannot exist!

This result is related to the anomaly: During the tunneling event, the axial charge of the vacuum changes, so instantons have to be accompanied by fermions. The tunneling amplitude is non-zero only in the presence of external quark sources, because zero modes in the denominator of the quark propagator can cancel against zero modes in the determinant. Consider the fermion propagator in the instanton field

$$S(x,y) = \frac{\psi_0(x)\psi_0^+(y)}{im} + \sum_{\lambda\neq 0} \frac{\psi_\lambda(x)\psi_\lambda^+(y)}{\lambda + im}, \tag{17}$$

where $i\not{D}\psi_\lambda = \lambda\psi_\lambda$. For N_f light quark flavors the instanton amplitude is proportional to m^{N_f}. Instead of the tunneling amplitude, let us calculate a $2N_f$-quark Green's function $\langle\prod_f \bar{\psi}_f(x_f)\Gamma\psi_f(y_f)\rangle$, containing one quark and one antiquark of each flavor. Performing the contractions, the amplitude involves N_f fermion propagators (17), so that the zero mode contribution involves a factor m^{N_f} in the denominator.

The result can be written in terms of an effective Lagrangian [11]. It is a non-local $2N_f$-fermion interaction, where the quarks are emitted or absorbed in zero mode wave functions. The result simplifies if we take the long wavelength limit (in reality, the interaction is cut off at momenta $k > \rho^{-1}$) and averaged over the instanton position and color orientation. For $N_f = 1$ the result is [11,14]

$$\mathcal{L}_{N_f=1} = \int d\rho\, n_0(\rho) \left(m\rho - \frac{4}{3}\pi^2\rho^3 \bar{q}_R q_L \right), \tag{18}$$

where $n_0(\rho)$ is the tunneling rate. Note that the zero mode contribution acts like a mass term. For $N_f = 1$, there is only one chiral $U(1)$ symmetry, which is anomalous. This means that the anomaly breaks chiral symmetry and gives a fermion mass term. This is not true for more than one flavor. For $N_f = 2$, the result is

$$\mathcal{L}_{N_f=2} = \int d\rho\, n_0(\rho) \Bigg[\prod_f \left(m\rho - \frac{4}{3}\pi^2\rho^3 \bar{q}_{f,R} q_{f,L} \right) \tag{19}$$
$$+ \frac{3}{32}\left(\frac{4}{3}\pi^2\rho^3\right)^2 \left(\bar{u}_R\lambda^a u_L \bar{d}_R\lambda^a d_L - \bar{u}_R\sigma_{\mu\nu}\lambda^a u_L \bar{d}_R\sigma_{\mu\nu}\lambda^a d_L \right) \Bigg].$$

One can easily check that the interaction is $SU(2) \times SU(2)$ invariant, but $U(1)_A$ is explicitly broken. This Lagrangian is of the type first studied by Nambu and Jona-Lasinio [12] and widely used as a model for chiral symmetry breaking and as an effective description for low energy chiral dynamics. It can be transformed to the form discussed above when we compared it to the NJL interaction.

2.5 The Quark Condensate in the Mean Field Approximation

We showed in the last section that in the presence of light fermions, tunneling can only take place if the tunneling event is accompanied by N_f fermions which change their chirality. But in the QCD vacuum, chiral symmetry is broken and the quark condensate $\langle\bar{q}q\rangle = \langle\bar{q}_L q_R + \bar{q}_R q_L\rangle$ is non-zero. This means that there is a finite amplitude for a quark to change its chirality and we expect the instanton density to be finite.

For a sufficiently dilute system of instantons, we can estimate the instanton density in full QCD from the expectation value of the $2N_f$ fermion operator in the effective Lagrangian (19). Using the factorization assumption [4], we find that the factor $\prod_f m_f$ in the instanton density should be replaced by $\prod_f m_f^*$, where the effective quark mass is given by

$$m_f^* = m_f - \frac{2}{3}\pi^2\rho^2\langle\bar{q}_f q_f\rangle. \tag{20}$$

This shows that if chiral symmetry is broken, the instanton density is finite in the chiral limit.

This obviously raises the question whether the quark condensate itself can be generated by instantons. This question can be addressed using several different techniques (for a review, see [1,2]). One possibility is to use the effective interaction (19) and to calculate the quark condensate in the mean field (Hartree-Fock) approximation. This corresponds to summing the contribution of all "cactus" diagrams to the full quark propagator. The result is a gap equation [13]

$$\int \frac{d^4k}{(2\pi)^4}\frac{M^2(k)}{k^2+M^2(k)} = \frac{N}{4N_c V}, \tag{21}$$

which determines the constituent quark mass $M(0)$ in terms of the instanton density (N/V). Here, $M(k) = M(0)k^2\varphi'^2(k)/(2\pi\rho)$ is the momentum dependent effective quark mass and $\varphi'(k)$ is the Fourier transform of the zero mode profile [13]. The quark condensate is given by

$$\langle\bar{q}q\rangle = -4N_c\int \frac{d^4k}{(2\pi)^4}\frac{M(k)}{M^2(k)+k^2}. \tag{22}$$

Using our standard parameters $(N/V) = 1\ \text{fm}^{-4}$ and $\rho = 1/3$ fm, one finds $\langle\bar{q}q\rangle \simeq -(255\,\text{MeV})^3$ and $M(0) = 320$ MeV. Parametrically, $\langle\bar{q}q\rangle \sim (N/V)^{1/2}\rho^{-1}$ and $M(0) \sim (N/V)^{1/2}\rho$. Note that both quantities are not proportional to (N/V) but rather to $(N/V)^{1/2}$. This is a reflection of the fact that spontaneous breaking of chiral symmetry is not a single instanton effect, but involves infinitely many instantons.

A very instructive way to study the mechanism for chiral symmetry breaking at a more microscopic level consists in considering the distribution of eigenvalues of the Dirac operator. A general relation that connects the spectral density $\rho(\lambda)$ of the Dirac operator to the quark condensate was given by the Banks-Casher relation

$$\langle \bar{q}q \rangle = -\pi\rho(0). \tag{23}$$

This result is analogous to the Kondo formula for the electrical conductivity. Just like the conductivity is given by the density of states at the Fermi surface, the quark condensate is determined by the level density at zero virtuality λ. For a disordered, random, system of instantons the zero modes interact and form a band around $\lambda = 0$. As a result, the eigenstates are delocalized and chiral symmetry is broken. On the other hand, if instantons are strongly correlated, for example bound into topologically neutral molecules, the eigenvalues are pushed away from zero, the eigenstates are localized and chiral symmetry is unbroken. As we will see below, which scenario is precisely realized depends on the parameters of the theory, like the number of light flavors and the temperature. Of course, for "real" QCD with two light flavors at $T = 0$, we expect chiral symmetry to be broken. This is supported by numerical simulations of the partition function of the instanton liquid, see [1].

2.6 The Qualitative Picture of the Instanton Ensemble

Using basically such expressions and the known value of the quark condensate, it was pointed out in [5] that all would be consistent only if the typical instanton size happened to be significantly smaller than their separation[3], $R = n^{-1/4} \approx 1$ fm, namely $\rho_{\max} \sim 1/3$ fm.

In Fig. 1 one can see lattice data on the instanton size distribution, obtained by cooling the original gauge fields. A similar distribution can also be obtained from lowest fermionic Dirac eigenmodes: in this case no "cooling" is needed.

Let me now show another evidence for this value of the instanton size, taken from the pion form factor calculated in the instanton model [15]. In Fig. 2, we show how the experimentally measured pion size correlates with the input mean instanton size: one can see that the value 0.35 fm is a clear winner. If so, the following qualitative picture of the QCD vacuum has emerged:

1. Since the instanton size is significantly smaller than the typical separation R between instantons, $\rho/R \sim 1/3$, the vacuum is fairly dilute. The fraction of spacetime occupied by strong fields is only a few percent.
2. The fields inside the instanton are very strong, $G_{\mu\nu} \gg \Lambda^2_{QCD}$. This means that the semi-classical approximation is valid, and the typical action is large

$$S_0 = 8\pi^2/g^2(\rho) \sim 10 - 15 \gg 1. \tag{24}$$

[3] Derived in turn from the gluon condensate and the topological susceptibility.

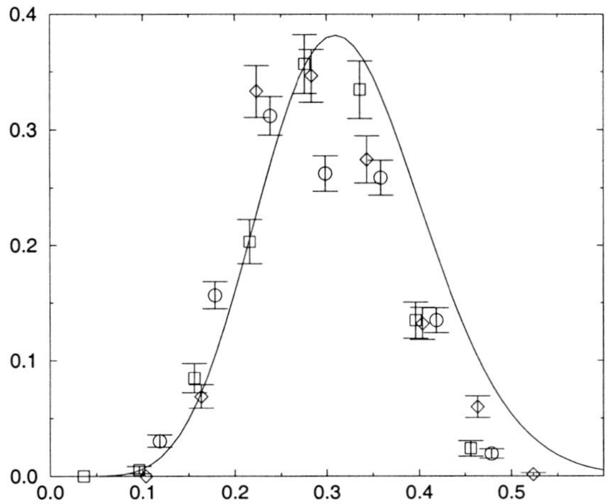

Fig. 1. The instanton density $dn/d\rho d^4z$ [fm^{-5}] versus its size ρ [fm]. The points are from the lattice work [10] for this theory, with $\beta = 5.85$ (diamonds), 6.0 (squares) and 6.1 (circles). Their comparison should demonstrate that the results are rather lattice-independent. The line corresponds to the proposed expression $\sim exp(-2\pi\sigma\rho^2)$, see the text

Higher order corrections are proportional to $1/S_0$ and presumably small.

3. Instantons retain their individuality and are not destroyed by interactions. From the dipole formula, one can estimate

$$|\delta S_{int}| \sim (2-3) \ll S_0. \quad (25)$$

4. Nevertheless, interactions are important for the structure of the instanton ensemble, since

$$\exp|\delta S_{int}| \sim 20 \gg 1. \quad (26)$$

This implies that interactions have a significant effect on correlations among instantons, and the instanton ensemble in QCD is not a dilute gas but an *interacting liquid.*

The aspects of the QCD vacuum for which instantons are most important are those related to light fermions. Their importance in the context of chiral symmetry breaking is related to the fact that the Dirac operator has a chiral zero mode in the field of an instanton. These zero modes are localized quark states around instantons, like atomic states of electrons around nuclei. At a finite density of instantons those states can become collective, like atomic

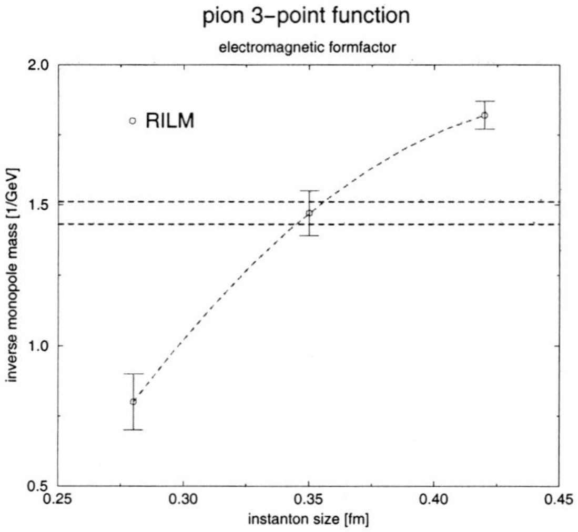

Fig. 2. The fitted parameter M of the pion form factor $f \sim M^2/(Q^2+M^2)$, versus the instanton size.

states in metals. The resulting delocalized state corresponds to the wave function of the quark condensate.

Direct tests of all these ideas on the lattice are possible. One may have a look at the lowest eigenmodes and see if they are related to instantons or something else (monopoles, vortices...) by identifying their shapes - 4d bumps, lines, or 2 d sheets, respectively. So far, only bumps representing instantons were seen.

One may also test the local chirality of the lowest eigenmodes. Just at this school I learned from one of its young participants, Christof Gattringer, about his version of chiral lattice fermions and the nice results he got. Among those, remarkably well defined, is the separation between instanton-based and perturbative-like lowest modes, revealed by the so-called participation ratios.

Let me now explain the *lowest QCD scale* generated by instantons, as mentioned above. The width of the *zero mode zone* of states is of the order of the root-mean-square matrix element of the Dirac operator $< I|\not{D}|J > \sim \rho^2/R^3$. Here, the states I,J are some instanton and anti-instanton zero modes, ρ is the instanton size and $R \sim n^{-1/4} \approx 1$ fm is the distance between their centers. Note the small factor $(\rho/R)^2 \sim 1/10$. The Dirac eigenvalues from the zone have similar magnitudes. Now, the eigenvalues enter together with quark mass m: only if this quark mass is smaller than this scale, we start seeing the physics of the zero mode zone. In particular, for *quenched* QCD (or an instanton liquid) there is no determinant and the zone states have a rather wrong spectrum. However, only if the quark mass is small compared to its width we start observing the difference. Only recently lattice practitioners were able to do so: Indeed, quenched QCD results at small m start deviating quite drastically from the correct answers.

2.7 Interacting Instantons

In the QCD partition function there are two types of fields, gluons and quarks, and so the first question one addresses is *which integral to take first.*

(i) One way is to eliminate *gluonic* degrees of freedom first. A physical motivation for this may be that gluonic states are heavy and an effective fermionic theory should be better suited to derive an effective low-energy fermionic theory. It is a well-trodden path and one can follow it to the development of a similar four-fermion theory, the NJL model. One can do simple mean-field or random-field approximation (RPA) diagrams, and find the mean condensate and properties of the Goldstone mesons [13]. The results for color superconductors at high density reported below are done with the same technique as well. But nevertheless, not much can really be done in such a NJL-like approach. In fact, multiple attacks during the last 40 years on the NJL model *beyond the mean field* basically failed. In particular, one might think that, if baryons are states with three quarks and if using quasi-local four-fermion Lagrangians for the three-body system is a solvable quantum mechanical problem, one could at least tell if nucleons are or are not bound in NJL. In fact one cannot: the results depend strongly on subtleties of how the local limit for the interaction is defined, and there is no clear answer to this question. Other notorious attempts to sum more complicated diagrams deal with the possible modification of the chiral condensate. Some works even claim that those diagrams destroy it *completely*!

Going from NJL to instantons improves the situation enormously: the shape of the form factor is no longer a guess (it is provided by the shape of zero modes) and one can in principle evaluate any particular diagram. However, *summing them all up* appears like an impossible task.

(ii) The solution to this problem was found. For that, one has to follow the opposite strategy and do the *fermion* integral first. The first step is simple and standard: fermions only enter quadratically, leading to a fermionic determinant. In the instanton approximation, it leads to the Interacting Instanton Liquid Model (IILM), defined by the following partition function

$$Z = \sum_{N_+, N_-} \frac{1}{N_+! N_-!} \int \prod_i^{N_+ + N_-} [d\Omega_i \, d(\rho_i)] \exp(-S_{\mathrm{int}}) \prod_f^{N_f} \det(\hat{D} + m_f) \quad (27)$$

describing a system of pseudo-particles interacting via the bosonic action and the fermionic determinant. Here, $d\Omega_i = dU_i \, d^4 z_i \, d\rho_i$ is the measure in color orientation, position, and size associated with single instantons, and $d(\rho)$ is the single instanton density $d(\rho) = dn_{I,\bar{I}}/d\rho dz$.

The gauge interaction between instantons is approximated by a sum of pure two-body interactions $S_{\text{int}} = \frac{1}{2}\sum_{I\neq J} S_{\text{int}}(\Omega_{IJ})$. Genuine three-body effects in the instanton interaction are not important as long as the ensemble is reasonably dilute. Implementation of this part of the interaction (quenched simulation) is quite analogous to usual statistical ensembles made of atoms.

As already mentioned, quark exchanges between instantons are included in the fermionic determinant. Finding a diagonal set of fermionic eigenstates of the Dirac operator is similar to what people are doing, e.g., in quantum chemistry when electron states for molecules are calculated. The difficulty of our problem is, however, much bigger, because this set of fermionic states should be determined for *all* configurations which appear during the Monte-Carlo process.

If the set of fermionic states is, however, limited to the subspace of instanton zero modes, the problem becomes tractable numerically. Typical calculations in the IILM involved up to N$\sim$ 100 instantons (and anti-instantons), which means that the determinants of $N \times N$ matrices are involved. Such determinants can be evaluated by an ordinary workstation (and even PC these days) so quickly that a straightforward Monte Carlo simulation of the IILM is possible in a few minutes. On the other hand, expanding the determinant in a sum of products of matrix elements, one can easily identify the sum of all closed loop diagrams up to order N in the 't Hooft interaction. Thus, in this way one can actually take care of about 100 factorial diagrams!

3 Hadronic Structure and the QCD Correlation Functions

3.1 Correlators as a Bridge Between Hadronic and Partonic Worlds

Consider two currents separated by a *space-like* distance x (which can be considered as the spatial distance, or an Euclidean time) and introduce correlation functions of the type

$$K(x) =< T(J(x)J(0)) > \tag{28}$$

with $J(x) = \bar{\psi}(x)\Gamma\psi(x)$. The matrix Γ contains γ_μ for vector currents, γ_5 for the pseudoscalar, or 1 for the scalars, etc. and also a flavor matrix, if needed.

We will start with isovector-vector and axial currents, and then discuss four scalar-pseudoscalar channels: π (P=-1, I=1), σ or f_0 (P=+1, I=0), η (P=-1, I=0), and δ or a_0 (P=+1, I=1).

In a (relativistic) field theory, correlation functions of gauge invariant local operators are the proper tool to study the spectrum of the theory. The correlation functions can be calculated either from the physical states (mesons, baryons, glueballs) or in terms of the fundamental fields (quarks and gluons) of the theory. In the latter case, we have a variety of techniques at our disposal, ranging from perturbative QCD, the OPE, to models of QCD and lattice simulations. For this reason, correlation functions provide a bridge between hadronic phenomenology on the one side and the underlying structure of the QCD vacuum on the other side.

Loosely speaking, hadronic correlation functions play the same role for understanding the forces between quarks as the NN scattering phase shifts did in the case of nuclear forces. In the case of quarks, however, confinement implies that we cannot define scattering amplitudes in the usual way. Instead, one has to focus on the behavior of gauge invariant correlation functions at short and intermediate distance scales. The available theoretical and phenomenological information about these functions was recently reviewed in [3].

In all cases at small x we expect $K(x) \approx K_0(x)$, where the latter corresponds to just *free* propagation of (about massless) light quarks. The zeroth order correlators are all just $K_0(x) = 12/(\pi^4 x^6)$, basically the square of the massless quark propagator.

The first deviations due to non-perturbative effects can be studied using the Wilson OPE in Ref. [4]. For all scalar and pseudoscalar channels the resulting first correction is

$$\frac{K(x)}{K_0(x)} = 1 + \frac{x^4}{384} < (gG)^2 > +... \tag{29}$$

The "gluon condensate" is assumed to be made out of a soft vacuum field, and therefore all arguments can be simply taken at the point $x = 0$. The so-called *standard* value of the "gluon condensate" appearing here was estimated previously from charmonium sum rules:

$$< (gG)^2 >_{SVZ} \approx 0.5\,\mathrm{GeV}^4. \tag{30}$$

Thus, the OPE suggests the following scale, at which the correction becomes equal to the first term:

$$x_{OPE} = (384/ < (gG)^2 >_{SVZ})^{1/4} \approx 1.0 \text{ fm}. \tag{31}$$

This seems to be completely consistent with the approximation used. However, as Novikov, Shifman, Vainshtein, and Zakharov soon noticed [6], this (and other OPE corrections) completely failed to describe all the $J^P = O^\pm$ channels. We return to this issue after we consider vectors and axials.

3.2 Vector and Axial Correlators

The information available on vector correlation functions from experimental data on $e^+e^- \to hadrons$, the OPE and other exact results was reviewed in [3]. Since then, however, new high statistics measurements of hadronic τ decays $\tau \to \nu_\tau + hadrons$ have been done. For definiteness, we use results of one of them, namely the ALEPH experiment at CERN [16,17].

The vector and axial-vector correlation functions are $\Pi_V(x) = \langle j_\mu^a(x) j_\mu^a(0)\rangle$ and $\Pi_A(x) = \langle j_\mu^{5\,a}(x) j_\mu^{5\,a}(0)\rangle$. Here, $j_\mu^a(x) = \bar{q}\gamma_\mu \frac{\tau^a}{2} q$, $j_\mu^{5\,a}(x) = \bar{q}\gamma_\mu\gamma_5 \frac{\tau^a}{2} q$ are the isotriplet vector and axial-vector currents. The Euclidean correlation functions have the spectral representation [3]

$$\Pi_{V,A}(x) = \int ds\, \rho_{V,A}(s) D(\sqrt{s}, x), \tag{32}$$

where $D(m,x) = m/(4\pi^2 x)K_1(mx)$ is the Euclidean coordinate-space propagator of a scalar particle with mass m. We shall focus on the linear combinations $\Pi_V + \Pi_A$ and $\Pi_V - \Pi_A$. These combinations allow for a clearer separation of different nonperturbative effects. The corresponding spectral functions $\rho_V \pm \rho_A$ measured by the ALEPH collaboration are shown in Fig. 3. The errors are a combination of statistical and systematic ones (below we use them, conservatively, as pure systematic): the main problem seems to be the separation into V and A of channels with Kaons, which may affect $V - A$ at $s > 2$ GeV at the 10% level. None of our conclusions are sensitive to it.

In QCD, the vector and axial-vector spectral functions must satisfy chiral sum rules. Assuming that $\rho_V - \rho_A = 0$ above $s > m_\tau^2$ and using ALEPH data below it, one finds that all four of the sum rules are satisfied within the experimental uncertainty, but the central values differ significantly from the chiral predictions [16]. In general, both functions are expected to have oscillations of decreasing amplitude, and putting $\rho_V - \rho_A$ to zero at an arbitrary point implies the appearance of spurious dimension $d = 2, 4$ operators in the correlation functions at small x. Therefore, we have decided to terminate the data above a specially tuned point, $s_0 = 2.5\,\mathrm{GeV}^2$, enforcing all four chiral sum rules. (The reader should, however, be aware of the fact that we have, in fact, slightly moved the data points in the small-x region within the error band.) Finally we add the pion-pole contribution (not shown in Fig. 3), which corresponds to an extra term $\Pi_A^\pi(x) = f_\pi^2 m_\pi^2 D(m_\pi, x)$. The resulting correlation functions $\Pi_V(x) \pm \Pi_A(x)$ are shown in Fig. 4.

We begin our analysis with the combination $\Pi_V - \Pi_A$. This combination is sensitive to chiral symmetry breaking, while perturbative diagrams as well as gluonic operators cancel out.

In Fig. 4, we compare the measured correlation functions with predictions from the instanton liquid model (in its simplest form, random instanton liquid with parameters n, ρ fixed in [5] and discussed above).

The agreement of the instanton prediction with the measured $V - A$ correlation is impressive: it extends all the way from short to large distances. At distances $x > 1.25$ fm both combinations are dominated by the pion contribution while at intermediate x the ρ, ρ', and a_1 resonances contribute.

We shall now focus our attention on the $V + A$ correlation function. The unique feature of this function is the fact that the correlator remains close to the free-field behavior for distances as large as 1 fm. This phenomenon was referred to as "super-duality" in [3]. The instanton model reproduces this

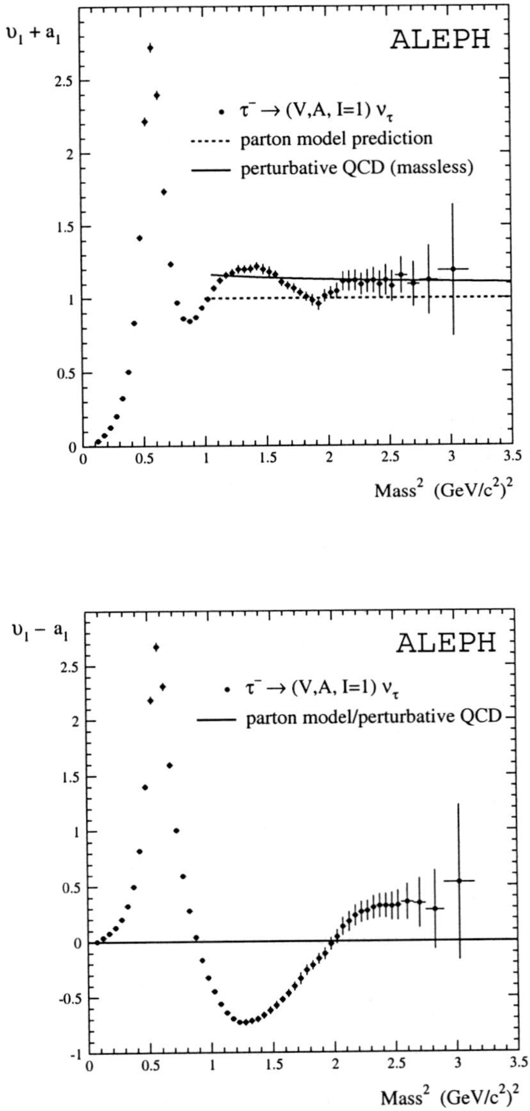

Fig. 3. Spectral functions $v(s) \pm a(s) = 4\pi^2(\rho_V(s) + \rho_A(s))$ extracted by the ALEPH collaboration from τ-lepton hadronic decays

feature of the $V + A$ correlator. We also notice that for small x the deviation of the correlator in the instanton model from the free-field behavior is small compared to the perturbative $O(\alpha_2/\pi)$ correction. This opens the possibility of precision studies of the pQCD contribution. But before we do so, let us

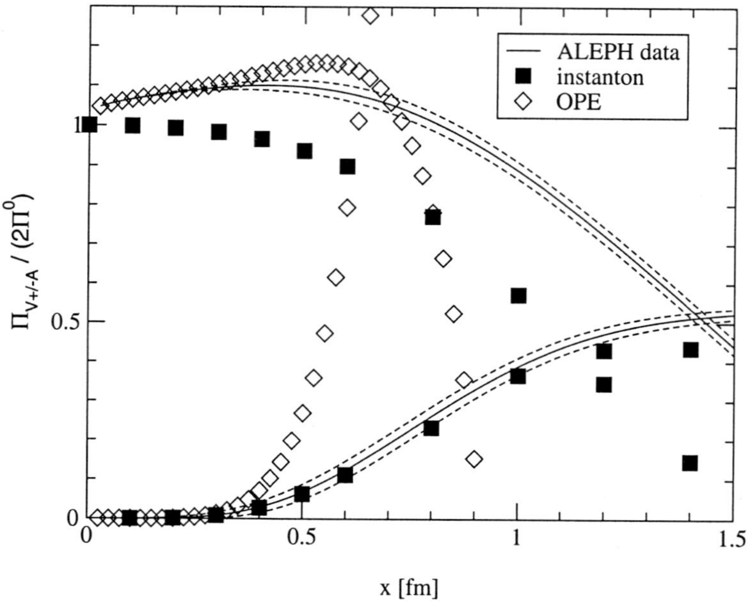

Fig. 4. Euclidean coordinate space correlation functions $\Pi_V(x) \pm \Pi_A(x)$ normalized to the free-field behavior. The solid lines show the correlation functions reconstructed from the ALEPH spectral functions and the dotted lines are the corresponding error bands. The squares show the result of a random instanton liquid model and the diamonds of the OPE fit described in the text

compare the correlation functions to the OPE prediction

$$\frac{\Pi_V(x) + \Pi_A(x)}{2\Pi_0(x)} = 1 + \frac{\alpha_s}{\pi} - \frac{1}{384}\langle g^2 G^2_{\mu\nu}\rangle x^4 - \frac{4\pi^3}{81}\alpha_s(x)\langle\bar{q}q\rangle \log(x^2) x^6 + \dots \quad (33)$$

Note that the perturbative correction is attractive while the power corrections of dimension $d = 4$ and $d = 6$ are repulsive. Direct instantons also induce an $O(x^4)$ correction $1 - \frac{\pi^2}{12}\left(\frac{N}{V}\right)x^4 + \dots$, which is consistent with the OPE because in a dilute instanton liquid we have $\langle g^2 G^2\rangle = 32\pi^2(N/V)$. This term can indeed be seen in the instanton calculation and causes the correlator to drop below 1 at small x. It is possible to extract the value of Λ_{QCD} (we find $\alpha_s(m_\tau) = 0.35$) and one obtains a clear indication of a running coupling. This is only possible because the non-perturbative corrections (represented by instantons) are basically cancelling each other to a very high degree in the $V + A$ channel.

Why is this happening? The first order in the 't Hooft interaction is indeed absent, due to chirality mismatch. There is no general theoretical reason why all non-perturbative terms of higher order should also do so: but the ALEPH data, when used wrongly, give a hint that they actually do so.

3.3 Spin-Zero Correlation Functions

Now we will look at cases that are completely opposite to those just considered: the instanton-induced effects will be large. Furthermore, the four channels actually show a completely different non-perturbative deviation from K_0 at small x: half of them (π, σ) deviate upwards, and another pair (η, δ) deviate downwards.

Let me first demonstrate, however, that the OPE scale determined above cannot be correct. All we have to do is to evaluate the strength of the pion contribution to the correlator in question

$$K_\pi(x) = \frac{\lambda_\pi^2}{4\pi^2 x^2} . \tag{34}$$

The coupling constant is defined as $\lambda_\pi =< 0|J(0)|\pi >$ and the rest is nothing more than the scalar massless propagator[4]. Because both the pion term and the gluon condensate correction happen to be $1/x^2$, let us compare the coefficients. Ideal matching would mean that they are about the same

$$\lambda_\pi^2 \approx \frac{< (gG)^2 >_{SVZ}}{8\pi^2} . \tag{35}$$

The r.h.s. is about 0.0063 GeV^4. However, phenomenology tells us that (unlike the better known coupling to the axial current f_π) the coupling λ_π is surprisingly large[5]. The l.h.s. of this relation is actually $\lambda_\pi^2 = (.48\text{GeV})^4 = 0.053\,\text{GeV}^4$, about *10 times larger* than the r.h.s. It means a much larger non-perturbative effect is needed to explain the deviation from the perturbative behavior.

Now, let us see why this is so. The instanton effects in spin-0 channels are in these cases much larger because the effect of the 't Hooft interaction appears in those cases in first order. Furthermore, since its flavor structure is non-diagonal $(\bar{u}u)(\bar{d}d)$ the correlator of the two π^0 currents $(\bar{u}\gamma_5 u - \bar{d}\gamma_5 d)$ has opposite sign as compared to the correlator of the η' currents $(\bar{u}\gamma_5 u + \bar{d}\gamma_5 d)$. What it means is that instantons are as attractive in the pion channel as they are repulsive in the η' case. The situation is reversed in the scalar channels: the isoscalar sigma is attractive and the isovector is repulsive.

[4] We can ignore the pion mass at the distances in question. We also ignore contributions of other states, which can only add positively to the correlator and make the disagreement only worse.

[5] The reason for that is that the pion is rather compact and also the wave function is concentrated at its center, so that its value at $r = 0$ is large. We return to this point in the discussion of the "instanton liquid" model.

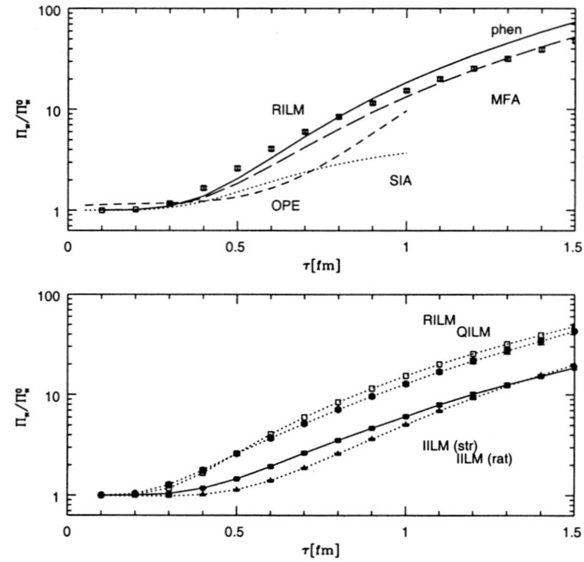

Fig. 5. Pion correlation function in various approximations and instanton ensembles. In the top figure we show the phenomenological expectation (solid), the OPE (dashed), the single instanton (dotted) and mean-field approximations (long-dashed) as well as data in the random instanton ensemble. In the bottom figure we compare different instanton ensembles, random (open squares), quenched (circles) and interacting (streamline: solid squares; ratio ansatz: solid triangles)

Full results from versions of the instanton liquid model for pion correlators are shown in Fig. 5. Different versions of the model (mentioned in the figure as IILM(rat) etc.) differ by a particular ansatz for the gauge field used, from which the interaction is calculated. Note, furthermore, that these figures contain also a curve marked "phen": this is what the correlator actually looks like according to phenomenology.

We simply show a few results of correlation functions in the different instanton ensembles (for original Refs. see [1]). Some of them (like vector and axial-vector ones) turned out to be easy: nearly any variant of the instanton model can reproduce well the (experimentally known!) correlators. Some of them are very much sensitive to details of the model: two such cases are shown in Figs. 5 and 6. The pion correlation functions in the different ensembles are qualitatively very similar. The differences are mostly due to different values of the quark condensate (and the physical quark mass) in the different ensembles. Using the Gell-Mann-Oaks-Renner relation, one can extrapolate the pion mass to the physical value of the quark masses. The results are consistent with the experimental value in the streamline ensemble (both quenched and unquenched), but they are clearly too small in the ratio

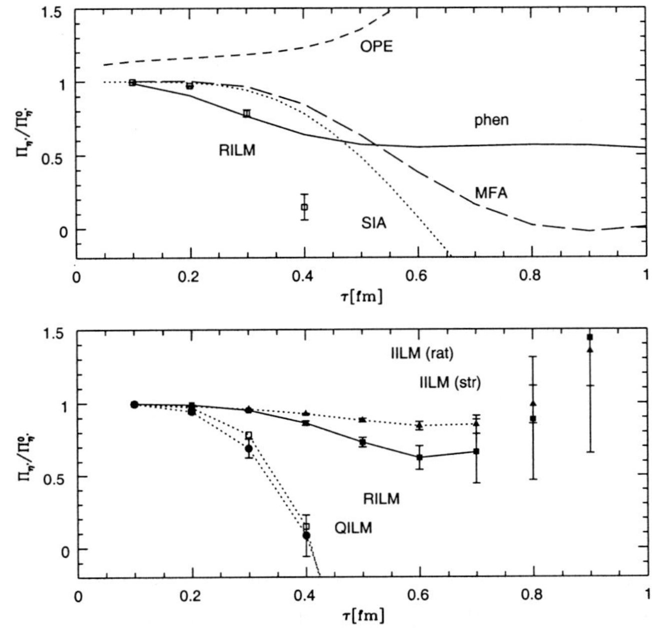

Fig. 6. η' meson correlation functions. The various curves and data sets are labeled as in Fig. 5. Note that the random instanton liquid model (RILM) and the quenched version (no fermionic determinant, only bosonic interactions) predict an η' correlator to be negative. The same unphysical behavior has been found on the lattice

ansatz ensemble. This is a reflection of the fact that the ratio ansatz ensemble is not sufficiently dilute.

The situation is drastically different in the η' channel. Among the ~ 40 correlation functions calculated in the random ensemble, only the η' and the isovector-scalar δ were found to be completely unacceptable. The correlation function decreases very rapidly and becomes *negative* at $x \sim 0.4$ fm. This behavior is incompatible even with a normal spectral representation. The interaction in the random ensemble is too repulsive and the model "over-explains" the $U(1)_A$ anomaly.

The results in the unquenched ensembles (closed and open points) significantly improve the situation. This is related to dynamical correlations between instantons and anti-instantons (topological charge screening). The single instanton contribution is repulsive, but the contribution from pairs is attractive. Only if correlations among instantons and anti-instantons are sufficiently strong the correlators are prevented from becoming negative. Quantitatively, the δ and η_{ns} masses in the streamline ensemble are still too heavy as compared to their experimental values. In the ratio ansatz, on the other hand, the correlation functions even show an enhancement at distances on the order of 1 fm, and the fitted masses are too light. This shows that the η' channel is very sensitive to the strength of correlations among instantons.

In summary, pion properties are mostly sensitive to global properties of the instanton ensemble, in particular its diluteness. Good phenomenology demands $\bar{\rho}^4 n \simeq 0.03$, as originally suggested in [5]. The properties of the ρ meson are essentially independent of the diluteness, but show sensitivity to $\bar{I}I$ correlations. These correlations become crucial in the η' channel.

3.4 Baryonic Correlation Functions

The existence of a strongly attractive interaction in the pseudoscalar quark-antiquark (pion) channel also implies an attractive interaction in the scalar quark-quark (diquark) channel. This interaction is phenomenologically very desirable, because it immediately explains why the nucleon is light, while the Δ (S=3/2, I=3/2) is heavy.

The so called Ioffe currents (with no derivatives and the minimum number of quark fields) are local operators which can excite states with nucleon quantum numbers. Those with positive parity and spin 1/2 can also be represented in terms of scalar and pseudoscalar diquarks

$$\eta_{1,2} = (2,4)\left\{\epsilon_{abc}(u^a C d^b)\gamma_5 u^c \mp \epsilon_{abc}(u^a C\gamma_5 d^b)u^c\right\}. \qquad (36)$$

Nucleon correlation functions are defined by $\Pi^N_{\alpha\beta}(x) = \langle\eta_\alpha(0)\bar{\eta}_\beta(x)\rangle$, where α, β are the Dirac indices of the nucleon currents. In total, there are six different nucleon correlators: the diagonal $\eta_1\bar{\eta}_1$, $\eta_2\bar{\eta}_2$ and the off-diagonal $\eta_1\bar{\eta}_2$ correlators, each contracted with either the identity or $\gamma\cdot x$. Let us focus on the first two of these correlation functions (for more detail, see [1] and references therein).

The correlation function Π^N_2 in the interacting ensemble is shown in Fig. 7. The fact that the nucleon in IILM is actually bound can also be demonstrated by comparing the full nucleon correlation function with that of three non-interacting quarks (the cube of the average propagator). The full correlator is significantly larger than the non-interacting one.

There is a significant enhancement over the perturbative contribution which is nicely described in terms of the nucleon contribution. Numerically, we find[6] $m_N = 1.019$ GeV. In the random ensemble, we have measured the nucleon mass at smaller quark masses and found $m_N = 0.96 \pm 0.03$ GeV. The nucleon mass is fairly insensitive to the instanton ensemble. However, the strength of the correlation function depends on the instanton ensemble. This is reflected by the value of the nucleon coupling constant, which is smaller in the IILM. In [18] we studied all six nucleon correlation functions. We showed

[6] Note that this value corresponds to a relatively large current quark mass $m = 30$ MeV.

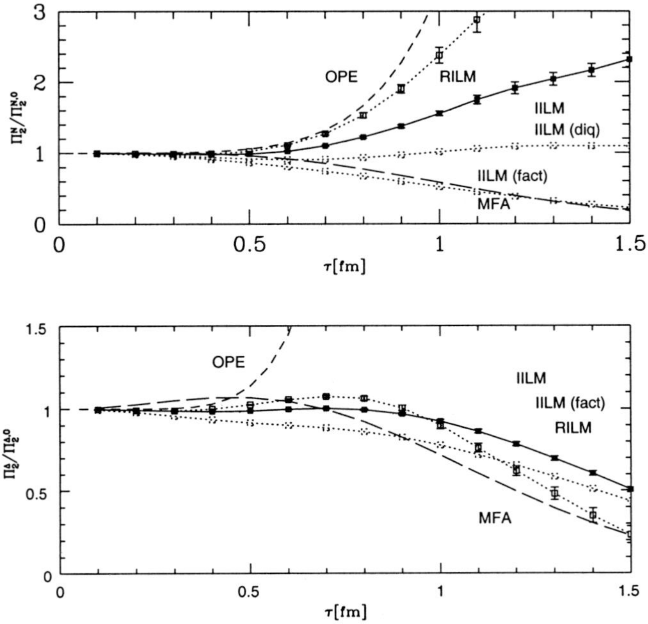

Fig. 7. Nucleon and Δ correlation functions Π_2^N and Π_2^Δ. Curves labeled as in Figs. 5 and 6 for mesonic correlators

that all correlation functions can be described with the same nucleon mass and coupling constants.

The fitted value of the threshold is $E_0 \simeq 1.8$ GeV, indicating that there is little strength in the "three-quark continuum" (dual to higher resonances in the nucleon channel). A significant part of this interaction was traced down to the strongly attractive *scalar diquark* channel. The nucleon (at least in IILM) is a strongly bound diquark plus a loosely bound third quark. The properties of this diquark picture of the nucleon continue to be disputed by phenomenologists. We will return to diquarks in the next section, where they will become Cooper pairs of color superconductors.

In the case of the Δ resonance, there exists only one independent Ioffe current, given (for the Δ^{++}) by

$$\eta_\mu^\Delta = \epsilon_{abc}(u^a C\gamma_\mu u^b)u^c. \quad (37)$$

However, the spin structure of the correlator $\Pi^\Delta_{\mu\nu;\alpha\beta}(x) = \langle \eta^\Delta_{\mu\alpha}(0)\bar{\eta}^\Delta_{\nu\beta}(x)\rangle$ is much richer. In general, there are ten independent tensor structures, but the Rarita-Schwinger constraint $\gamma^\mu\eta_\mu^\Delta = 0$ reduces this number to four.

The mass of the Δ resonance is too large in the random model, but closer to experiment in the unquenched ensemble. Note that, similar to the nucleon, part of this discrepancy is due to the value of the current quark mass. Nevertheless, the Δ-N mass splitting in the unquenched ensemble is $m_\Delta - m_N = 409$ MeV, i.e. larger but still comparable to the experimental value of 297 MeV. It mostly comes from the *absent scalar diquarks* in the Δ channel.

4 The Phases of QCD

4.1 The Phase Diagram

In this section we discuss QCD in extreme conditions, such as finite temperature/density. Let me first emphasize why it is interesting and instructive to do so. It is not simply to practice once again the semi-classical or perturbative methods similar to what has been done before in the vacuum. What we are looking for here are *new phases* of QCD (and related theories), namely new self-consistent solutions that differ qualitatively from what we have in the QCD vacuum.

One such phase occurs at high enough temperature $T > T_c$: it is known as Quark-Gluon Plasma (QGP). It is a phase understandable in terms of basic quark and gluon-like excitations [39], without confinement and with unbroken chiral symmetry in the massless limit[7]. One of the main goals of the heavy-ion program, especially at the new dedicated Brookhaven facility RHIC, is to study transitions to this phase.

Another one, which has been getting much attention recently, is the direction of finite density. Very robust color superconductivity was found to be the case here. Let me also mention one more frontier, which has not yet attracted sufficient attention: namely a transition (or many transitions?) as the number of light flavors N_f grows. The minimal scenario includes a transition from the usual hadronic phase to a more unusual QCD phase, the *conformal* one, in which there are no particle-like excitations and correlators are power-like in the infrared. Even the position of the critical point is unknown. The main driving force of these studies is the intellectual challenge it provides.

The QCD phase diagram as we understand it now is shown in Fig 8, in the baryonic chemical potential μ (normalized per quark, not per baryon) and the temperature T plane. Some part of it is old: it has the hadronic phase at small values of both parameters, and a QGP phase at large T and μ.

The phase transition line separating them most probably does not really start at $T = T_c, \mu = 0$ but at an "endpoint" E, a remnant of the so-called QCD tricritical point, which QCD has in the chiral (all quarks are massless) limit. Although we do not know where it is[8], we hope to find it one day in experiment. The proposed ideas rotate around the fact that the order parameter, the VEV of the sigma meson, is at this point truly massless, and creates a kind of "critical opalescence". Similar phenomena were predicted

[7] It does not mean though, that it is a simple issue to understand even the high-T limit of QCD, related to non-perturbative 3d dynamics.

[8] Its position is very sensitive to the precise value of the strange quark mass m_s.

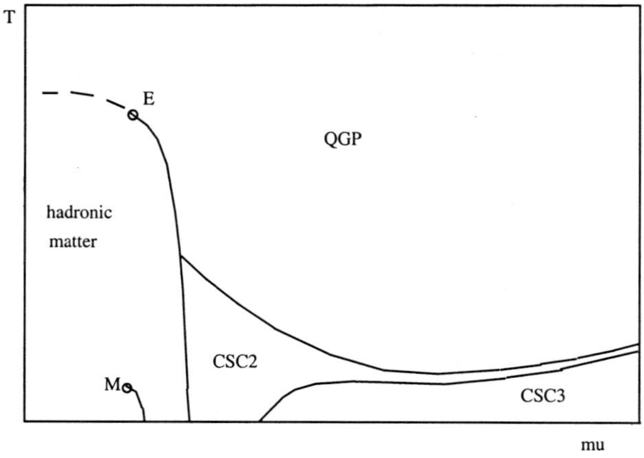

Fig. 8. Schematic phase diagram of QCD, in the plane temperature T and baryon chemical potential μ. E and M show critical endpoints of first-order transitions: M (from multi-fragmentation) is that for a liquid-gas transition in nuclear matter. The color superconducting phases, CSC2 and CSC3, are explained in the text

and then indeed observed at the endpoint of another line (called M from multi-fragmentation), separating liquid nuclear matter from the nuclear gas phase.

The large-density (and low-T) region looks rather different from what was shown at conferences just a year ago: two new color superconducting phases appear there. Unfortunately heavy-ion collisions do not cross this part of the phase diagrams and so it belongs to neutron star physics.

Above I mentioned an approach to high density starting from the vacuum. One can also work in the opposite direction, starting from very large densities and going down. Since the electric part of one-gluon exchange is screened, the Cooper pairs appear due to magnetic forces. It is interesting by itself, as a rare example: one has to take care of *time delay effects* of the interaction. The results are indefinitely growing gaps at large $\mu > 10$ GeV, as $\Delta \sim \mu exp(-\frac{3\pi^2}{\sqrt{2}g(\mu)})$ [35].

4.2 Finite-Temperature Transition and Large Number of Flavors

There is no place here to discuss in detail the rather extensive lattice data available now, and I only mention some results related to instantons. In the vacuum a quasi-random set of instantons leads to chiral-symmetry breaking and quasi-zero modes: but how, in the same terms, does the high-T phase look like?

The simplest solution would be just a *suppression* of instantons at $T > T_c$, and at some early time people thought this is what actually happens. However, it should not be like this because the Debye screening, which is killing them, only appears at $T = T_c$. Lattice QCD works have also found no depletion of the instanton density up to $T = T_c$.

On the other hand, the absence of the condensate and quasi-zero modes implies that the "liquid" is now broken into finite pieces. The simplest of them are pairs, or instanton-anti-instanton molecules. This is precisely what instanton simulations have found [1], see Fig. 9. Whether it is indeed so on the lattice is not yet clear: nice molecules were located, but the evidence for the molecular mechanism of chiral restoration is still far from being convincing. (No alternatives I am aware of have been proposed so far, though.)

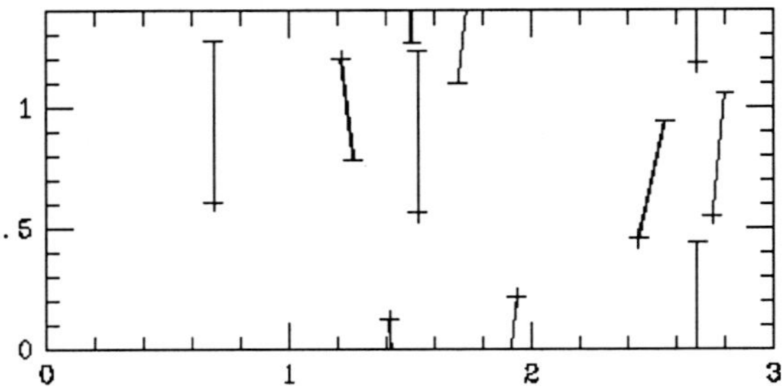

Fig. 9. Typical configuration from an instanton liquid simulation, at $T > T_c$. Lines indicate the direction in which quark propagators are the largest. Clear pairing of instantons and anti-instantons are observed: the pairs tend to have the same spatial position and are separated mostly by Euclidean time

The results of IILM simulations with a variable number of flavors $N_f = 2, 3, 5$[9] flavors with equal masses can be summarized as follows. For $N_f = 2$ there is a second-order phase transition which turns into a line of first-order transitions in the $m - T$ plane for $N_f > 2$. If the system is in the chirally restored phase ($T > T_c$) at $m = 0$, we find a discontinuity in the chiral order parameter if the mass is increased beyond some critical value. Qualitatively, the reason for this behavior is clear. Increasing the temperature raises the role of correlations caused by the fermion determinant, increasing the quark mass has the opposite effect. We also observe that increasing the number of flavors lowers the transition temperature. Again, increasing the number of flavors means that the determinant is raised to a higher power, so fermion induced correlations become stronger. For $N_f = 5$ we find that the transition temperature drops to zero and the instanton liquid has a chirally symmetric ground state, provided the dynamical quark mass is less than some critical value. Studying the instanton ensemble in more detail shows that in this case all instantons are bound into molecules.

[9] The case $N_f = 4$ is omitted because in this case it is very hard to determine whether the phase transition happens at $T > 0$.

Unfortunately, little is known about QCD with a large numbers of flavors from lattice simulations. There are data by the Columbia group for $N_f = 4$. The most important result is that chiral symmetry breaking effects were found to be drastically smaller as compared to $N_f = 0, 2$. In particular, the mass splittings between chiral partners such as $\pi - \sigma$, $\rho - a_1$, $N(\frac{1}{2}^+) - N(\frac{1}{2}^-)$, extrapolated to $m = 0$ were found to be 4-5 times smaller. This agrees well with what was found in the interacting instanton model: more work in this direction is certainly needed.

4.3 High Density and Color Superconductivity

This topic is covered much in detail by the article of M. Alford in this volume, so I only add a few remarks here.

Although the idea of color superconductivity originates from the 70's, the field of high-density QCD was in a dormant state for a long time till two papers [20,21] (posted on the same day) in 1998 have claimed gaps about 100 times larger than previously thought. The field is booming since then, as one can see from about 250 citations those papers got in 2 years.

The, then, Princeton group (Alford-Rajagopal-Wilczek) has been thinking about different pairings from a theoretical perspective, but our (Stony Brook) team (Rapp-Schäfer-Shuryak-Velkovsky) had started from the impressive qq pairing phenomenon found theoretically [18] in the instanton liquid model *inside the nucleon.* As explained above, we have found it to be, roughly speaking, a small drop of CS matter, made of one Cooper pair of a certain sort (the *ud scalar diquark*) and one massive quark[10]. Schäfer heroically attempted numerical simulations of the instanton liquid model at finite μ: although he was not very successful[11], he found out strange "polymers" made of instantons connecting by two quark lines going through. It took us some time to realize that we see paths of condensed diquarks! It was like finding superconductivity by watching electrons moving on the computer screen.

The main point I would like to emphasize here is that the qq pairing of such diquarks has in fact deep dynamical roots: it follows from the same basic dynamics as the "superconductivity" of the QCD vacuum, the chiral symmetry breaking. These spin-isospin-zero diquarks are related to pions, as we will see below.

The most straightforward argument for deeply bound diquarks came from the bi-color ($N_c = 2$) theory: in it the scalar diquark is degenerate with pions. By continuation from $N_c = 2$ to 3 a trace of it should exist in real QCD[12].

[10] This is contrary to Δ (decuplet) baryons, which is a small drop of "normal" quark matter without scalar diquarks.

[11] For the same reason as lattice people cannot do it: the fermionic determinant is not real.

[12] The instanton-induced interaction strength in a diquark channel is $1/(N_c - 1)$ of that for the $\bar{q}\gamma_5 q$ one. It is the same at $N_c = 2$, zero for large N_c, and is exactly in between for $N_c = 3$.

Instantons create the following amusing *triality*: there are three attractive channels which compete: (i) The instanton-induced attraction in the $\bar{q}q$ channel leading to chiral symmetry breaking. (ii) The instanton-induced attraction in qq, which leads to color superconductivity. (iii) The *light-quark-induced* attraction of $\bar{I}I$, which leads to pairing of instantons into "molecules" and a QGP phase without *any* condensates.

At very high density we also can find an *arbitrarily dilute instanton liquid*, as shown recently in [36]. The reason it cannot exist in the vacuum or at high T is that if the instanton density goes below some critical value, there cannot be any condensate. (The system then breaks into instanton molecules or other clusters and chiral symmetry is restored.) However, at high density the superconducting condensate can be created perturbatively as well (we mentioned it above) and there is no problem. The dilute instantons interact by exchanging very light η' (which would be massless without instantons): one can calculate an effective Lagrangian, a theta angle dependence etc.

Bi-color QCD: A Very Special Theory. One reason why it is special (well known to the lattice community): its fermionic determinant is *real* even for non-zero μ, which makes simulations possible. However, the major interest in this theory is related the so-called *Pauli-Gursey symmetry*. We have argued above that pions and diquarks appear at the same one-instanton level, and are, so to say, brothers. In bi-color QCD they become identical twins: due to the additional symmetry mentioned the diquarks are *degenerate* with mesons.

In particular, chiral symmetry breaking is done like this: $SU(2N_f) \to Sp(2N_f)$, and for $N_f = 2$ the coset $K = SU(4)/Sp(4) = SO(6)/SO(5) = S^5$. Those 5 massless modes are pions plus the scalar diquark S and its anti-particle $\bar{S}$.

Vector diquarks are degenerate with vector mesons, etc. Therefore, the scalar-vector splitting is in this case about twice the constituent quark mass, or about 800 MeV. It should be compared to the binding in the "real" $N_c = 3$ QCD of only 200-300 MeV, and to zero binding in the large-N_c limit.

The corresponding sigma model describing this chiral symmetry breaking was worked out in [20]; for the further development, see [24]. As argued in [20], in this theory the critical value of the transition to color superconductivity is simply $\mu = m_\pi/2$, or zero in the chiral limit. The diquark condensate is just a rotated $< \bar{q}q >$ one, and the gap is the constituent quark mass. Recent lattice works [27] display it in great detail, building confidence for other cases.

New Studies Reveal Possible New Crystalline Phases. These phases still have a somewhat debatable status, so I have not indicated them on the phase diagram.

Once again, there were two papers submitted by chance on the same day. The "Stony Brook" team [25] has found that a "chiral crystal" with oscillating $< \bar{q}q(x) >$ (similar to Overhouser spin waves in solid-state physics) can compete with the BCS 2-flavor superconductor at its onset, or $\mu \approx 400$ MeV.

The proper position of this phase is somewhere in between the hadronic phase (with constant $< \bar{q}q >$) and a color superconductor.

The "MIT group"[26] has looked at the oscillating superconducting condensate $< qq(x) >$, following earlier works on the so called LOFF phase in usual superconductors. One has found that it is appearing when the difference between Fermi momenta of different quark flavors become comparable to the gap. The natural place for it on the phase diagram is close to the line at which color superconductivity disappears because the gap goes to zero.

5 High-Energy Collisions of Hadrons and Heavy Ions

5.1 The Little Bang: AGS, SPS and Now the RHIC Era

Let me start with a brief comparison of these two magnificent explosions: the Big Bang versus the Little Bang, as we call heavy-ion collisions.

The expansion law is roughly the Hubble law in both, $v(r) \sim r$, although strongly anisotropic in the Little Bang. The Hubble constant tells us the expansion rate today: similarly the radial flow tells us the final magnitude of the transverse velocity. The acceleration history is not really well measured. For the Big Bang people use the distance of supernovae, we use Ω^- which does not participate at the late stages to learn *what was the velocity earlier.* Both show small dipole (quadrupole or elliptic for Little Bang) components, what has some physics, and who knows maybe we will see higher-harmonics fluctuations later on, like in the universe. As we will discuss below, in both cases the major puzzle is how this large entropy has been actually produced, and why it happened so early.

The major lessons we learned from AGS experiments ($E_{LAB} = 2 - 12$ AGeV) are:

(i) Strangeness enhancement over simple multiple NN collisions appears from very low energies, and heavy-ion collisions quickly approach a nearly ideal chemical equilibrium of strangeness.

(ii) "Flows" of different species, in their radial directed and elliptical form, are in this energy domain driven by collective potentials and absorptions: they are not really flows in a hydrodynamic sense. All of them strongly diminish by the high end of the AGS region, demonstrating the onset of "softness" of the EoS. Probably it is some precursor of the QCD phase transition.

Several important lessons came so far from CERN SPS data:

(i) Much more particle ratios have been measured there. Overall those show a surprisingly good degree of chemical equilibration: the chemical freeze-out parameters are tantalizingly close to the QGP phase boundary.

(ii) Dileptons show that the radiation spectral density is very different in dense matter compared to an ideal hadronic gas. The most intriguing data are the CERES finding of "melting of the ρ", which seem to be transformed into a wide continuum reaching down to invariant masses as low as 400 MeV.

It puts in doubt the "resonance gas" view of hadronic matter at these conditions. Intermediate mass dileptons studied by NA50 can be well described by thermal radiation with QGP rates.
(iii) The impact parameter of J/ψ and ψ' suppression in Pb-Pb collisions studied by the NA50 collaboration shows a rather non-trivial behavior. More studies are needed, including especially measurements of the open-charm yields, to understand the origin and magnitude of the suppression.

However, during the last several months those discussions have been overshadowed by a list of news from RHIC. It had its first run in summer 2000 and first results were reported recently at the Quark Matter 2001 conference [28]. The present status is discussed by M. Gyulassy in this volume.

A brief summary can be as follows. The results so far have shown that heavy-ion collisions (AA) at these energies significantly differ *both* from the pp collisions at high energies and the AA collisions at lower (SPS/AGS) energies. The main features of these data are quite consistent with the QGP (or Little Bang) scenario, in which entropy is produced promptly and the subsequent expansion is close to an adiabatic expansion of an equilibrated hot medium.

Let me mention here two other pictures of the heavy-ion production, discussed prior to the appearance of these data. One is the *string picture*, used in event generators like RQMD and UrQMD: they predicted effectively a very soft EoS and an elliptic flow decreasing with energy. The other one is a *pure minijet scenario*, in which most secondaries would come from independently fragmenting minijets. If so, there are basically no collective phenomena whatsoever.

Already the very first multiplicity measurements reported by the PHOBOS collaboration [53] have shown that the particle production per participating nucleon is no longer constant, as was the case at lower (SPS/AGS) energies. This new component may be due to long-anticipated *pQCD* processes, leading to perturbative production of new partons. Unlike high-p_t processes resulting in visible jets, those must be undetectable *"mini-jets"* with momenta $\sim 1 - 2$ GeV. Production and decay of such *mini-jets* was discussed in Ref. [54], also this scenario is the basis of the widely used event generator HIJING [50]. Its crucial parameter is the *cutoff scale* p_{min}: if fitted from *pp* data to be 1.5-2 GeV, it leads to the predicted mini-jet multiplicity $dN_g/dy \sim 200$ for central Au-Au collisions at $\sqrt(s) = 130\,\mathrm{AGeV}$. If those fragment independently into hadrons, and are supplemented by a "soft" string-decay component, the predicted total multiplicity was found to be in good agreement with the first RHIC multiplicity data. Because partons interact perturbatively, with their scattering and radiation being strongly peaked at small angles, their equilibration is expected to be relatively long [55]. However, the new set of RHIC data reported in [28] have provided serious arguments *against* the mini-jet scenario, and point toward a quite rapid entropy production rate and early QGP formation.

(i) If most of the mini-jets fragment independently, there are no *collective phenomena* such as a transverse flow related with the QGP pressure. However, it was found that those effects are very strong at RHIC. Furthermore, the STAR collaboration has observed a very robust *elliptic flow* [38], which is in perfect agreement with predictions of a hydrodynamic model [45,44] assuming an equilibrated QGP with its full pressure $p \approx \epsilon/3$ above the QCD phase transition. This agreement persists to rather peripheral collisions, in which the overlap almond-shaped region of two nuclei is only a couple of fm thick. STAR and PHENIX data on spectra of identified particles, especially $p, \bar{p}$, indicate a spectacular radial expansion, also in agreement with hydrodynamic calculations [45,44].

(ii) Spectra of hadrons at large p_t, especially the π^0 spectra, agree well with HIJING for peripheral collisions, but show much smaller yields for central ones, with rather different (exponential-shaped) spectra. It means the long-anticipated *"jet quenching"* at large p_t is seen for the first time, with a surprisingly large suppression factor $\sim 1/5$. Keeping in mind that jets originating from the surface outward cannot be quenched, the effect seems to be as large as it can possibly be. For that to happen, the outgoing high-p_t jets should propagate through matter with a parton population larger than the above-mentioned minijet density predicted by HIJING.

(iii) The curious interplay between collective and jet effects has also been studied by the STAR collaboration, in form of the elliptic asymmetry parameter $v_2(p_t)$. At large transverse momenta $p_t > 2$ GeV the data depart from hydrodynamic predictions and level off. When compared to predictions of jet quenching models worked out in [56], they also indicate a gluon multiplicity several times larger than the HIJING prediction, and are even consistent with its maximal possible value evaluated from the final entropy at freeze-out, $(dN/dy)_\pi \sim 1000$.

5.2 Collective Flows and EoS

If we indeed have produced excited matter (rather than just a bunch of partons which fly away and fragment independently), we expect to see certain collective phenomena. Ideally, those should be quantitatively reproduced by relativistic hydrodynamics, which is basically just local energy-momentum conservation plus the EoS we know from the lattice and from models.

The role of the QCD phase transition in matter expansion is significant. QCD lattice simulations [42] show approximately a first-order transition. Over a wide range of energy densities $e = 0.5 - 1.4$ GeV/fm^3, where the temperature T and the pressure p are nearly constant. So the ratio of pressure to energy density, p/e, decreases till a minimum at a particular energy density $e_{sp} \approx 1.4$ GeV/fm^3, known as the *softest point* [43]. Near e_{sp} a small pressure gradient cannot effectively accelerate the matter and the evolution stagnates. However, when the initial energy density is well above the QCD phase transition region, $p/e \approx 1/3$, and this pressure drives the collective motion. The energy densities reached at time ~ 1 fm/c at SPS($\sqrt{s}_{NN} = 17$ GeV) and RHIC ($\sqrt{s}_{NN} = 130$ GeV) are about 4 and 8 GeV/fm^3, respectively. We found that at RHIC conditions we are in the latter regime, and matter accelerates to $v \sim 0.2c$ *before* entering the soft domain. Therefore by freeze-out this motion changes the spatial distribution of matter dramatically: e.g., as shown in [37], the initial almond-shape distribution 10 fm/c later looks like two separated shells, with a little "nut" in between.

The simplest way to see a hydrodynamic expansion is in the spectra of particles: on top of chaotic thermal distributions $\sim exp(-m_t/T), m_t^2 = p_t^2 + m^2$ one expects to see an additional broadening due to a hydrodynamic outward motion. This effect is especially large if particles are heavy, since a flow with velocity v adds a momentum mv.

Derek Teaney [45] has developed a comprehensive Hydro-to-Hadrons (H2H) model which combines the hydrodynamic description of the initial QGP/mixed phase ($e > 0.5$ GeV/fm^3) stages, where hadrons are not appropriate degrees of freedom, with a hadronic cascade RQMD for the hadronic stage. In this way, we can include different EoS displaying properties of the phase transition, and also incorporate complicated final-state interactions at freeze-out. The set of EoS used is shown in Fig. 10.

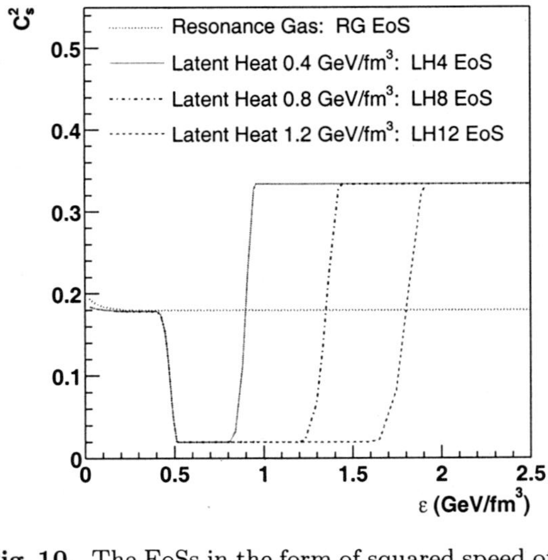

Fig. 10. The EoSs in the form of squared speed of sound $C_s^2 = dp/d\epsilon$ with variable latent heats 0.4 GeV/fm^3, 0.8 GeV/fm^3,.. labeled as LH4, LH8,..versus the energy density

The radial flow is usually characterized by the slope parameter T: Each particle spectrum is fitted to the form $dN/dp_t^2dy \sim exp(-m_t/T), m_t^2 = p_t^2 + m^2$. Although we denoted the slope by T, it is *not* the temperature: it incorporates random thermal motion and collective transverse velocity. The SPS NA49 slope parameters for pions and protons are shown in Fig. 11(a). The parameter T grows with the particle multiplicity due to an increased velocity of the radial flow. Furthermore, the rate of growth depends on the EoS: the softer it is, the less growth. The SPS NA49 data (corresponding to two data points of our fits to the spectra) favor the (relatively stiff) LH8 EoS. (Details of the fit, a discussion of the b-dependence etc. can be found in [45].) It is very important to get these parameters for RHIC, especially for heavy secondaries like nucleons and hyperons.

For *non-central* collisions the overlap region in the transverse plane has an elliptic, "almond", shape, and a larger pressure gradient forces matter to expand preferentially in the direction of the impact parameter [40]. Compared to the radial flow, the elliptic flow is formed earlier, and therefore it measures the early pressure. The *elliptic flow* is quantified experimentally by measuring the azimuthal distributions of the produced particles and calculating the elliptic flow parameter $V_2 = \langle cos(2\phi)\rangle$, where the angle ϕ is measured with respect to the impact parameter direction around the beam axis. It appears due to the elliptic *spatial* deformation of the overlap region in the nucleus-nucleus collision, quantified by its eccentricity $\epsilon_2 =< y^2 - x^2 > / < x^2 + y^2 >$, usually calculated in the Glauber model. Since the effect (v_2) is proportional to the cause (ϵ_2), the ratio v_2/ϵ_2 does not have a strong dependence on the impact parameters b, and this ratio is often used for comparison. (We will not do that below, in the detailed comparison to data, because $\epsilon_2(b)$ is not directly measured.)

In Fig. 11(b) the elliptic flow of the system is plotted as a function of the charged particle multiplicity at an impact parameter of 6 fm. Before discussing the energy dependence, let us quantify the magnitude of the elliptic flow at the SPS. Ideal relativistic hydrodynamics used in earlier works [40,44] generally over-predicts the elliptic flow by about a factor 2. An example of such kind is indicated by a star in Fig. 11(b): it is our hydro result (with the LH8 EoS) which has been followed hydrodynamically till very late stages, the freeze-out temperature is $T_f = 120$ MeV. By switching to the hadronic cascade at late stages, we have a more appropriate treatment of the resonance decays and re-scattering rate, and so one can see that it significantly reduces V_2, to the range much closer to the data points.

One might think that one can also do that by simply taking a *softer* EoS, e.g., by increasing the latent heat. However, it only happens till LH16 and then V_2 starts to even slightly increase again. The explanation of this non-monotonous behavior is the interplay of the initial "QGP push" for a stiffer EoS, with a longer time for the hadronic stage available for a softer EoS. We cannot show here details, but it turns out that a given (experimental)

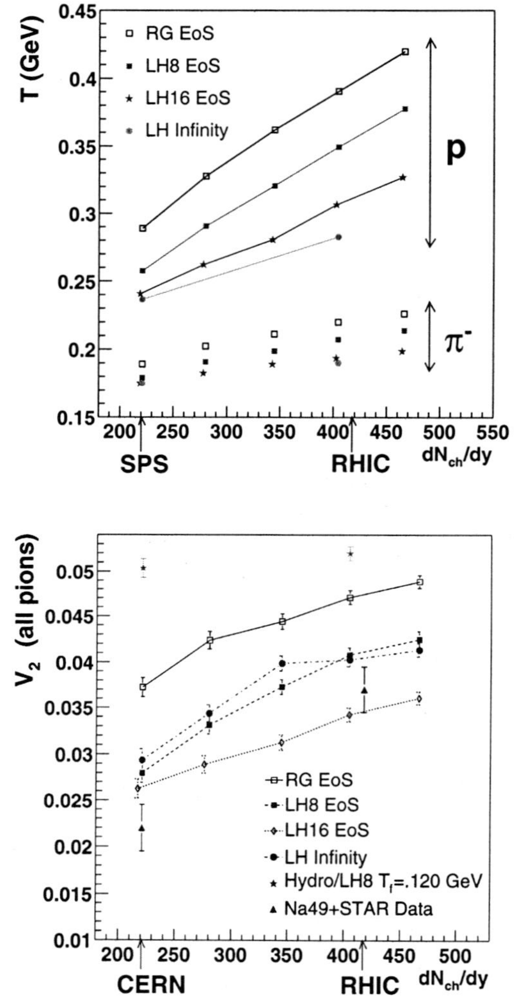

Fig. 11. The transverse mass slope T (a) and elliptic flow parameter V_2 (b) versus the midrapidity (y=0) charged particle multiplicity, for AuAu collisions with b=6

V_2 value can correspond to *two different solutions*, one with an earlier push and another with the later expansion dominating. Coincidentally, the STAR data points happen to be right at the onset of such a bifurcation, close to LH16. The *multiplicity dependence* of V_2 appears simple from Figure 11(b): all curves show growth with about the same rate. Note, however, that such a growth of V_2 from SPS to RHIC (first predicted in [48] where our first preliminary results have been shown) is not shared by most other models. In particular, *string-based* models like UrQMD predict a decrease by a factor of ≈ 2 [49]. It happens because strings produce no transverse pressure and so the effective EoS is super-soft at high energies. Models based on *independent parton scattering and decay* (such as HIJING) also predict a basically vanishing (or slightly negative) [50] V_2.

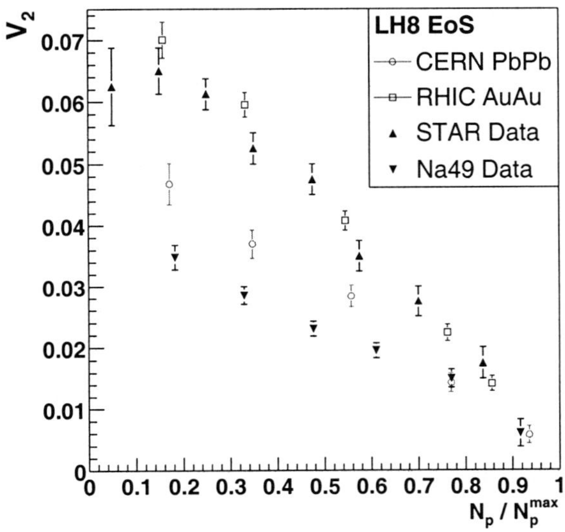

Fig. 12. V_2 versus impact parameter b, described experimentally by the number of participant nucleons, for the RHIC STAR and SPS NA49 experiments. Both are compared to our results for the LH8 EoS

In Fig. 12 we show how our results compare with data as a function of the impact parameter. One can see that the agreement becomes much better at RHIC. Furthermore, one may notice that the deviation from a linear dependence, which we predict, becomes visible at SPS for more peripheral collisions with $N_p/N_p^{max} < 0.6$ or so, while at RHIC only the most peripheral point, with $N_p/N_p^{max} = 0.05$ shows such a deviation. This clearly indicates that the hydrodynamic regime in general works much better at RHIC.

In summary, the flow phenomena observed at RHIC are stronger than at SPS. It is in complete agreement with the QGP scenario. All data on elliptic and radial flow can be nicely reproduced by the H2H model. Furthermore, we are able to restrict the EoS to those with the latent heat about 0.8 GeV/fm^3.

5.3 How QGP Happened to Be Produced/Equilibrated So Early?

One possible solution to the puzzle outlined above can be a *significantly lower cutoff scale* in AA collisions, as compared to $p_{min} = 1.5 - 2$ GeV fitted from the *pp* data. That increases perturbative cross sections, both due to a smaller momentum transfer and a larger coupling constant. As I argued over the years, the QGP is a new phase of QCD which is *qualitatively different* from the QCD vacuum: therefore the cut-off of pQCD may have entirely different values and be determined by different phenomena. Furthermore, since QGP is a plasma-like phase, which screens itself perturbatively [39], one may think of a cut-off to be determined *self-consistently* from a resummation of perturbative effects. These ideas known as *self-screening* or *initial-state saturation* were discussed in Refs. [55]. Although the scale in question grows with temperature or density, *just above* T_c it may actually be *smaller* than the value 1.5-2 GeV we observe in the vacuum. Its first experimental manifestation may be a dropping of the so-called "duality scale" in the observed dilepton spectrum, see the discussion in [59].

Another alternative to explain the large gluon population at RHIC would be an existence of more rapid multi-gluon production processes. Let us consider an alternative $non-perturbative$ scenario based entirely on non-perturbative processes involving *instantons* and *sphalerons* [58]. But before we do that, we have to take a look at hadronic collisions and briefly review a few recent papers on the subject.

At $s > 10^3$ GeV^2 hadronic cross sections as $\bar{p}p, pp, \pi p, Kp$, γN, and even $\gamma\gamma$ grow slowly with the collision energy s. This behavior can be parameterized well by a *soft Pomeron* phenomenology, but we will only use its logarithmically growing part

$$\sigma_{hh'}(s) = \sigma_{hh'}(s_0) + log(s/s_0) X_{hh'} \Delta + ... \tag{38}$$

ignoring both the higher powers of $log(s)$ and decreasing Regge terms. We will use those two parameters from the PDG-2000 recent fits, the intercept and its coefficient in $pp, \bar{p}p$ collisions, $\Delta = \alpha(0) - 1 = 0.093(2)$, $X_{NN} = 18.951(27)$ mb. Note a qualitative difference between constant and logarithmically growing parts of the cross section. The former can be explained by prompt color *exchanges*, as suggested by Low and Nussinov long ago. It nicely correlates with a flux tube picture of the final state. The *growing* part of the cross section cannot be generated by t-channel color exchanges and is associated with processes promptly producing some objects, with $log(s)$ coming from the longitudinal phase space. In pQCD it is *gluon* production, by processes like the one shown in Fig. 13(a). If iterated in the t-channel in ladder-type fashion, the result is approximately a BFKL pole [60]. Although the predicted power is much larger than Δ mentioned before, it seems to be consistent with much stronger growth seen in hard processes at HERA: thus it is therefore sometimes called the "hard pomeron". The physical origin of the cross section growth remains an outstanding open problem: neither the perturbative resummations nor many non-perturbative models are really quantitative. It is hardly surprising, since the scale at which a soft Pomeron operates (as seen, e.g., from the Pomeron slope $\alpha'(0) \approx 1/(2\ \text{GeV})^2$) is also the "substructure scale" mentioned above.

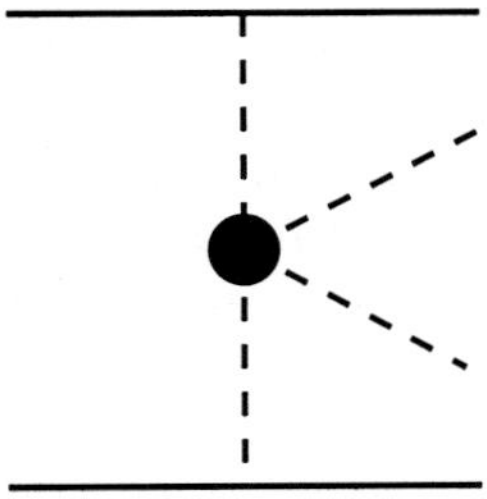

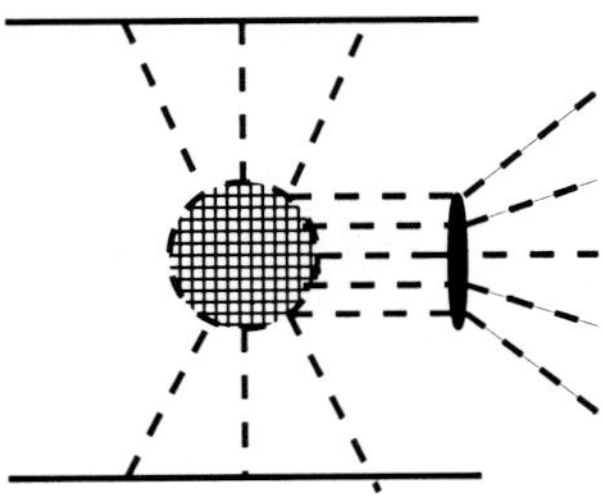

Fig. 13. (a) A typical inelastic perturbative process: two t-channel gluons collide, producing a pair of gluons; (b) An instanton-induced inelastic process incorporates collisions of multiple t-channel gluons with the instanton (the shaded circle), resulting in a multi-gluon production. The intermediate stage of the process, indicated by the horizontal dashed lines, corresponds to a time when outgoing glue is in the form of a coherent field configuration - the *sphaleron*. Since this part of the process corresponds to a motion above the barrier, it does not enter the calculation of the cross section, but is only needed for the prediction of the inclusive spectra, multiplicities etc.

A recent application of the instanton-induced dynamics to this problem has been discussed in several papers [61]. Especially relevant for this topic are two last works, which use insights obtained a decade ago in the discussion of instanton-induced processes in electroweak theory [34], and the growing parts of the hh cross sections were ascribed to multi-gluon production via instantons, see Fig. 13(b). Among qualitative features of this theory an explanation is given why no odderon appears (instantons are SU(2) objects, in which quarks and antiquarks are not really distinct); furthermore an explanation of the small power Δ (it is proportional to the "instanton diluteness parameter" $n\rho^4$ as mentioned above) and the small size of the soft Pomeron (governed simply by the small size of instantons $\rho \sim 1/3$ fm) is offered. Although instanton-induced amplitudes contain a small "diluteness" factor, there is no extra penalty for the production of new gluons: thus one should expect instanton effects to beat perturbative amplitudes of sufficiently high order. This generic idea is also behind the present work, dealing with prompt multi-gluon production.

A technical description of the process can be split into two stages. The first one (at which one evaluates the probability) is the motion *under the barrier*, and it is described by Euclidean paths approximated by instantons. Their interaction with the high-energy colliding partons results in some energy deposition and subsequent motion *over the barrier*. At this second stage the action is real, and the factor $exp(iS)$ does not affect the probability, and we only need to consider it for final-state distributions. The relevant Minkowski paths start with configurations close to the QCD analogues of electroweak *sphalerons* [62], static spherically symmetric clusters of the gluomagnetic field which satisfy the Yang-Mills equations. (Those can be obtained from known electroweak solutions in the limit of infinitely large Higgs self-coupling.) Their mass in QCD is

$$M_{sph} \approx \frac{30}{g^2(\rho)\rho} \sim 2.5\,\text{GeV}. \tag{39}$$

Since those field configurations are close to a classically unstable saddle point at the top of the barrier, they roll downhill and develop gluoelectric fields. When both become weak enough, the solution can be decomposed into perturbative gluons. This part of the process can also be studied directly from the classical Yang-Mills equation. For electroweak sphalerons it has been done in Refs. [63], a calculation for its QCD version is in progress [64]. While rolling, the configurations tend to forget the initial imperfections (such as a non-spherical shape) since there is only one basic instability path downward: so the resulting fields should be nearly perfect spherical expanding shells. Electroweak sphalerons decay into approximately 51 W, Z, H quanta, of which only about 10% are Higgses, which carry only 4% of the energy. Ignoring those, one can estimate a mean gluon multiplicity per sphaleron decay, by a simple re-scaling of the coupling constants: the result gives 3-4 gluons. Although this number is not large, it is important to keep in mind that they appear as a coherent expanding shell of a strong gluonic field.

It has been suggested in [58] that if sphaleron-type objects are copiously produced, with or instead of $p \sim 1$ GeV minijets, they may significantly increase the entropy produced and speed up the equilibration process.

Acknowledgments. It is a pleasure to thank the organizers of the School for their invitation: it was indeed a very interesting and successful one. This work was supported in parts by the US-DOE grant DE-FG-88ER40388.

References

1. T. Schäfer, E.V. Shuryak: Rev. Mod. Phys. **70**, 323 (1998); hep-ph/9610451
2. D. Diakonov: 'Chiral Symmetry Breaking by Instantons'. In: *Varenna 1995, *Selected topics in nonperturbative QCD** 397-432; hep-ph/9602375

3. E.V. Shuryak: Rev. Mod. Phys. **65**, 1 (1993)
4. M.A. Shifman, A.I. Vainshtein, V.I. Zakharov: Phys. Lett. **76B**, 471 (1978)
5. E.V. Shuryak: Nucl. Phys. **B203**, 93,116,140 (1982)
6. V.A. Novikov, M.A. Shifman, A.I. Vainshtein, V.I. Zakharov: Nucl. Phys. **B191**, 301 (1981)
7. A.A Belavin, A.M. Polyakov, A.S. Schvartz, Yu.S. Tyupkin: Phys. Lett. **B59**, 85-87 (1975)
8. S.L. Adler: Phys. Rev. **177**, 2426 (1969)
9. J.S. Bell, R. Jackiw: Nuovo Cim. **A60**, 47 (1969)
10. A. Hasenfratz, C. Nieter: hep-lat/9806026
11. G. 't Hooft': Phys. Rev. **D14**, 3432 (1976)
12. Y. Nambu, G. Jona-Lasinio: Phys. Rev. **122**, 345 (1961)
13. D. Diakonov, V.Yu. Petrov: Nucl. Phys. **B272**, 457 (1986)
14. M.A. Shifman, A.I. Vainshtein, V.I. Zakharov: Nucl. Phys. **B165**, 45 (1980)
15. A. Blotz, E.V. Shuryak: Phys. Rev. **D55**, 4055-4065 (1997)
16. R. Barate *et al.* [ALEPH Collaboration]: Z. Phys. **C76**, 15 (1997)
17. R. Barate *et al.* [ALEPH Collaboration]: Eur. Phys. J. **C4**, 409 (1998)
18. T. Schäfer, E.V. Shuryak., J. Verbaarschot: Nucl. Phys. **B412**, 143 (1994)
19. M. Anselmino et al.: Rev. Mod. Phys. **65**, 1199 (1993)
20. R. Rapp, T. Schäfer, E.V. Shuryak, M. Velkovsky: Phys. Rev. Lett. **81**, 53 (1998)
21. M. Alford, K. Rajagopal, F. Wilczek: Phys. Lett. **B422**, 247 (1998)
22. M. Alford, K. Rajagopal, F. Wilczek: Nucl.Phys. **B537**, 443 (1999) [hep-ph/9804403]
23. R. Rapp, T. Schäfer, E.V. Shuryak, M. Velkovsky: Ann. Phys. **280**, 35 (2000) [hep-ph 9904353]
24. J.B. Kogut, M.A. Stephanov, D. Toublan: Phys.Lett. **B464**, 183 (1999) [hep-ph/9906346]
25. R. Rapp, E.V. Shuryak, I. Zahed: Phys. Rev. **D63**, 034008 (2001) [hep-ph/0008207]
26. M. Alford, J. Bowers, K. Rajagopal: Phys. Rev. **D63**, 074016 (2001) [hep-ph/0008208]
27. S. Hands, I. Montvay, M. Oevers, L. Scorzato, J. Skullerud: Nucl. Phys. Proc. Suppl. **94**, 461 (2001) [hep-lat/0010085]
28. Proceedings of *Quark Matter 2001*, Nucl. Phys. A, in press
29. E.V. Shuryak, I. Zahed: hep-ph/0005152; Phys. Rev. D, in press
30. M. Nowak, E.V. Shuryak, I. Zahed: in progress
31. D. Kharzeev, E. Levin: BNL-NT-99-8; hep-ph/991221
32. E. Shuryak: hep-ph 0001189
33. D. Kharzeev, Y. Kovchegov, E. Levin: hep-ph/0007182
34. A. Ringwald: Nucl. Phys. **B330**, 1 (1990); O. Espinosa: Nucl.Phys. **B343**, 310 (1990); V.V. Khoze, A. Ringwald: Phys. Lett. **B259**, 106-112 (1991); V.I. Zakharov: Nucl. Phys. **B353**, 683 (1991); M. Maggiore, M. Shifman: Phys.Rev. **D46**, 3550-3564 (1992)
35. D.T. Son: Phys. Rev. **D59**, 094019 (1999) [hep-ph/9812287]
36. D.T. Son, M.A. Stephanov, A.R. Zhitnitsky: hep-ph/0103099
37. D. Teaney, E.V. Shuryak: Phys. Rev. Lett. **83**, 4951 (1999) [nucl-th/9904006]
38. K.H. Ackermann et al.: [STAR Collaboration] nucl-ex/0009011
39. E. Shuryak: Phys. Rep. **61**, 71 (1980); Phys. Lett. **78B**, 150 (1978); Sov. J. Nucl. Phys. **28**, 408 (1978)

40. J.Y. Ollitrault: Phys. Rev. **D46**, 229 (1992); Phys. Rev. **D48**, 1132 (1993)
41. H. Sorge: Phys. Rev. Lett. **78**, 2309 (1997)
42. F. Karsch, E. Laermann, A. Peikert: Phys. Lett. **B478**, 447 (2000) [hep-lat/0002003]
43. C.M. Hung, E. Shuryak: Phys. Rev. **C57**, 1891 (1998)
44. P.F. Kolb, J. Sollfrank, U. Heinz: hep-ph/0006129
45. D. Teaney, J. Lauret, E.V. Shuryak: in progress
46. S. Bass, A. Dumitru et al.: nucl-th/9902062
47. H. Sorge: Phys. Rev. **C52**, 3291(1995)
48. E.V. Shuryak: *Invited talk at 14th International Conference on Ultrarelativistic Nucleus-Nucleus Collisions (QM 99)*, Torino, Italy, 10-15 May 1999, Nucl.Phys. **A661**, 119 (1999) [hep-ph/9906443]
49. M. Bleicher, H. Stocker: hep-ph/0006147
50. X.-N. Wang, M. Gyulassy: Phys. Rev. **D44**, 3501 (1991)
51. H. Heiselberg, A. Levy: Phys. Rev. **C59**, 2716 (1999) [nucl-th/9812034]
52. D. Kharzeev, M. Nardi: nucl-th/0012025
53. B.B. Back et al. [PHOBOS Collaboration]: Phys. Rev. Lett. **85**, 3100 (2000) [hep-ex/0007036]
54. J.P. Blaizot, A.H. Mueller: Nucl. Phys. **B289**, 847 (1987); K. Kajantie, P.V. Landshoff, J. Lindfors: Phys. Rev. Lett. **59**, 2527 (1987)
55. T.S. Biro, E. van Doorn, B. Muller, M.H. Thoma, X.N. Wang: Phys. Rev. **C48**, 1275 (1993) [nucl-th/9303004]
    L. Xiong, E. Shuryak, Phys. Rev. **C49**, 2203 (1994) [hep-ph/9309333]
    R. Baier, A.H. Mueller, D. Schiff, D.T. Son: hep-ph/0009237
56. M. Gyulassy, I. Vitev , X.N. Wang: Phys. Rev. Lett. **86**, 2537 (2001) [nucl-th/0012092]
57. K.H. Ackermann et al. [STAR Collaboration]: nucl-ex/0009011
58. E.V. Shuryak: hep-ph/0101269
59. R. Rapp, J. Wambach: nucl-th/0001014
60. E. Kuraev, L. Lipatov, V. Fadin: Sov. Phys. JETP **45**, 199 (1977); I. Balitsky, L. Lipatov: Sov. J. Nucl. Phys. **28**, 822 (1978); L. Lipatov, Sov. Phys. JETP **63**, 904 (1986)
61. D. Kharzeev, Y. Kovchegov, E. Levin: hep-ph/0007182; E. Shuryak, I. Zahed: Phys. Rev. **D62**, 085014 (2000); [hep-ph/0005152]; M. A. Nowak, E. V. Shuryak, I. Zahed: hep-ph/0012232; Phys. Rev. D, in press
62. N. Manton: Phys. Rev. **D28**, 2019 (1983); F.R. Klinkhamer, N. Manton: Phys. Rev. **D30**, 2212 (1984)
63. J. Zadrozny: Phys. Lett. **B284**, 88 (1992)
    M. Hellmund, J. Kripfganz: Nucl. Phys. **B373**, 749 (1992)
64. G.W. Carter, E.V. Shuryak: Sphaleron decays, in progress

The Color Glass Condensate and Small-x Physics

Larry McLerran

Nuclear Theory Group, Brookhaven National Laboratory
Upton, NY 11793, USA

Summary. The Color Glass Condensate is a state of high-density gluonic matter which controls the high-energy limit of hadronic matter. The article begins with a discussion of general problems of high-energy strong interactions. The infinite-momentum-frame description of a single hadron at very small x is developed, and this picture is applied to the description of ultrarelativistic nuclear collisions. Recent developments in the renormalization group description of the Color Glass Condensate are reviewed.

1 General Considerations

1.1 Introduction

QCD is the correct theory of hadronic physics. It has been tested in various experiments. For high-energy short-distance phenomena, perturbative QCD computations successfully confront experiment. In lattice Monte-Carlo computations, one gets a successful semi-quantitative description of hadronic spectra, and perhaps in the not too distant future one will obtain precise quantitative agreement.

At present, however, all analytic computations and all precise QCD tests are limited to a small class of problems which correspond to short-distance physics, or to semi-quantitative comparisons with the results of lattice gauge theory numerical computations. For the short-distance phenomena, there is some characteristic energy transfer scale E, and one uses asymptotic freedom,

$$\alpha_S(E) \rightarrow 0 \tag{1}$$

as $E \rightarrow \infty$. For example, in Fig. 1, two hadrons collide to make a pair of jets. If the transverse momenta of the jets is large, the strong coupling strength which controls this production is evaluated at the p_T of the jet. If $p_T >> \Lambda_{QCD}$, then the coupling is weak and this process can be computed in perturbation theory. QCD has also been extensively tested in deep inelastic scattering. In Fig. 2, an electron exchanges a virtual photon with a hadronic target. If the virtual photon momentum transfer Q is large, then one can use weak coupling methods.

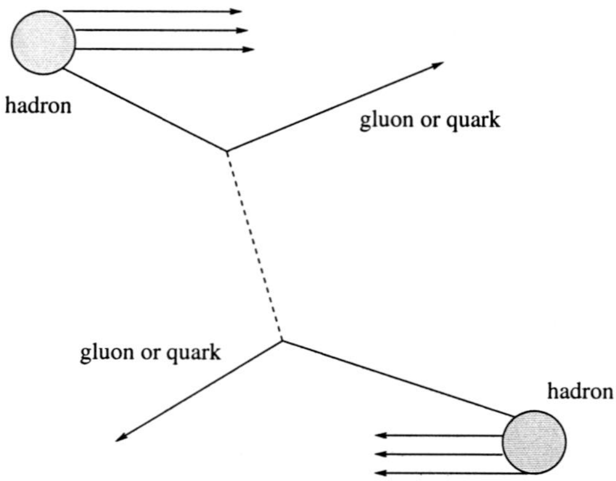

Fig. 1. Hadron-hadron scattering to produce a pair of jets

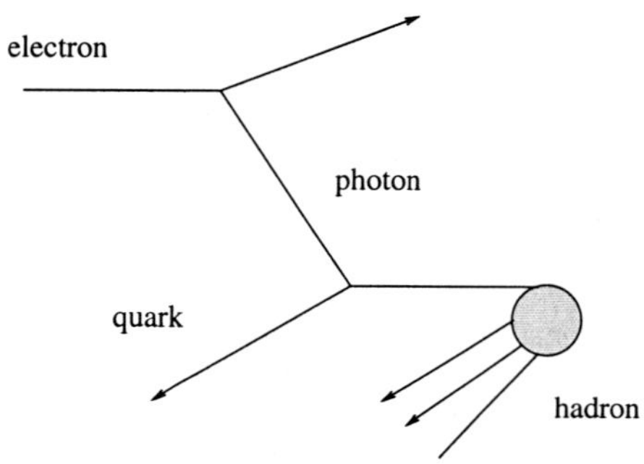

Fig. 2. Deep inelastic scattering of an electron on a hadron

One question which we might ask is whether there are non-perturbative "simple phenomena" which arise from QCD which are worthy of further effort. The questions I would ask before I would become interested in understanding such phenomena are

- Is the phenomenon simple in structure?
- Is the phenomenon pervasive?
- Is it reasonably plausible that one can understand the phenomenon from first principles, and compute how it would appear in nature?

I will argue that *gross* or *typical* processes in QCD, which by their very nature are pervasive, appear to follow simple patterns. The main content of this first chapter is to show some of these processes, and pose some simple questions about their nature which we do not yet understand.

My goal is to convince you that much of these average phenomena of strong interactions at extremely high energies is controlled by a new form of hadronic matter, a dense condensate of gluons. This is called the Color Glass Condensate since

- Color: The gluons are colored.
- Glass: We shall see that the fields associated with the glass evolve very slowly relative to natural time scales, and are disordered. This is like a glass which is disordered and is a liquid on long time scales but seems to be a solid on short time scales.
- Condensate: There is a very high density of massless gluons. These gluons can be packed until their phase space density is so high that interactions prevent more gluon occupation. This forces at increasingly high density the gluons to occupy higher momenta, and the coupling becomes weak. The density saturates at $dN/d^2p_Td^2r_T \sim 1/\alpha_s >> 1$, and is a condensate.

In this article, I will try to explain why the above is very plausible.

1.2 Total Cross-Sections at Asymptotic Energy

Computing total cross sections as $E \to \infty$ is one of the great unsolved problems of QCD. Unlike for processes which are computed in perturbation theory, it is not required that any energy transfer becomes large as the total collision energy $E \to \infty$. Computing a total cross section for hadronic scattering therefore appears to be intrinsically non-perturbative. In the 60's and early 70's, Regge theory was extensively developed in an attempt to understand the total cross section. The results of this analysis were to my mind inconclusive, and certainly can not be claimed to be a first-principles understanding from QCD.

The total cross section for pp and $\overline{p}p$ collisions is shown in Fig. 3. Typically, it is assumed that the total cross section grows as ln^2E as $E \to \infty$. This is the so called Froissart bound which corresponds to the maximal growth allowed by unitarity of the S matrix. Is this correct? Is the coefficient of ln^2E universal for all hadronic precesses? Why is the unitarity limit saturated? Can we understand the total cross section from first principles in QCD? Is it understandable in weakly coupled QCD, or is it an intrinsically non-perturbative phenomenon?

1.3 How Are Particles Produced in High-Energy Collisions?

In Fig. 4, I plot the multiplicity of produced particles in pp and in $\overline{p}p$ collisions. The last six points correspond to the $\overline{p}p$ collisions. The three upper points are the multiplicity in $p\overline{p}$ collisions, and the bottom three have the mutliplicity at zero energy subtracted. The remaining points correspond to pp. Notice that the pp points and those for $p\overline{p}$ with zero energy multiplicity subtracted fall on the same curve. The implication is that whatever is causing the increase in multiplicity in these collisions may be from the same mechanism. Can we compute $N(E)$, the total multiplicity of produced particles as a function of energy?

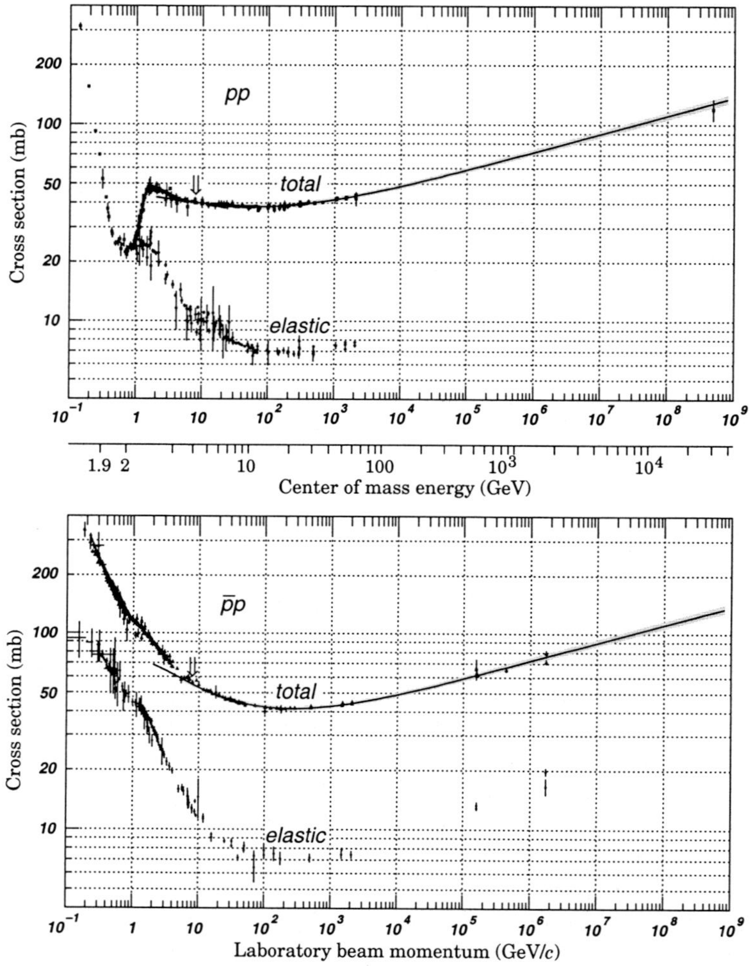

Fig. 3. The cross sections for pp and $p\overline{p}$ scattering

1.4 Some Useful Variables

At this point it is useful to develop some mathematical tools. I will introduce kinematic variables: light cone coordinates. Let the light cone longitudinal momenta be

$$p^{\pm} = \frac{1}{\sqrt{2}}(E \pm p_z). \tag{2}$$

Note that the invariant dot product is given by

$$p \cdot q = p_t \cdot q_t - p^+ q^- - p^- q^+ \tag{3}$$

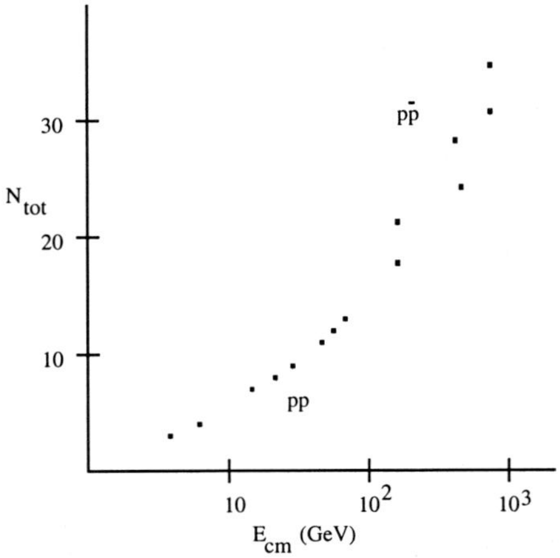

Fig. 4. The total multiplicity in pp and $p\overline{p}$ collisions

and that

$$p^+p^- = \frac{1}{2}(E^2 - p_z^2) = \frac{1}{2}(p_T^2 + m^2) = \frac{1}{2}m_T^2. \tag{4}$$

This equation defines the transverse mass m_T. (Please note that my metric is the negative of that conventionally used in particle physics.)

Consider a collision in the center of mass frame as shown in Fig. 5. In this figure, we have assumed that the colliding particles are large compared to the size of the produced particles. This is true for nuclei, or if the typical transverse momenta of the produced particles are large compared to Λ_{QCD}, since the corresponding size will be much smaller than a fm. We have also assumed that the colliding particles have an energy which is large enough so that they pass through one another and produce mesons in their wake. This is known to happen experimentally: the particles which carry the quantum numbers of the colliding particles typically lose only some finite fraction of their momenta in the collision.

The right moving particle which initiates the collision shown in Fig. 5 has $p_1^+ \sim \sqrt{2} \mid p_z \mid$ and $p_1^- \sim \frac{1}{2\sqrt{2}} m_T^2 / \mid p_z \mid$. For the colliding particles $m_T = m_{projectile}$ (because the transverse momentum is zero, the transverse mass equals the particle mass). For particle 2, we have $p_2^+ = p_1^-$ and $p_2^- = p_1^+$.

If we define the Feynman x of a produced pion as

$$x = p_\pi^+ / p_1^+, \tag{5}$$

then $0 \leq x \leq 1$. (This definition agrees with Feynman's original one if the energy of a particle in the center of mass frame is large and the momentum is positive. We will use this definition as a generalization of the original one of Feynman since it is invariant under longitudinal Lorentz boosts.) The rapidity

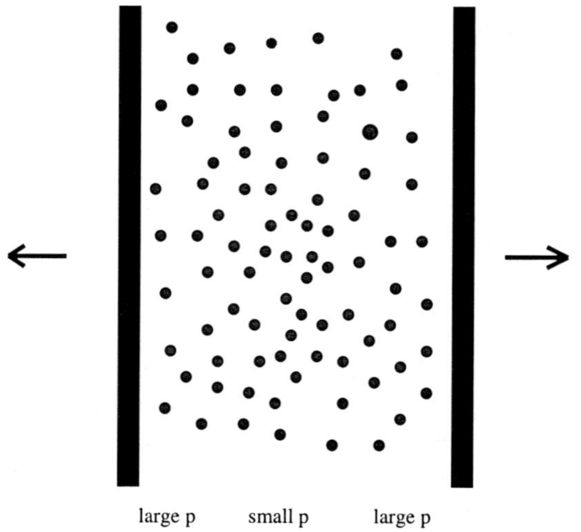

Fig. 5. A hadron-hadron collision. The produced particles are shown as circles

of a pion is defined to be

$$y = \frac{1}{2} ln(p_\pi^+/p_\pi^-) = \frac{1}{2} ln(2p^{+2}/m_T^2). \tag{6}$$

For pions, the transverse mass includes the transverse momentum of the pion.

The pion rapidity is always in the range $-y_{CM} \leq y \leq y_{CM}$ where $y_{CM} = ln(p^+/m_{projectile})$. All the pions are produced in a distribution of rapidities within this range.

A distribution of produced particles in a hadronic collision is shown in Fig. 6. The leading particles are clustered around the projectile and target rapidities. For example, in a heavy ion collision, this is where the nucleons would be. The higher curve shows the distribution of produced mesons.

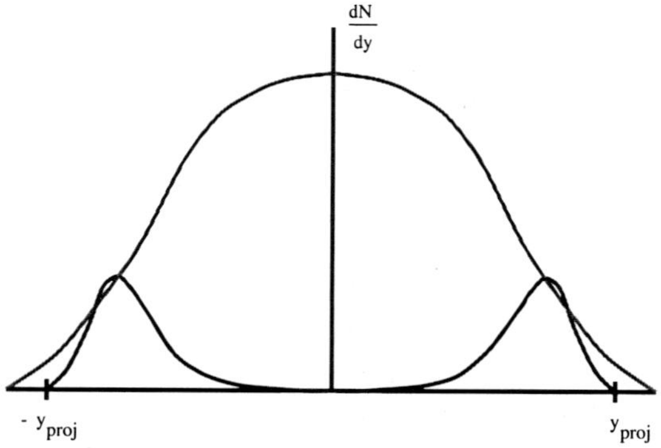

Fig. 6. The rapidity distribution of particles produced in a hadronic collision

These definitions are useful, among other reasons, because of their simple properties under longitudinal Lorentz boosts: $p^\pm \to \kappa^{\pm 1} p^\pm$ where κ is a constant. Under boosts, the rapidity just changes by a constant.

The distribution of mesons, largely pions, shown in Fig. 6 are conveniently thought about in the center of mass frame. Here we imagine the positive rapidity mesons as somehow related to the right moving particle and the negative rapidity particles as related to the left moving particles. We define $x = p^+/p^+_{projectile}$ and $x' = p^-/p^-_{projectile}$ and use x for positive rapidity pions and x' for negative rapidity pions.

Several theoretical issues arise in multiparticle production. Can we compute dN/dy? Or even dN/dy at $y = 0$? How does the average transverse momentum of produced particles $< p_T >$ behave with energy? What is the ratio of produced strange/nonstrange mesons, and corresponding ratios of charm, top, bottom etc. at $y = 0$ as the center of mass energy approaches infinity? Does multiparticle production as $E \to \infty$ at $y = 0$ become simple, understandable and computable?

There is a remarkable feature of rapidity distributions of produced hadrons, which we shall refer to as Feynman scaling. If we plot rapidity distributions of produced hadrons at different energies, then as function of the distance from the fragmentation region, the rapidity distributions are to a good approximation independent of energy. This is illustrated in Fig. 7. This means that as we go to higher and higher energies, the new physics is associated with the additional degrees of freedom at small rapidities in the center of mass frame (small-x degrees of freedom). The large-x degrees of freedom do not change much. This suggests that there may be some sort of renormalization group description in rapidity where the degrees of freedom at larger x are held fixed as we go to smaller values of x. We shall see that in fact these large-x degrees of freedom act as sources for the small-x degrees of freedom, and the renormalization group is generated by integrating out low-x degrees of freedom to generate these sources.

1.5 Deep Inelastic Scattering

In Fig. 2, deep inelastic scattering is shown. Here an electron emits a virtual photon which scatters from a quark in a hadron. The momentum and energy transfer of the electron is measured, and the results of the break-up are not. I will not develop the theory of deep inelastic scattering in this article. Suffice it to say, that this measurement is sufficient at large momenta transfer Q^2 to measure the distributions of quarks in a hadron.

To describe the quark distributions, it is convenient to work in a reference frame where the hadron has a large longitudinal momentum p^+_{hadron}. The corresponding light cone momentum of the constituent is $p^+_{constituent}$. We define $x = p^+_{constituent}/p^+_{hadron}$. This x variable is equal to the Bjorken x variable, which can be defined in a frame independent way. In this frame-independent definition, $x = Q^2/2p \cdot Q$, where p is the momentum of the

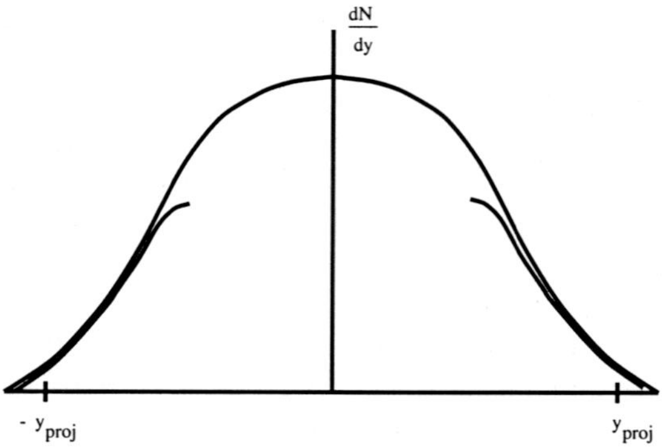

Fig. 7. Feynman scaling of rapidity distributions. The two different shapes correspond to rapidity distributions at different energies

hadronic target and Q is the momentum of the virtual photon. The cross section which one extracts in deep inelastic scattering can be related to the distributions of quarks inside a hadron, dN/dx.

It is useful to think about the distributions as a function of rapidity. We define this for deep inelastic scattering as

$$y = y_{hadron} - ln(1/x) \tag{7}$$

and the invariant rapidity distribution as

$$dN/dy = xdN/dx. \tag{8}$$

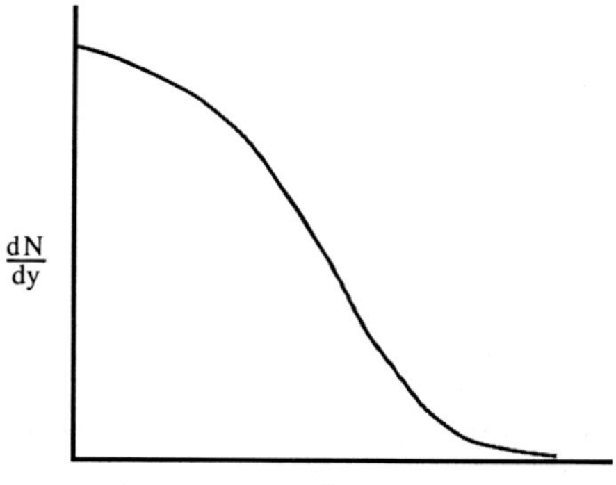

Fig. 8. The rapidity distribution of gluons inside of a hadron

In Fig. 8, a typical dN/dy distribution for constituent gluons of a hadron is shown. This plot is similar to the rapidity distribution of produced particles in deep inelastic scattering. The main difference is that we have only half of the plot, corresponding to the left moving hadron in a collision in the center of mass frame.

We shall later argue that there is in fact a relationship between the structure functions as measured in deep inelastic scattering and the rapidity distributions for particle production. We will argue that the gluon distribution function is in fact proportional to the pion rapidity distribution.

The small-x problem is that in experiments at HERA the rapidity distribution function for quarks grows as the rapidity difference between the quark and the hadron grows. This growth appears to be more rapid than simply $\mid y_{proj} - y \mid$ or $(y_{proj} - y)^2$, and various theoretical models based on the original considerations of Lipatov and colleagues suggest it may grow as an exponential in $\mid y_{proj} - y \mid$.[1] (Consistency of the BFKL approach with the more established DGLAP evolution equations remains an outstanding theoretical problem [2].) If the rapidity distribution grew at most as y^2, then there would be no small-x problem. We shall try to explain the reasons for this later in this article.

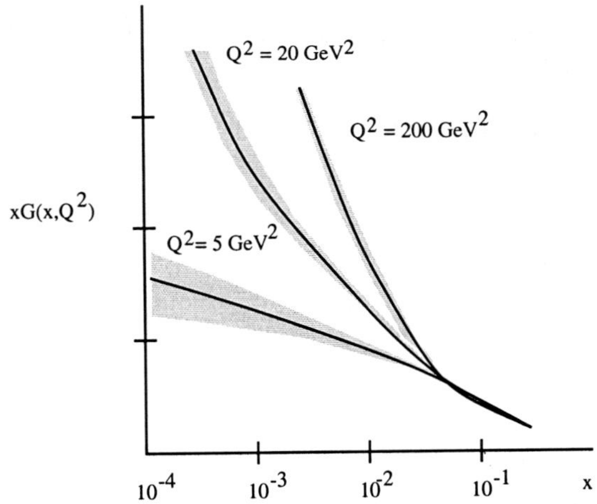

Fig. 9. The ZEUS data for the gluon structure functions

In Fig. 9, the ZEUS data for the gluon structure function is shown [3]. I have plotted the structure function for $Q^2 = 5$ GeV^2, 20 GeV^2, and 200 GeV^2. The structure function depends upon the resolution of the probe, that is Q^2. Note the rise of $xg(x)$ at small x, this is the small-x problem. If one had plotted the total multiplicity of produced particles in pp and $\overline{p}p$ collisions on the same plot, one would have found rough agreement in the shape of the curves. Here I would use $y = log(E_{cm}/1\ GeV)$ for the pion production data. This is approximately the maximal value of rapidity difference between centrally produced pions and the projectile rapidity. The total multiplicity would be rescaled so that at small-x it matches the gluon structure functions. This demonstrates the qualitative similarity between the gluon structure function and the total multiplicity.

Why is the small-x rise in the gluon distribution a problem? Consider Fig. 10, where we view hadron head on [4]-[5]. The constituents are the valence quarks, gluons, and sea quarks. As we add more and more constituents, the hadron becomes more and more crowded. If we were to try to measure these constituents with say an elementary photon probe, as we do in deep inelastic scattering, we might expect that the hadron would become so crowded that we could not ignore the shadowing effects of constituents as we make the measurement. (Shadowing means that some of the partons are obscured by virtue of having another parton in front of them. For hard spheres, for example, this would result in a decrease of the scattering cross section relative to what is expected from incoherent independent scattering.)

In fact, in deep inelastic scattering, we are measuring the cross section for a virtual photon γ^* and a hadron, $\sigma_{\gamma^* hadron}$. Making x smaller corresponds to increasing the energy of the interaction (at fixed Q^2). An exponential growth in the rapidity corresponds to a power law growth in $1/x$, which in turn implies a power law growth with energy. This growth, if it continues forever, violates unitarity. The Froissart bound will allow at most $ln^2(1/x)$. (The Froissart bound is a limit on how rapidly a total cross section can rise. It follows from the unitarity of the scattering matrix.)

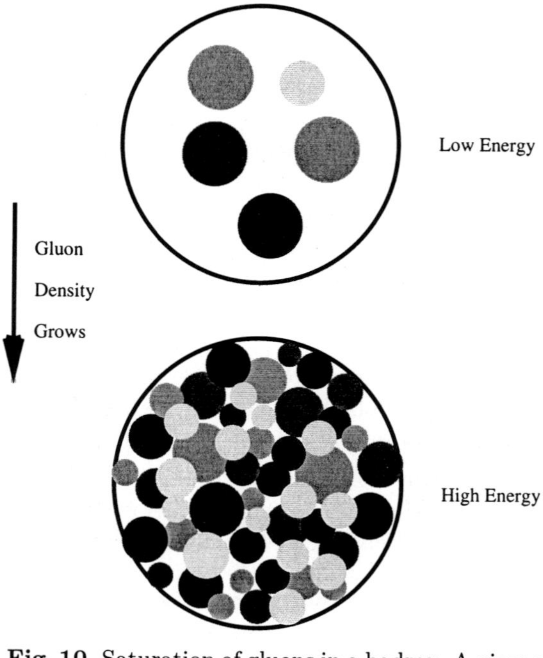

Fig. 10. Saturation of gluons in a hadron. A view of a hadron head on as x decreases

We shall later argue that in fact the distribution functions at fixed Q^2 do saturate and cease growing so rapidly at high energy. The total number of gluons, however, demands a resolution scale, and we will see that the natural intrinsic scale is growing at smaller values of x, so that effectively, the total number of gluons within this intrinsic scale is always increasing. The quantity

$$\Lambda^2 = \frac{1}{\pi R^2} \frac{dN}{dy} \tag{9}$$

defines this intrinsic scale. Here πR^2 is the cross section for hadronic scattering from the hadron. For a nucleus, this is well defined. For a hadron, this is less certain, but certainly if the wave lengths of probes are small compared to R, this should be well defined. If

$$\Lambda^2 >> \Lambda^2_{QCD} \tag{10}$$

as the HERA data suggest, then we are dealing with weakly coupled QCD since $\alpha_S(\Lambda) << 1$.

Even though QCD may be weakly coupled at small x, that does not mean the physics is perturbative. There are many examples of nonperturbative physics at weak coupling. An example is instantons in electroweak theory, which lead to the violation of baryon number. Another example is the atomic physics of highly charged nuclei. The electron propagates in the background of a strong nuclear Coulomb field, but on the other hand, the theory is weakly coupled and there is a systematic weak coupling expansion which allows for computation of the properties of high Z atoms (Z is the charge of the nucleus).

We call this assortment of gluons a Color Glass Condensate. The name follows from the fact that the gluons are colored, and we have seen that they are very dense. For massless particles we expect that the high density limit will be a Bose condensate. The phase space density will be limited by repulsive gluon interactions, and be of order $1/\alpha_s >> 1$. The glass nature follows because these fields are produced by partons at higher rapidity, and in the center of mass frame, they are Lorentz time dilated. Therefore the induced fields at smaller rapidity evolve slowly compared to natural time scales. These fields are also disordered. These two properties are similar to that of a glass which is a disodered material that is a liquid on long time scales and a solid on short ones.

If the theory is local in rapidity, then the only parameter which can determine the physics at that rapidity is Λ^2. Locality in rapidity means that there are not long-range correlations in the hadronic wave function as a function of rapidity. In pion production, it is known that except for overall global conserved quantities such as energy and total charge, such correlations are of short range. Note that if only Λ^2 determines the physics, then in an approximately scale invariant theory such as QCD, a typical transverse momentum of a constituent will also be of order Λ^2. If $\Lambda^2 >> 1/R^2$, where R is the radius of the hadron, then the finite size of the hadron becomes irrelevant. Therefore at small enough x, all hadrons become the same. The physics should only be controlled by Λ^2.

There should therefore be some equivalence between nuclei and, say, protons. When their Λ^2 values are the same, their physics should be the same. We can take an empirical parameterization of the gluon structure functions as

$$\frac{1}{\pi R^2}\frac{dN}{dy} \sim \frac{A^{1/3}}{x^\delta}, \tag{11}$$

where $\delta \sim 0.2 - 0.3$. This suggests that there should be the following correspondences:

- RHIC with nuclei $\sim$ HERA with protons
- LHC with nuclei $\sim$ HERA with nuclei

Estimates of the parameter Λ for nuclei at RHIC energies give $\sim 1 - 2$ GeV, and at LHC $2 - 3$ GeV.

Since the physics of high gluon density is weak coupling we have the hope that we might be able to do a first principle calculation of

- the gluon distribution function,
- the quark and heavy quark distribution functions,
- the intrinsic p_T distributions of quarks and gluons.

We can also suggest a simple escape from unitarity arguments which suggest that the gluon distribution function must not grow at arbitrarily small x. The point is that at smaller x, we have larger Λ and correspondingly larger p_T. A typical parton added to the hadron has a size of order $1/p_T$. Therefore although we are increasing the number of gluons, we do it by adding in more gluons of smaller and smaller size. A probe of size resolution $\Delta x \geq 1/p_T$ at fixed Q will not see partons smaller than this resolution size. They therefore do not contribute to the fixed Q^2 cross section, and there is no contradiction with unitarity.

1.6 Heavy-Ion Collisions

In Fig. 11, the standard light cone cartoon of heavy-ion collisions is shown [6]. To understand the figure, imagine we have two Lorentz contracted nuclei approaching one another at the speed of light. Since they are well localized, they can be thought of as sitting at $x^\pm = 0$, that is along the light cone, for $t < 0$. At $x^\pm = 0$, the nuclei collide. To analyze this problem for $t \geq 0$, it is convenient to introduce a time variable which is Lorentz covariant under longitudinal boosts

$$\tau = \sqrt{t^2 - z^2} \tag{12}$$

and a space-time rapidity variable

$$\eta = \frac{1}{2} ln \left(\frac{t - z}{t + z} \right) . \tag{13}$$

For free streaming particles

$$z = vt = \frac{p_z}{E} t \tag{14}$$

we see that the space-time rapidity equals the momentum space rapidity

$$\eta = y. \tag{15}$$

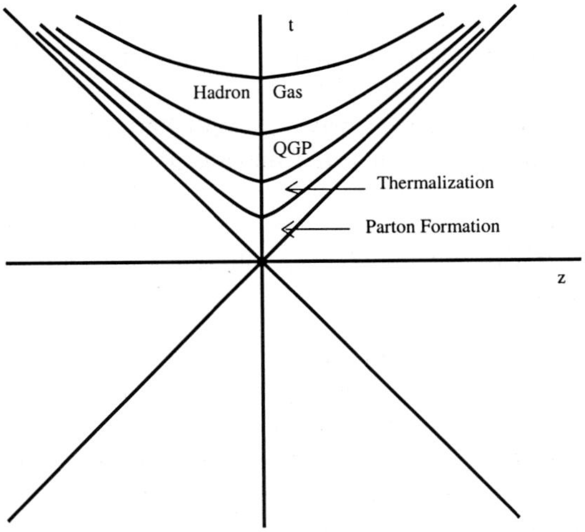

Fig. 11. A space-time figure for ultrarelativistic heavy-ion collisions

If we have distributions of particles which are slowly varying in rapidity, it should be a good approximation to take the distributions to be rapidity invariant. This should be valid at very high energies in the central region. By the correspondence between space-time and momentum space rapidity, it is plausible therefore to assume that distributions are independent of η. Therefore distributions are the same on lines of constant τ, which is as shown in Fig. 11. At $z = 0$, $\tau = t$, so that τ is a longitudinally Lorentz invariant time variable.

We expect that at very late times, we have a free streaming gas of hadrons. These are the hadrons which eventually arrive at our detector. At some earlier time, these particles decouple from a dense gas of strongly interacting hadrons. As we proceed earlier in time, at some time there is a transition between a gas of hadrons and a plasma of quarks and gluons. This may be through a first-order phase transition where the system might exist in a mixed phase for some length of time, or perhaps there is a continuous change in the properties of the system.

At some earlier time, the quarks and gluons of the quark-gluon plasma are formed. This is at RHIC energies, a time of the order of a fm, perhaps as small as 0.1 fm. As they form, the particles scatter from one another, and this can be described using the methods of transport theory. At some later time they have thermalized, and the system can be approximately described using the methods of perfect fluid hydrodynamics.

In the time between that for which the quarks and gluons have been formed and $\tau = 0$, the particles are being formed. This is where the initial conditions for a hydrodynamic description are made.

In various levels of sophistication, one can compute the properties of matter made in heavy-ion collisions at times later than the formation time. The problems are understood in principle for $\tau \geq \tau_{formation}$ if perhaps not in fact. Very little is known about the initial conditions.

In principal, understanding the initial conditions should be the simplest part of the problem. At the initial time, the degrees of freedom are most energetic and therefore one has the best chance to understand them using weak coupling methods in QCD.

There are two separate classes of problems one has to understand for the initial conditions. First the two nuclei which are colliding are in single quantum mechanical states. Therefore for some early time, the degrees of freedom must be quantum mechanical. This means that

$$\Delta z \Delta p_z \geq 1. \tag{16}$$

Therefore classical transport theory cannot describe the particle down to $\tau = 0$ since classical transport theory assumes we know a distribution function $f(\mathbf{p}, \mathbf{x}, t)$, which is a simultaneous function of momenta and coordinates. This can also be understood as a consequence of entropy. An initial quantum state has zero entropy. Once one describes things by classical distribution functions, entropy has been produced. Where did it come from?

Another problem which must be understood is classical charge coherence. At very early time, we have a tremendously large number of particles packed into a longitudinal size scale of less than a fm. This is due to the Lorentz contraction of the nuclei. We know that the particles cannot interact incoherently. For example, if we measure the field due to two opposite charges at a distance scale r large compared to their separation, we know the field falls as $1/r^2$, not $1/r$. On the other hand, in cascade theory, interactions are taken into account by cross sections which involve matrix elements squared. There is no room for classical charge coherence.

There are a whole variety of problems one can address in heavy-ion collisions such as:

- What is the equation of state of strongly interacting matter?

- Is there a first-order QCD phase transition?

These issues and others would take us beyond the scope of this article. The issues that I would like to address are related to the determination of the initial conditions, a problem which can hopefully be addressed using weak coupling methods in QCD.

1.7 Universality

There are two separate formulations of universality that are important in understanding small-x physics.

The first is a weak universality. This is the statement that physics should only depend upon the variable [7]

$$\Lambda^2 = \frac{1}{\pi R^2} \frac{dN}{dy}. \tag{17}$$

As discussed above, this universality has immediate experimental consequences which can be directly tested.

The second is a strong universality which is meant in a statistical mechanical sense. At first sight it appears to be a formal idea with little relation to experiment. If it is, however, true, its consequences are very powerful and far reaching. What we shall mean by strong universality is that the effective action which describes a small-x distribution function is critical and at a fixed point of some renormalization group. This means that the behavior of correlation functions is given by universal critical exponents, and these universal critical exponents depend only on general properties of the theory such as the symmetries and dimensionality.

Since the correlation functions determine the physics, this statement says that the physics is not determined by the details of the interactions, only by very general properties of the underlying theory!

2 A Very-High-Energy Nucleus

In this chapter, I will consider the properties of a single nucleus [7]-[9]. I will develop the theory of the small x part of the nucleus, the components most relevant in the high energy limit. I will begin with some general considerations. This will largely be done to develop approximations which will be useful later, and leads directly to the Color Glass description. I then present a brief review of light cone quantization. Finally, I turn to a computation of the color fields which describe the nuclear wave function at small x. I show that in a simple approximation for the Color Glass, one recovers saturation, and that the phase space density of the fields is of order $1/\alpha_s$ which is typical of a Bose condensate.

2.1 Approximations and the Color Glass

In the previous chapter, I argued that when we go to small x

$$\Lambda^2 = \frac{1}{\pi R^2}\frac{dN}{dy} >> \Lambda^2_{QCD}, \tag{18}$$

the theory is weakly coupled, $\alpha_s(\Lambda) << 1$. The typical transverse momentum scale of constituents of this low-x part of the hadron wave function is

$$p_T^2 \sim \Lambda^2 >> 1/R^2_{had}. \tag{19}$$

This equation means that the scale of transverse variation of the hadron is over much larger sizes than the transverse de Broglie wave length. I can therefore treat the hadron as having a well-defined size and collisions will have a well-defined impact parameter.

For our purposes, it is sufficient to treat the hadron as a thin sheet of infinite transverse extent. The transverse variation in radius can be reinserted in an almost trivial generalization of these considerations. The thinness of the sheet follows because I shall assume that the sources for the fields at small x come from partons at much larger x which are Lorentz contracted to size scales much smaller than can be resolved. In Fig. 12, a nucleus in the infinite momentum frame is shown, within the approximations described above.

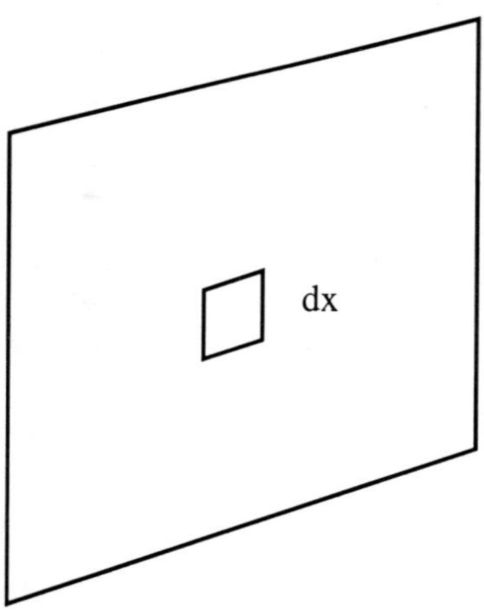

Fig. 12. A single nucleus in the infinite momentum frame as seen by a small x probe

Recall from the first chapter, we introduced rapidities associated with produced particles in hadron-hadron collisions,

$$\begin{aligned} y &= \frac{1}{2}ln(p^+/p^-) = ln(\sqrt{2}p^+/M_T) \\ &= ln(\sqrt{2}p^+_{had}/M_T) + ln(p^+/p^+_{had}) \sim y_{had} - ln(1/x). \end{aligned} \tag{20}$$

This expression shows that the rapidity of produced hadrons can be written in the form used to describe the rapidity of constituents of a hadron. If we were to think of both the constituents and produced partons as pions, they would be the same, or alternatively if we think of both the produced and constituent partons as gluons. We can convert to spacetime rapidity using the uncertainty relation $p^{\pm}x^{\mp} \sim 1$ and get

$$y \sim \frac{1}{2} ln(x^-/x^+) \sim y_{had} - ln(x^- p^+_{had}). \tag{21}$$

We have assumed in deriving this relationship that the typical values of the proper time $\tau = \sqrt{x^+x^-}$ are not large compared to natural scales such as a transverse mass. These relations argue that all rapidities, up to shifts of order one, are the same. We can identify all momentum-space and space-time rapidities! This has the profound consequence that at high energies momentum space and space-time are intrinsically correlated, and particles which arise from a localized region of momentum space rapidity also arise from a localized region of space-time rapidity.

Now we illustrate a high-energy hadron in terms of space-time rapidity. This has the effect of spreading out the thin sheet shown in Fig. 12, as shown in Fig. 13. Note that the partons which are shown have an ordering in momentum space rapidity which corresponds to their coordinate space rapidity. Fast partons are to the left. In this Figure, I have drawn a tube of transverse extent dx which goes through the nucleus. I take $dx << 1$ fm so that one is resolving the constituents of ordinary hadrons. Notice that when $dx \to 0$, the longitudinal separation between hadrons which intersect the tube becomes large. If I also require that the energy is high enough so that there is always a large number of partons which intersect the box (which are longitudinally well separated), then one can think of a source associated with the charge inside the box, and this charge has a random distribution over the transverse area of the box. (At what scale the source becomes random is not entirely clear from this discussion. This will be resolved more carefully later.)

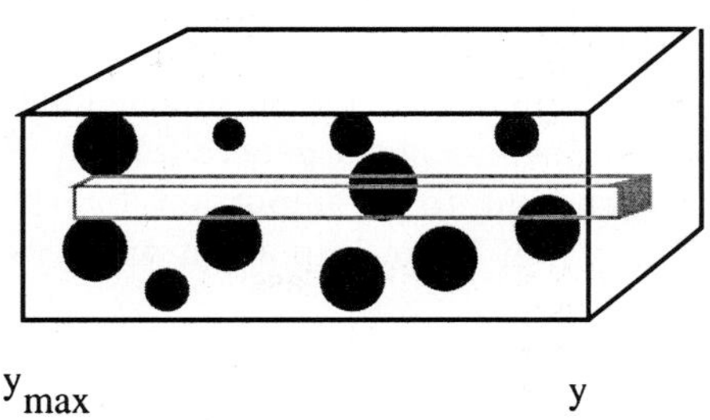

Fig. 13. A single nucleus shown in terms of the space-time rapidity. The full circles indicate partons

In the limit where $1/\Lambda << dx << 1$ fm, there are many charges inside the box of dimension dx. The charge should go over to a classical charge on this resolution scale because we can ignore commutators of charges

$$| [Q^a, Q^b] | = | if^{abc} Q^c | << Q^2. \tag{22}$$

We can define a current associated with this charge which is localized in the sheet as

$$J^\mu_a = \delta^{\mu +} \delta(x^-) q^a(x_T). \tag{23}$$

The + component of the current is the only important one because the sheet is traveling near the speed of light. The source $q^a(x_T)$ is a c-number color charge density which is a random variable on the sheet. It is only defined on scales $1/\Lambda << dx << 1$ fm. The δ-function of x^- expresses the fact that the source is on a thin sheet. In fact, for many applications, we will have to relax the δ-function assumption, and work with a charge density which includes the effect of distribution in x^- as

$$q^a(x_T) = \int dx^- \ \rho^a(x^-, x_T) \tag{24}$$

and where for many purposes

$$\rho^a(x^-, x_T) \sim \delta(x^-) q^a(x_T). \tag{25}$$

We now know how to write down a theory. It is a theory where one computes the classical gluon field in stationary phase approximation and then integrates over a random source function. Its measure is

$$Z = \int [dA][d\rho] exp \left\{ iS[A] + iJ^+ A^- - \frac{1}{2} \int dx^- d^2 x_T \frac{\rho^2(x^-, x_T)}{\mu^2(x^-)} \right\}. \tag{26}$$

In this theory, we have assumed that the sources are randomly distributed as a Gaussian. This turns out to be an approximation valid in a particular range of resolution dx, and can be fixed up for a wider range. This will be discussed when we do the renormalization group. The sources and fields are coupled together in the standard $J \cdot A$ form. This results in the problem that the extended current conservation law $D_\mu J^\mu = 0$ makes J not an independent function. This problem can be avoided by introducing a generalization of the $J \cdot A$ coupling. This generalization turns out to be important for the renormalization group analysis of this theory, but is not important when we compute the classical field associated with these sources.

The above theory implicitly has cutoffs in it. We have discussed the range of dx for which this effective theory is valid. Implicit in the analysis is that the fields we are computing have p^+ values much less than those of the sources. This implies there is an upper p^+ cutoff in the fields A considered.

If we were to Fourier analyze the sources ρ, they would have their support for $| p^+ |$ which is greater than that of the cutoff. This cutoff, is of course, entirely arbitrary, and the lack of dependence of physical quantities upon this cutoff forms the basis of the renormalization group.

Notice that this theory, in spite of having a gauge dependent source, is gauge invariant on account of the integration over all sources. This computation of classical fields associated with sources and then averaging over sources is similar to the mathematics of glasses. The physical origin of this similarity is the Lorentz time dilation of the source for the fields and the disorder of the gluon field. The Lorentz time dilation is of course an approximation, and if one were to observe these classical fields over long enough time scales they would evolve, as do the atoms in a glass.

Notice that

$$< \rho^a(x)\rho^b(y) >= \delta^{ab}\delta^{(3)}(x-y)\mu^2(x^-) \tag{27}$$

so that μ^2 is the charge squared per unit transverse area per unit x^- scaled by $1/(N_c^2-1)$.

2.2 Light Cone Quantization

Before discussing the properties of classical fields associated with these sources, it is useful to review some properties of light cone quantization [10]. This will allow us to pick out physical observables, such as the gluon density, from expectation values of gluon field operators.

Light cone coordinates are

$$x^\pm = \frac{1}{\sqrt{2}}(x^0 \pm x^3) \tag{28}$$

and momenta

$$p^\pm = \frac{1}{\sqrt{2}}(p^0 \pm p^3). \tag{29}$$

The invariant dot product is

$$p \cdot x = p_t \cdot x_t - p^+x^- - p^-x^+, \tag{30}$$

where p_t and x_t are transverse coordinates. This implies that in this basis the metric is $g^{+-} = g^{-+} = -1$, $g^{ij} = \delta^{ij}$ where i,j refer to transverse coordinates. All other elements of the metric vanish.

An advantage of light cone coordinates is that if we do a Lorentz boost along the longitudinal direction with Lorentz gamma factor $\gamma = cosh(y)$ then $p^\pm \rightarrow e^{\pm y}p^\pm$.

If we let x^+ be a time variable, we see that the variable p^- is to be interpreted as an energy. Therefore, when we have a field theory, the component of the momentum operator P^- will be interpreted as the Hamiltonian. The remaining variables are to be thought of as momenta and spatial coordinates. In Fig. 14, there is a plot of the z, t plane. The line $x^+ = 0$ provides a surface where initial data might be specified. Time evolution is in the direction normal to this surface.

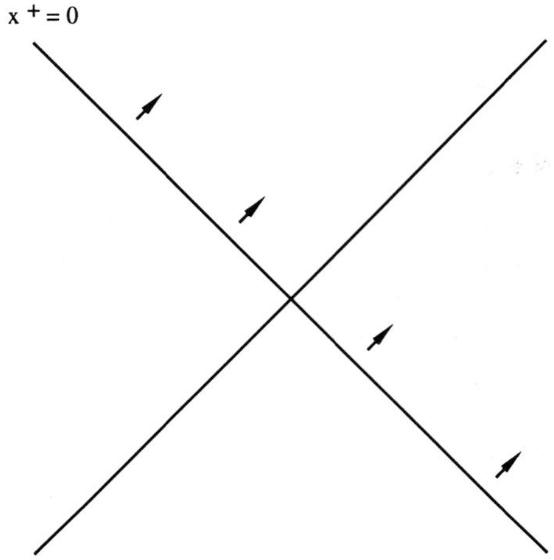

Fig. 14. The initial value problem in light cone coordinates

We see that an elementary wave equation

$$(p^2 + M^2)\phi = 0 \tag{31}$$

holds and is particularly simple in light cone gauge. Since $p^2 = p_t^2 - 2p^+p^-$ this equation is of the form

$$p^-\phi = \frac{p_t^2 + M^2}{2p^+}\phi \tag{32}$$

and is first order in time. In light cone coordinates, the dynamics looks similar to that of the Schrödinger equation. The initial data to be specified is only the value of the field on the initial surface.

In the conventional treatment of the Klein-Gordon field, one must specify the field and its first derivative (the momentum) on the initial surface. In light cone coordinates, the field is sufficient and the field momentum is redundant. This means that the field momentum will not commute with the field on the initial time surface!

Lets us work all this out with the example of the Klein-Gordon field. The action for this theory is

$$S = -\int d^4x \left\{ \frac{1}{2}(\partial\phi)^2 + \frac{1}{2}M^2\phi^2 \right\}. \tag{33}$$

The field momentum is

$$\Pi(x_t, x^-) = \frac{\delta S}{\delta \partial_+ \phi} = \partial_- \phi = \frac{\partial}{\partial x^-}\phi. \tag{34}$$

Note that Π is a derivative of ϕ on the initial time surface. It is therefore not an independent variable, as would be the case in the standard canonical quantization of the scalar field.

We postulate the equal-time commutation relation

$$[\Pi(x_t, x^-), \phi(y_t, y^-)] = -\frac{i}{2}\delta^{(3)}(x-y) \tag{35}$$

(The factor of two in the above expression is subtle and comes from a careful reduction of constrained Dirac bracket quantization for the classical theory to quantum field theory. It can be checked by verifying that we get the correct result for the Hamiltonian.) Here the time is $x^+ = y^+ = 0$ in both the field and field momentum. We see therefore that

$$\partial_- [\phi(\mathbf{x}), \phi(\mathbf{y})] = -\frac{i}{2}\delta^{(3)}(x-y) \tag{36}$$

or

$$[\phi(x), \phi(y)] = -\frac{i}{2}\epsilon(x^- - y^-)\delta^{(2)}(x-y). \tag{37}$$

Here $\epsilon(v)$ is $1/2$ for $v > 0$ and $-1/2$ for $v < 0$.

These commutation relations may be realized by the field

$$\begin{aligned} \phi(x) &= \int \frac{d^3p}{(2\pi)^3 2\sqrt{2}p^+} e^{ipx} a(p) \\ &= \int_{p^+>0} \frac{d^3p}{(2\pi)^3 2\sqrt{2}p^+} \left\{ e^{ipx} a(p) + e^{-ipx} a^\dagger(p) \right\}. \end{aligned} \tag{38}$$

Using

$$[a(p), a^\dagger(q)] = 2p^+(2\pi)^3\delta^{(3)}(p-q) \tag{39}$$

one can verify that the equal-time commutation relations for the field are satisfied.

The quantity $1/p^+$ in the expression for the field in terms of creation and annihilation operators is singular when $p^+ = 0$. When we use a principle value prescription, we reproduce the form of the commutation relations postulated above with the factor of $\epsilon(x^- - y^-)$. Different prescriptions correspond to different choices for the inversion of $\frac{1}{\partial^-}$. One possible prescription is the Leibbrandt-Mandelstam prescription $1/p^+ = p^-/(p^+p^- + i\epsilon)$. This prescription has some advantages relative to the principle value prescription in that it maintains causality at intermediate stages of computations, while the principle value prescription does not. In the end, for physical quantities, the choice of prescription cannot result in different results. Of course, in some schemes the computations may become prohibitively difficult.

The light cone Hamiltonian is

$$P^- = \int_{p^+>0} \frac{d^3p}{(2\pi)^3 2p^+} \frac{p_t^2 + M^2}{2p^+} a^\dagger(p) a(p) \tag{40}$$

with obvious physical interpretation.

In a general interacting theory, the Hamiltonian will, of course, be more complicated. The representation for the fields in terms of creation and annihilation operators will be the same as above. Note that all particles created by a creation operator have positive P^+. Therefore, since the vacuum has $P^+ = 0$, there can be no particle content to the vacuum. It is a trivial state. Of course, this must be wrong since the physical vacuum must contain condensates such as the one responsible for chiral symmetry restoration. It can be shown that such non-perturbative condensates arise in the $P^+ = 0$ modes of the theory. We have not been careful in treating such modes. For perturbation theory, presumably to all orders, the above treatment is sufficient for our purposes.

2.3 Light Cone Gauge QCD

In QCD we have a vector field A_a^μ. This can be decomposed into longitudinal and transverse parts as

$$A_a^\pm = \frac{1}{\sqrt{2}}(A_a^0 \pm A_a^z) \tag{41}$$

and the transverse as lying in the two-dimensional plane orthogonal to the beam z axis. Light cone gauge is

$$A_a^+ = 0. \tag{42}$$

In this gauge, the equation of motion

$$D_\mu F^{\mu\nu} = 0 \tag{43}$$

is for the + component

$$D_i F^{i+} - D^+ F^{-+} = 0, \tag{44}$$

which allows one to compute A^- in terms of A^i as

$$A^- = \frac{1}{\partial^{+2}} D^i \partial^+ A^i. \tag{45}$$

This equation says that we can express the longitudinal field entirely in terms of the transverse degrees of freedom which are specified by the transverse fields entirely and explicitly. These degrees of freedom correspond to the two polarization states of the gluons.

We therefore have

$$A^i_a(x) = \int_{p^+>0} \frac{d^3p}{(2\pi)^3 2p^+} \left(e^{ipx} a^i_a(p) + e^{-ipx} a^{i\dagger}_a(p)\right), \tag{46}$$

with

$$[a^i_a(p), a^{j\dagger}_b(q)] = 2p^+ \delta_{ab} \delta^{ij} (2\pi)^3 \delta^{(3)}(p-q), \tag{47}$$

where the commutator is at equal light cone time x^+.

2.4 Distribution Functions

We would like to explore some hadronic properties using light cone field operators. For example, suppose we have a hadron and ask what is the gluon content of that hadron. Then we would compute

$$\frac{dN_{gluon}}{d^3p} =< h \mid a^\dagger(p) a(p) \mid h > . \tag{48}$$

If we express this in terms of the gluon field, we find

$$\frac{dN_{gluon}}{d^3p} = \frac{2p^+}{(2\pi)^3} < h \mid A^{ia}(\mathbf{p}, x^+) A^{ia}(-\mathbf{p}, x^+) \mid h > \tag{49}$$

which can be related to the gluon propagator. The quark distribution for quarks of flavor i (for the sum of quarks and antiquarks) would be given in terms of creation and annihilation operators for quarks as

$$\frac{dN_i}{d^3p} =< h \mid \{b^\dagger_i(p) b_i(p) + d^\dagger_i(p) d_i(p)\} \mid h >, \tag{50}$$

where b corresponds to quarks and d to antiquarks. The creation and annihilation operators for quarks and gluons can be related to the quark coordinate space field operators by techniques similar to those above [11].

2.5 The Classical Gluon Field

To compute the gluon distribution function, we need the expectation value of the gluon field. To lowest order in weak coupling, this is given by computing the classical gluon field and then averaging over sources. The classical equation of motion is

$$D_\mu F^{\mu\nu} = \delta^{\nu +} \rho(x^-, x_T). \tag{51}$$

To solve this equation we shall work in the gauge $\mathcal{A}^- = 0$, and then gauge rotate the solution back to light cone gauge $A^+ = 0$. The solution in $\mathcal{A}^- = 0$ gauge is:

$$\mathcal{A}^i = 0,$$
$$-\nabla_T^2 \mathcal{A}^+ = \overline{\rho}. \tag{52}$$

Here $\overline{\rho} = U^\dagger(x)\rho U(x)$ is the source which has been gauge rotated to this new gauge. Since the measure for integration over sources is gauge invariant, we do not have to distinguish between these sources since we can rotate one into the other.

To rotate back to light cone gauge we use

$$\mathcal{A}^\mu = U^\dagger A^\mu U + \frac{i}{g} U^\dagger \partial^\mu U \tag{53}$$

so that the gauge rotation matrix U is

$$\partial^+ U = -igU\mathcal{A}^+, \tag{54}$$

where

$$\mathcal{A}^+ = \alpha = \frac{1}{-\nabla_T^2} \nabla \overline{\rho}. \tag{55}$$

The solution is [9]-[12]

$$U^\dagger = Pexp\left\{ ig \int_{x_0^-}^{x^-} dz^- \alpha(z^-, x_T) \right\}. \tag{56}$$

There is a choice of boundary condition here associated with x_0^+. The ambiguity with this choice is associated with a residual gauge freedom. We shall resolve this by choosing retarded boundary conditions, $x_0^- \to -\infty$. This boundary condition lets us construct the solution for U at some x_1^- knowing only information about α for $x^- < x_1^-$.

The solution in light cone gauge is therefore:

$$A^+ = A^- = 0,$$
$$A^i = \frac{i}{g} U \nabla^i U^\dagger. \tag{57}$$

If x^- is outside of the range of support of the source ρ, this can be written as

$$A^i = \theta(x^-) \frac{i}{g} V \nabla^i V^\dagger, \tag{58}$$

where

$$V^\dagger(x_T) = Pexp\left(ig \int_{-\infty}^{\infty} dz^- \alpha(z^-, x_T) \right). \tag{59}$$

We now have an explicit expression for the gluon field in terms of the sources [9],[12]. For our Gaussian weight function, we can now compute the expectation value of the gluon fields which gives the gluon distribution function. The details of such a computation are given in Ref. [12]. It is a straightforward computation to perform: One can expand the exponentials and compute term by term in the expansion. The series exponentiates. One subtlety occurs due to logarithmic infrared infinity which is regulated on a scale of order of a fm, where transverse charge correlations go to zero since all hadrons are color singlets [13]. The result is

$$< A^i_a(\mathbf{x}, x^+) A^i_a(\mathbf{0}, x^+) > = \frac{N_c^2 - 1}{\pi \alpha_s N_c} \frac{1}{x_T^2} \times \\ \left(1 - exp\{x_T^2 Q_s^2 ln(x_T^2 \Lambda_{QCD}^2)/4\}\right) . \quad (60)$$

In this equation, the saturation momentum is defined as

$$Q_s^2 = 2\pi N_c \alpha_s^2 \int dx^- \mu^2 \quad (61)$$

and is of the order α_s^2 times the charge squared per unit area.

This expression can be Fourier transformed to produce the gluon distribution function, with a result as shown in Fig. 15. We can understand this plot from the properties of the coordinate space distribution function. We notice that the dominant scale factor in the problem is Q_s, so to a first approximation everything scales in terms of this quantity. Large p_T corresponds to small x_T, and the coordinate space distribution behaves as $ln(x_T^2)$ which corresponds to $1/p_T^2$. This is typical of a bremstrahlung spectrum. At larger x_T, distribution is of order $1/x_T^2$, which Fourier transforms into $ln(p_T^2)$ at small p_T. The softer x_T dependence at large x_T can be traced to a dipole cancellation of the fields. The monopole charge field, seen at short distances, is $ln(x_T^2)$ and the dipole cancellation should set in at large distances when one cannot resolve individual charges, and reduce this by two powers of x_T.

The overall scale of the curve is $1/\alpha_s$. The quantity we are plotting is in fact the phase space density of gluons. At small α_s, this density becomes large, and the Color Glass becomes a condensate. Hence the name, Color Glass Condensate.

This form of the gluon distribution function illustrates how the problems with unitarity can be solved. Let us assume that the saturation momentum Q_s is rapidly increasing as $x \to 0$. If we start with an x so that $Q >> Q_s$, then, as x decreases, the number of gluons which can be seen in scattering rises like Q_s^2, see Fig. 15. Eventually, Q_s becomes larger than Q, in which case, the number of gluons rises slowly, like $ln(Q_s)$. At this point the cross section saturates since the number of gluons which can be resolved stops growing, and we are consistent with unitarity constraints.

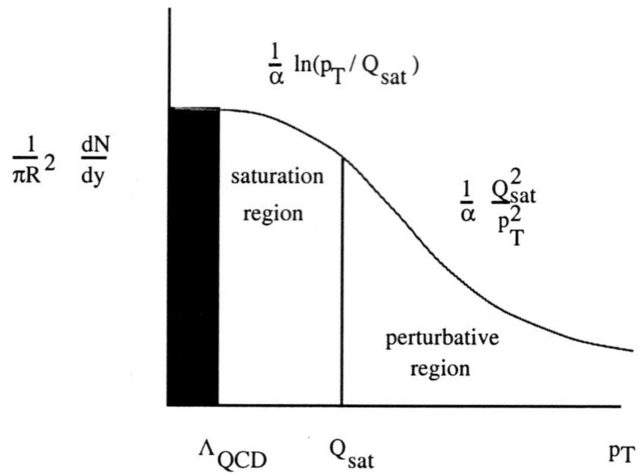

Fig. 15. The gluon distribution function

The gluon distribution function is defined to be

$$xG(x, Q^2) = \int_0^{Q^2} d^2p_T \frac{dN}{d^2p_T dy}. \tag{62}$$

This behaves in the saturation region as $\pi R^2 Q^2$, and in the large-Q region as $\pi R^2 Q_s^2$. We expect that $Q_s^2 \sim charge^2/area$ and due to the random nature of the way charges add, $Q_s^2 \sim R$. Therefore in the saturation region, the gluon distribution function is proportional to the surface area of the hadron, that is the gluons can only be seen which are on the surface of the hadron. In the large-Q region, one sees gluons from the entire hadron, that is, the hadron has become transparent.

2.6 The Structure of the Gluon Field

The gluon field arises from a charge density which is essentially a δ-function in x^-. In order to solve the equations of motion, the field must have a discontinuity at $x^- = 0$. This can be achieved with a field which is a two-dimensional gauge transform of zero field strength on one side of the sheet and a different gauge transform of zero on the other side. The field strength $F^{\mu\nu}$ is therefore zero if μ and ν are both in the two-dimensional transverse space. If either index is $-$, it also vanishes since there is no change in the x^+ direction. The only non-vanishing component is therefore F^{i+}, and this is a δ-function in the x^- direction. Since $F^{i\pm} = E^i \pm \epsilon^{ij} B^j$, we see that

$$\mathbf{E} \perp \mathbf{B} \perp \mathbf{z}. \tag{63}$$

The fields are therefore transversely polarized to the direction of motion and live in the two-dimensional sheet where the charges sit. These are the non-Abelian generalizations of the Lienard-Wiechert potentials of electrodynamics. The density of these fields is of order $1/\alpha_s$. A picture of the Color Glass Condensate is shown in Fig. 16.

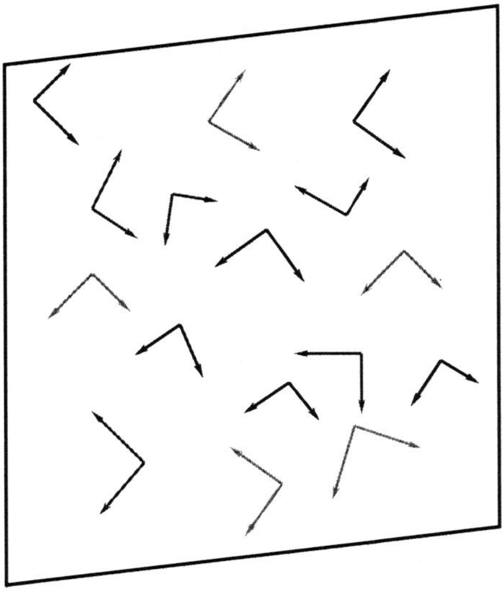

Fig. 16. The non-Abelian Lienard-Wiechert potentials that form the Color Glass Condensate

3 Hadron-Hadron Collisions and the Initial Conditions for Heavy-Ion Collisions

3.1 Phenomenology of Mini-Jets

In the last chapter, we argued that at small x, the typical gluon constituent of a hadron acquires a transverse momentum of order Q_s and that this can grow as $x \to 0$. This leads us to hope that in hadron-hadron collisions, this will be the typical momentum scale of particle production. If true, then the processes are weakly coupled and computable using weak coupling methods.

This is reminiscent of past attempts to compute particle production by mini-jets [14]-[15]. On dimensional grounds, the cross section for jet production is $d\sigma/dyd^2p_T \sim \alpha_s^2/p_T^4$. If we attempt to compute the total cross section for jet production

$$\frac{d\sigma}{dy} \sim \alpha_s^2 \int_{\Lambda_{QCD}^2} \frac{d^2p_T}{p_T^4}, \tag{64}$$

the result is infrared sensitive and presumably would be cut off at Λ_{QCD}. In early computations, one introduced an ad-hoc cutoff which was fixed, and hopefully large enough so that one could compute the minijet component. This, of course, left unanswered many questions about the origin of this cutoff, and the effects of particles produced below the cutoff scale.

In this chapter, we will argue that the Color Glass Condensate cuts off the integral at a scale of order Q_s, the saturation momentum. At large p_T, dimensional arguments tell us that the density of produced particles has the form

$$\frac{1}{\pi R^2} \frac{dN}{d^2 p_T dy} = \kappa \frac{1}{\alpha_s} \frac{Q_s^4}{p_T^4}. \tag{65}$$

The factor of $1/p_T^4$ comes about because the high-p_T tail is controlled by perturbation theory. The $1/\alpha_s$ arises because of the large density of gluons in the condensate. In fact, if we can successfully formulate the particle production problem classically, we expect that in general

$$\frac{1}{\pi R^2} \frac{dN}{d^2 p_T dy} = \frac{1}{\alpha_s} F(Q_s^2/p_T^2). \tag{66}$$

At large $p_T >> Q_s$, $F \sim Q_s^4/p_T^4$ and for small $p_T << Q_s$, F should be slowly varying (logarithmic) or a constant. A plot is shown in Fig. 17.

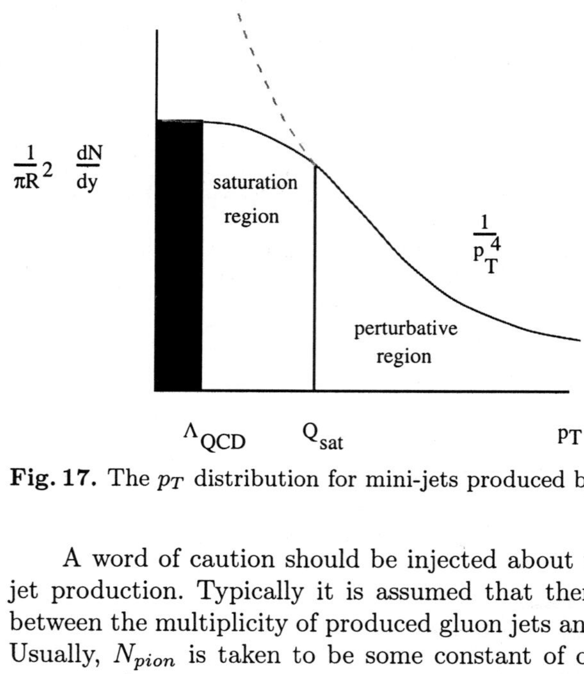

Fig. 17. The p_T distribution for mini-jets produced by a Color Glass Condensate

A word of caution should be injected about the interpretation of mini-jet production. Typically it is assumed that there is a simple relationship between the multiplicity of produced gluon jets and the multiplicity of pions. Usually, N_{pion} is taken to be some constant of order one times N_{gluon}. In our considerations, we can only talk about the gluon mini-jet production, and it is beyond the scope of these article to relate this to the final-state multiplicity. Suffice it to say that the situation is controversial, particularly in heavy-ion collisions where there can be much final-state interaction [16].

Recall that in heavy-ion collisions, we expect that $Q_s^2 \sim A^{1/3}$. At large p_T, Eqn. 66 predicts that

$$\frac{dN}{d^2 p_T dy} \sim \pi R^2 \frac{Q_s^4}{p_T^4} \sim \frac{A^{4/3}}{p_T^4}. \tag{67}$$

This result is consistent with hard incoherent scattering. At small p_T,

$$\frac{dN}{d^2p_T dy} \sim \pi R^2 \sim A^{2/3} \tag{68}$$

which is consistent with much shadowing, and the gluons are produced from the surface of the nuclei.

The total multiplicity per unit rapidity

$$\frac{dN}{dy} \sim R^2 \int_{Q_s^2} d^2p_T \frac{Q_s^4}{p_T^4} \sim R^2 Q_s^2 \sim A \tag{69}$$

is proportional to A, just as in color string models! This is because for the Color Glass Condensate, the cutoff in transverse momentum depends on A. (If one was careful with the factors of α_s in the above equation, one would predict mild logarithmic modifications of the linear dependence on A.) In addition to the A dependence, there is also a correlation between the energy dependence of the gluon distribution function at saturation and the multiplicity of minijets since

$$\pi R^2 Q_s^2 = \int_x^1 dx' G(x', Q_s), \tag{70}$$

a relationship which follows from the last chapter.

We can be a little more careful with the numerical factors which determine the saturation momentum [17]. Using the results of the last chapter,

$$Q_s^2 = \frac{2\pi N_c \alpha_s^2}{\pi R^2 (N_c^2 - 1)} Q_{color}^2. \tag{71}$$

Here Q_{color}^2 is the color charge squared of all quarks and gluons at larger x values than that of interest. For a quark,

$$Q_{quark}^2 = \frac{1}{N_c(N_c^2 - 1)} tr\ \tau_a^2 = \frac{1}{2N_c}, \tag{72}$$

and for a gluon

$$Q_{gluon}^2 = \frac{N_c}{(N_c^2 - 1)}. \tag{73}$$

We find that

$$Q_{color}^2 = \frac{N_{quark}}{2N_c} + \frac{N_c N_{gluon}}{(N_c^2 - 1)}. \tag{74}$$

If we plug in numbers, at RHIC energies corresponding to $x \sim 10^{-2}$, $Q_s \sim 1 - 2$ GeV

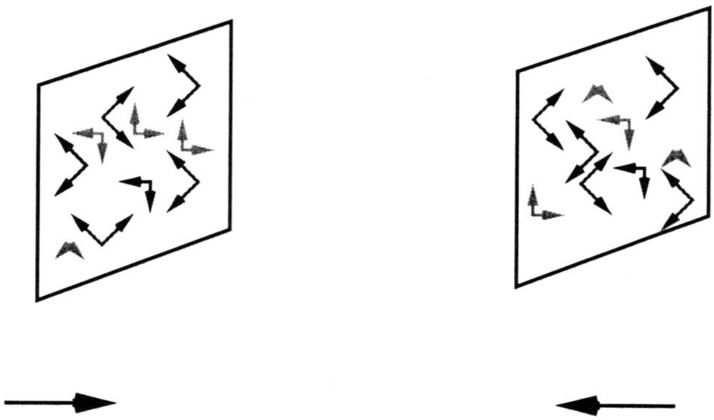

Fig. 18. A collision of two ultra-relativistic hadrons

3.2 Classical Description of Hadron Collisions

We want to describe the collision of two ultra-relativistic hadrons. A collision is shown in Fig. 18. The hadrons have been Lorentz contracted to thin sheets and a Color Glass Condensate sits in the planes of both sheets.

Before the collision the non-zero fields are for the right moving nucleus, $F^{i+} \sim \delta(x^-)$ and for the left moving nucleus $F^{i-} \sim \delta(x^+)$. Before the nuclei pass through one another, nothing happens and the fields in each sheet are static. When they pass through one another, the sum of these two fields is not a solution of the equations of motion, unlike the case in electrodynamics, and this induces a time evolution of the fields [18].

One can understand this from the vector potentials. In Fig. 19 a space-time diagram is shown for the scattering. In the backward light cone, Region I, the field is a pure two-dimensional gauge transform of zero field. In crossing into Regions II and III, the fields must have a discontinuity to match the charge on the surfaces of the light cone. This requires the vector potential to be different gauge transforms of zero field strength, G_2 and G_3 in these regions. Now in going to Region IV, one could solve either for the sources on the left edge of the forward light cone with a gauge transform of zero or the right edge of the forward light cone with a different gauge transform of zero. One cannot satisfy the equations of motion for the fields in the presence of the sources on both edges of the light cones with the same gauge transform of zero field strength. One must produce a field in the forward light cone which is not a gauge, and therefore matter is produced.

The situation in QCD is completely different than that in electrodynamics. In electrodynamics, one must produce pairs of charged particles to make matter in the forward light cone. This arises from a quantum correction to the equations of motion. In QCD, matter is produced classically.

The procedure for solving this problem is now straightforward, in principle. One solves the classical equations of motion in the forward light cone with boundary conditions at the edges of the forward light cone. We will

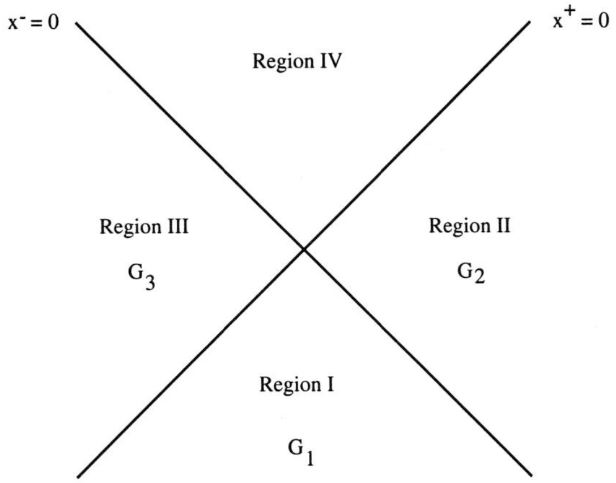

Fig. 19. A space-time diagram for the vector potentials in hadron-hadron scattering

shortly determine these boundary conditions and the form of the solution in the forward light cone. Then one evolves the equations of motion into the far future. At some time, the energy density becomes dilute, and the field equations should linearize in some gauge. One can then identify the quanta of the linearized fields in the standard way that one does classical radiation theory in electrodynamics.

3.3 The Form of the Classical Field

Before the collision, the form of the classical field can be taken as

$$\begin{aligned} A^+ &= A^- = 0, \\ A^i &= \theta(x^-)\theta(-x^+)\alpha_1^i(x_T) + \theta(-x^-)\theta(x^+)\alpha_2^i(x_T), \end{aligned} \tag{75}$$

where the α^i are two-dimensional gauge transforms of zero field. We will consider the collision of identical hadrons. The solution in the forward light cone is therefore expected to be boost invariant. After the collision, a boost invariant solution is:

$$\begin{aligned} A^+ &= x^+\alpha(\tau, x_T), \\ A^- &= x^-\beta(\tau, x_T), \\ A^i &= \alpha_3^i(\tau, x_T). \end{aligned} \tag{76}$$

We can choose the gauge

$$x^+A^- + x^-A^+ = 0 \tag{77}$$

so that

$$\alpha(\tau, x_T) = -\beta(\tau, x_T). \tag{78}$$

In the forward light cone, the equations of motion are

$$\frac{1}{\tau^3}\partial_\tau \tau^3 \partial_\tau \alpha - [D^i,[D^i,\alpha]] = 0 \tag{79}$$

and

$$\frac{1}{\tau}\partial_\tau \tau \partial_\tau \alpha_3^i - ig\tau^2[\alpha,[D^i,\alpha]] - [D^j,F^{ji}] = 0. \tag{80}$$

The boundary conditions are determined by matching the solution in Regions II and III to that in the forward light cone. The result is that α and α_3^i must both be regular as $\tau \to 0$ and

$$\begin{aligned} \alpha_3^i(0,x_T) &= \alpha_1^i(x_T) + \alpha_2^i(x_T), \\ \alpha(0,x_T) &= \frac{-ig}{2}[\alpha_1^i(x_T),\alpha_2^i(x_T)]. \end{aligned} \tag{81}$$

The problem is now well defined, and these equations may be numerically solved.

The behaviour of these solution at large τ can be extracted. With $V(x_T)$ an element of the group, the solution is a small fluctuation field up to a possible large gauge transformation

$$\begin{aligned} \alpha(\tau,x_T) &= V\epsilon(\tau,x_T)V^\dagger, \\ \alpha_3^i(\tau,x_T) &= V(\epsilon_3^i(\tau,x_T) + \frac{i}{g}\partial^i)V^\dagger. \end{aligned} \tag{82}$$

The small fluctuation fields ϵ and ϵ^i solve the equations

$$\frac{1}{\tau^3}\partial_\tau \tau^3 \partial_\tau \epsilon - \nabla_T^2 \epsilon = 0 \tag{83}$$

and

$$\frac{1}{\tau}\partial_\tau \tau \partial_\tau \epsilon^i - (\nabla_T^2 \delta^{ij} - \nabla^i \nabla^j)\epsilon^j = 0. \tag{84}$$

At large τ, these linear equations can be Fourier analyzed with the result

$$\epsilon^a(\tau,x_T) = \int \frac{d^2k_T}{(2\pi)^2\sqrt{2\omega}} \frac{1}{\tau^{3/2}} \left(a_1^a(k_T)e^{-ik\cdot x} + c.c.\right) \tag{85}$$

and

$$\epsilon^{ai}(\tau,x_T) = \int \frac{d^2k_T}{(2\pi)^2\sqrt{2\omega}} \epsilon^{ij}\frac{k^j}{\omega}\frac{1}{\tau^{1/2}} \left(a_2^a(k_T)e^{-ik\cdot x} + c.c.\right). \tag{86}$$

In these equations $\omega =| k_T |$.

One can compute the energy distribution associated with these fields as

$$\frac{dE}{dyd^2k_T} = \frac{\omega}{(2\pi)^2} \sum_{ia} \mid a_i^a(k_T) \mid^2 \tag{87}$$

and the multiplicity distribution is given by dividing this by ω, that is

$$\frac{dN}{dyd^2k_T} = \frac{1}{(2\pi)^2} \sum_{ia} \mid a_i^a(k_T) \mid^2 . \tag{88}$$

These last two formulae correspond to those of free quantum field theory when we replace $a(p), a^*(p)$ by the creation and annihilation operators $a(p), a^\dagger(p)$. The $a(p)$ are the classical quantities which correspond to the quantum creation and annihilation operators. These formulae show how to use the classical solutions to compute distributions of produced minijets.

3.4 Numerical Results for Mini-Jet Production

Krasnitz and Venugopalan have numerically solved the classical equations for mini-jet production.[19] This involves finding a gauge invariant discretization of the classical equations of motion. One then solves the classical equations for a fixed ρ_1 and ρ_2, and extracts the produced radiation. An ensemble of sources is produced with the Gaussian weight of the Color Glass, which then produces an ensemble of radiation fields. These fields are then averaged to generate the mini-jet distributions.

In Fig. 20, the form of the numerical results for mini-jet production is illustrated. At large p_T, the results of analytic studies are reproduced which up to logarithms is $\sim 1/p_T^4$. At $p_T \leq Q_s$, the distribution flattens out [17]-[18]. To good numerical accuracy, the result in this region can be fit to a two-dimensional Bose-Einstein distribution,

$$\frac{1}{\pi R^2} \frac{dN}{d^2p_T dy} \sim \frac{1}{\alpha_s} \frac{1}{e^{\kappa p_T/Q_s} - 1} \tag{89}$$

where κ is a constant of order 1.

The result at large p_T can be computed analytically by expanding the equations in powers of the gluon field. At high p_T, the phase space is not so heavily occupied, so a field strength expansion makes sense. At small p_T, it is not at all certain that the result is in fact an exact two-dimensional Bose-Einstein distribution [19]-[21]. In any case, the origins of this distribution have nothing at all to do with thermodynamics, and it is a useful example of the traps one can fall into if one assumes that exponential distribution corresponds to a temperature and thermalization.

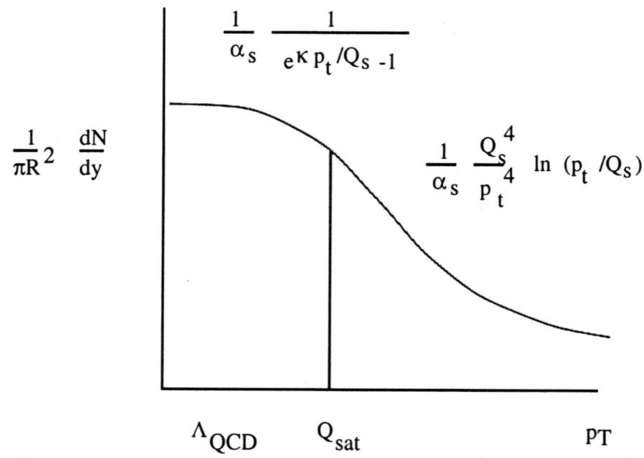

Fig. 20. An illustration of the results generated by numerical simulation of the classical equations for mini-jet production

3.5 pA Scattering

An interesting example of minijet production is given by the collision of two hadrons of different size [22]-[24]. We will generically refer to this as pA scattering, although most of our considerations could be generalized to $A'A$ nuclear collisions. In Fig. 21, the transverse momentum distribution for pA scattering is shown.

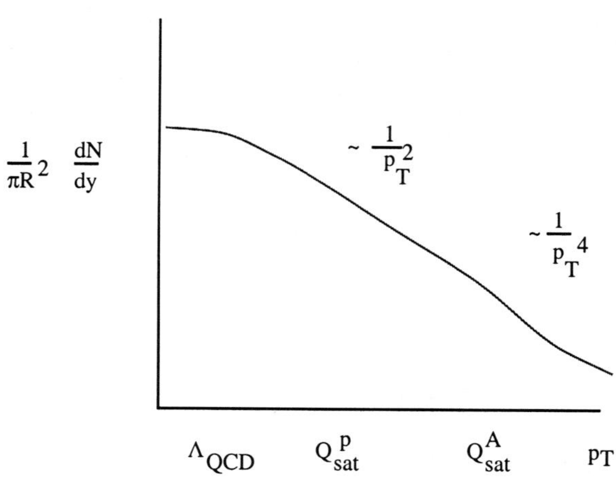

Fig. 21. The p_T distribution for particles produced in a pA collision

There are three distinct regions, which follow from the fact that there are two saturation scales, Q_s^A and Q_s^p, and $Q_s^p << Q_s^A$ since $(Q_s^A)^2 \sim A^{1/3}$. At very large p_T where the fields from both nuclei are small, the distribution can be computed from perturbation theory, and the distribution falls as $1/p_T^4$

and is proportional to $(Q_s^A Q_s^p)^2$. This first region is for $p_T >> Q_s^A$. An intermediate region follows where the field from the nucleus is strong but the field from the proton is weak and can be treated perturbatively. This intermediate region is for $Q_s^p << p_T << Q_s^A$. There is finally the region where $p_T << Q_s^p$ and both fields are strong.

We expect that in the intermediate region, the transverse momentum dependence will be in between the flat behaviour at small p_T and the $1/p_T^4$ behaviour characteristic of large p_T. The naive expectation is $1/p_T^2$ in the intermediate region. The total multiplicity can be computed if one understands this intermediate region since the dominant contribution arises here. The strength in this intermediate region should involve the total charge squared from the proton, but that from the nucleus should go like p_T^2 so that when combined with the $1/p_T^4$, one gets a distribution proportional to $1/p_T^2$. This softer behaviour of the distribution function follows since we are inside the region where we expect coherence from the field of the nucleus, and since the distribution should extrapolate between $1/p_T^4$ at very large p_T and a constant, up to logarithms, at very small p_T.

In fact, it is possible to compute the behaviour in this intermediate region. The equations for the classical production can be analytically solved for any $p_T >> Q_s^A$. The solution in the forward light cone are plane waves which are gauge transformed by the field of the large nucleus. The boundary conditions determine the strength of these waves.

For the total multiplicity, in the large-p_T region $dN/dyd^2p_T \sim A^{1/3}$. We expect that as we interpolate between the proton fragmentation region and that of the nucleus, we go between $O(1)$ and $O(A^{1/3})$ as shown in Fig. 22. For p_T in the intermediate region, we expect that dN/dyd^2p_T is of order 1 except for a small region of rapidity around the fragmentation region of the nucleus. The total integrated multiplicity arises from this latter region so we expect that $dN/dy \sim O(1)$.

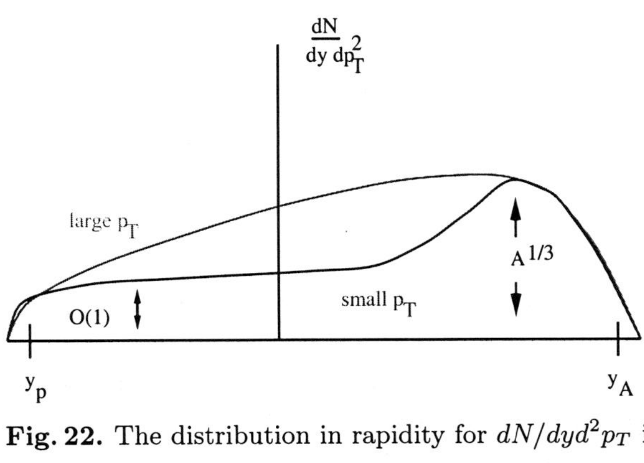

Fig. 22. The distribution in rapidity for dN/dyd^2p_T in a pA collision

3.6 Thermalization

After the gluons are produced in hadron-hadron collisions, they may rescatter from one another [16]. If one goes to very small x so that the density of gluons becomes very large, one expects that the gluons will eventually thermalize. Due to the very large typical p_T, $\alpha_s << 1$, and this takes a time $\tau \sim 1/(\alpha_s^2 Q_s)$ which is longer by a factor of $1/\alpha_s^2$ than the natural time scale. The system therefore becomes dilute relative to its natural scale.

In the first diagram of Fig. 23, there is ordinary Coulomb scattering. When all processes which populate and depopulate phase space are summed, this diagram is only naively logarithmically divergent, and is cut off by the density dependent Debye mass, $\rho_{gluons} << p_T^3$. In the second diagram, there is no such cancellation, and the diagram is of order $1/(\alpha_s \sqrt{\rho_{gluons}})$. At a time of order $1/(\alpha_s^2 Q_s)$ for a density decreasing like $1/\tau$ as we expect for ultrarelativistic nuclear collisions, the diagram is enhanced by a factor of $1/\alpha_s$. This cancels the extra factor of α_s coming from the diagram being higher order in perturbation theory [16].

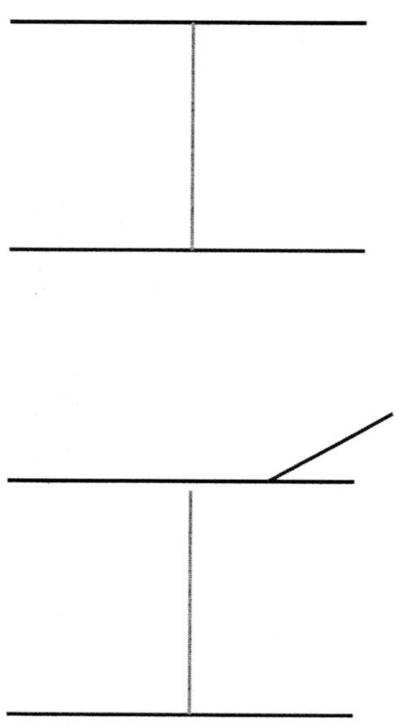

Fig. 23. The diagrams for gluon scattering which lead to thermalization

What appears to happen is that as the system gets more dilute, it thermalizes due to multigluon production. This will modify the relationship between the number of gluons produced as mini-jets and the pion multiplicity. How this actually works is not yet fully understood.

4 The Renormalization Group

The effective action for the theory we have described must be gauge invariant and properly describe the dynamics in the presence of external sources. For the theory which we have written down in the past chapters with the $J \cdot A$ coupling of source to field, gauge invariance is only retained if we impose

$$D_\mu J^\mu = 0. \tag{90}$$

This equation presents a problem in our formulation since it implies that the source cannot be independently specified from the field. This did not present a problem for the classical theory since one could find a solution which solved the constraint. When we compute quantum corrections and proceed to a renormalization group treatment, we must be more careful.

In a clever series of papers [25], it was shown that one can generalize the $J \cdot A$ coupling. This led Jalilian-Marian, Kovner, Leonidov and Weigert to propose the action

$$\begin{aligned} S = & -\frac{1}{4}\int d^4x F^a_{\mu\nu} F^{\mu\nu}_a + \frac{i}{N_c}\int d^2x_t dx^- \delta(x^-) \\ & \times \rho^a(x_t) tr T^a exp\left\{ i \int_{-\infty}^{\infty} dx^+ T \cdot A^-(x) \right\}. \end{aligned} \tag{91}$$

In this equation, the matrix T is in the adjoint representation of the gauge group. This is required for reality of the action. When this action is extremized to get the Yang-Mills equations, one can identify the current and show that the current is covariantly conserved. This action is invariant under gauge transformations which are identified at $x^+ = \pm\infty$. (Even this can be corrected to get a fully invariant theory if one generalizes even further to complex time Keldysh contours. As shown in Ref. [27], this further generalization does not affect the renormalization group in lowest non-trivial order.) This is a consequence of the gauge invariance of the measure of integration over the sources ρ. It will be taken as a boundary condition on the theory. In general, if we had not integrated over sources, one could not define a gauge invariant theory with a source, as gauge rotations would change the definition of the source. Here because the source is integrated over in a gauge invariant way, the problem does not arise.

The most general gauge invariant theory which we can write down is generated from

$$Z = \int [d\rho] e^{-F[\rho]} \int [dA] e^{iS[A,\rho]}. \tag{92}$$

This is a generalization of the Gaussian ansatz described in the previous lecture. It allows for a slightly more complicated structure of stochastic variation of the sources. The Gaussian ansatz can be shown to be valid when evaluating structure functions at large transverse momenta

$$F_{Gaussian}[\rho] = \frac{1}{2} \int dx^- d^2x_t \, \frac{\rho^2(x_t)}{\mu^2(x^-)}. \tag{93}$$

This theory is an effective theory valid only in a limited range of rapidity much less than the rapidity of the source. The sources for this theory sit at higher rapidity. This happens because as we go to lower values of rapidity, the fluctuations in the field are integrated out and are replaced by sources and an integral over fluctuations in the source. The renormalization group equations which we will describe are what make the theory independent of this cutoff. To fully determine F in the above equation demands a full solution of these renormalization group equations. This has yet to be done, although there are now approximate solutions for small and large transverse momentum of the fields [27]-[26].

We can understand this a little better by imagining what happens when we compute a quantum correction to the classical theory. This quantum correction will generate terms proportional to $\alpha_s ln(\Lambda^+/p^+)$ where Λ^+ is the p^+ cutoff for our effective theory. Clearly these corrections are small and sensible only if $e^{-1/\alpha_s}\Lambda^+ << p^+ << \Lambda^+$. If we want to generate a good effective theory at smaller values of p^+, we need to break the theory into intervals of p^+ with each interval sufficiently small so that the quantum fluctuations are small and computable. The relation between one interval and the next is the renormalization group.

The remarkable thing that happens when one integrates out the fluctuations interval is that only the function F which controls the source strength is modified! The functional form of F is modified so that this equation is of the form

$$\frac{d}{dy} Z = -H(\rho, \delta/\delta\rho) Z, \tag{94}$$

where

$$y = ln(\Lambda_i^+/\Lambda_f^+), \tag{95}$$

and $\Lambda_{i,f}^+$ are the cutoffs at the initial and final values, and

$$Z = e^{-F}. \tag{96}$$

This equation is of the form of the time evolution for a two-spatial-dimension quantum field theory where the coordinates are ρ and the momenta are $d/d\rho$.

4.1 How to Compute the RG Effective Hamiltonian

In the Eqn. 94, the renormalization group Hamiltonian H was introduced. I will here outline how it is computed. We first take the theory defined for

$p^+ < \Lambda_i^+$. We integrate out the quantum fluctuations. In particular, the two point function is

$$\mathcal{G}^{ij}(x,y) =< (A^i(x) + \delta A^i(x))(A^i(y) + \delta A^i(y)) > . \quad (97)$$

In this equation, A is the classical background field and δA is the small fluctuation. At the momentum scales which will be of interest for $p^+ < \Lambda_f^+$, it is sufficient to consider the equal-time limit of this correlation function. We now identify

$$\begin{aligned} < \delta A \delta A > &= G < \delta\rho\delta\rho > G \\ &= G\chi G, \end{aligned} \quad (98)$$

where G is the Greens function in the classical background field A. We also identify

$$\begin{aligned} < \delta A > &= G < \delta\rho > \\ &= \sigma. \end{aligned} \quad (99)$$

We can get exactly the same result by modifying the weight function so that we reproduce χ and σ and move the cutoff to Λ_f^+ so that there are no longer quantum fluctuations to integrate out. This is the origin of the form of Eqn. 94.

Some technical comments about the computations are required. One must be extremely careful of gauge. The gauge prescriptions of retarded or advanced for $1/k^+$ singularities are used. We were not able to effectively use either Leibbrandt-Mandelstam or principle value prescriptions although this may be possible in principle. When one computes propagators in background fields, one gets analytic expressions in terms of line ordered phases of the source ρ. It is most convenient to compute these in $\delta A^- = 0$ gauge and express things in terms of the source in $A^- = 0$ gauge, and then rotate results back to light cone gauge. This can be carefully done only when the $1/k^+$ singularity is properly regularized.

If we change the variables to space-time rapidity, we can define

$$\alpha(y, x_T) = \frac{1}{-\nabla_T^2}\rho(y, x_T) \quad (100)$$

and

$$V^\dagger(y, x_T) = Pexp\left(ig\int_{-\infty}^{y} dy'\alpha(y', x_T)\right). \quad (101)$$

After much work, one finds

$$H = \frac{\alpha_s}{2}\int d^2x_T J^{ia}(x_T)J^{ia}(x_T), \quad (102)$$

where

$$J^{ia}(x_T) = \int \frac{d^2 z_T}{\pi} \frac{(x-z)^i}{(x-z)^2} (1 - V^\dagger(y, x_T) V(y, z_T))^{ab} \frac{1}{i} \frac{\delta}{\delta \alpha^b(y, z_T)}. \quad (103)$$

The Hamiltonian is positive definite and looks like a pure kinetic energy term (up to the multiplicative non-linearities) with no potential [27].

The renormalization group above can also be seen to be a consequence of equations for correlation functions of $V(y, x_T)$ [28]. In Ref. [29] it was shown that these equations for correlation functions were almost the same as those of the renormalization group. There was an error in this analysis associated with the subtleties of gauge fixing, and when repaired gives that these equations are precisely equivalent [27]. Meanwhile, Weigert showed that the equations for the correlation functions could be summarized as a Hamiltonian equation of the form above [30], which was also shown to be precisely the equations for the renormalization group Hamiltonian [27].

4.2 Quantum Diffusion

The Hamiltonian presented in the previous section is analogous to that without a potential. If we were to ignore the non-linearities associated with the matrices V, this would be the Hamiltonian for a free theory with only momenta and no potential.

If there was a potential in the Hamiltonian, then at large times the solution of the above renormalization group equations would be trivial

$$Z \sim exp(-yE_o), \quad (104)$$

where E_o is the ground state energy. All expectation values would become rapidity independent and the solution to the small-x problem would be trivial: x independence.

The solution to the above equation is more complicated. One can see this by studying a one-dimensional quantum mechanics problem:

$$\frac{d}{dy} Z = -\frac{p^2}{2} Z \quad (105)$$

with solution

$$Z = \frac{1}{\sqrt{2\pi y}} exp(-x^2/2y). \quad (106)$$

This equation describes diffusion. The width of the Gaussian in x grows with time. This is unlike solving

$$\frac{d}{dy} Z = -\left(\frac{p^2}{2} + V(x) \right) Z. \quad (107)$$

In this latter case, the coordinate x evolves towards the minimum of V, and then does undergo small fluctuations around this minimum.

We see therefore that the non-triviality of the small-x problem in QCD arises because of the quantum diffusive nature of the renormalization group equations.

4.3 Some Generic Features of the Renormalization Group Equation

If we compute the correlation function of two sources using Eq. (94), we find that

$$\frac{d}{dy} < \rho(x)\rho(y) >= - < \rho(x)\rho(y)H > . \tag{108}$$

At large k_T when the fields are linear, the gluon structure function is the same as the source-source correlation function up to a trivial factor of $1/k_T^2$. (The momentum k_T is conjugate to the coordinate $x_T - y_T$.) If we ignore the non-linearities in H, keeping the lowest-order non-vanishing terms, and if we integrate by parts the factors of $\delta/\delta\alpha(y, x_T)$, we get a closed linear equation for the correlation function. This is precisely the BFKL equation.

In fact, in the region where the equations are linear, one is in the high-k_T limit, and this also reduces to the DGLAP and BFKL equations, which are known to be equivalent if one computes distribution functions to leading order in $\alpha_s ln(1/x) ln(Q^2)$, where Q is some typical momentum for the correlation function, $Q \sim k_T$. When the non-linearities are important, the non-linearities of this equation cannot be ignored.

The situation is as shown in Fig. 24. In the linear region, one can choose to evolve using linear equations. In the $ln(Q^2)$ direction, the equation is the DGLAP equation and in the $ln(1/x)$ direction, it is the BFKL equation. There is a boundary region in the $ln(1/x)$-$ln(Q^2)$ plane. Within this boundary region, there is a high density of glue and the evolution becomes non-linear. One always collides with this region if one decreases x and holds Q^2 fixed or decreases Q^2 holding x fixed.

4.4 Some Limiting Solutions of the Renormalization Group Equations

In the small-k_T region, we expect that correlation functions such as $< V(x)V^\dagger(y) >$ are very small, since we are probing the theory at distance scales long compared to natural correlation lengths. In this limit, one might be able to ignore the non-linearities in the renormalization group equations. Using that $x^i/x^2 = \nabla^i/\nabla^2$, we have then

$$\int d^2x_T\ J^2(x_T) \sim \int d^2z d^2z' < z \mid \frac{1}{-\nabla_T^2} \mid z' > \frac{\delta}{\delta\alpha(y,z)} \frac{\delta}{\delta\alpha(y,z')} . \tag{109}$$

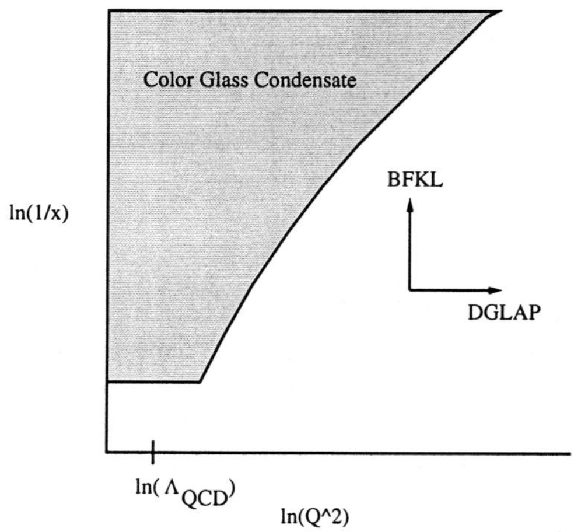

Fig. 24. The various regions of evolution for structure functions in the $ln(1/x)$-$ln(Q^2)$ plane

The solution for Eq. (94) is

$$F = \frac{\kappa}{2\alpha_s} \int dy d^2x_T \nabla_T^i \alpha(y, x_T) \nabla_T^i \alpha(y, x_T). \tag{110}$$

The small-k_T functional F is a pure scale-invariant Gaussian. It is universal and independent of initial conditions.

In the large-p_T region, we perform a mean-field analysis. The result is that discussed in chapter 2. For details of the analysis leading to the results of this section, the interested reader is referred to [26].

4.5 Some Speculative Remarks

The form of the renormalization group equation appears to be simple. It looks like it might even be possible to find exact solutions. In remarkable works [28],[31], Balitsky and Kovchegov have shown that the equation for the two-line correlation function $W(x)W^\dagger(y)$, where W is in the fundamental representation becomes a closed non-linear equation in the large-N_{color} limit. This means that at large N_{color} one can compute this correlation function at arbitrarily small x including all the non-linearities associated with small x.

Although Kovchegov's original derivation was for large nuclei, the result can be shown to follow directly from the renormalization group Eq. (94). This is done by taking the expectation value of $< W(x)W^\dagger(y) >$, using the form of the Hamiltonian and a factorization property of expectation values true in large N_{color}. A derivation is presented in Ref. [26] for the interested reader.

This result is interesting in itself since it means that all of the saturation effects for $F_2(x, Q^2)$ may be computed at small x. The Balitsky-Kovchegov equation has been solved numerically [32]. More important, it suggests that perhaps, at least in large N_{color}, the full renormalization group equations may be solved for F.

5 Concluding Comments

In this article, constraints of space have forced me to not mention many of the exciting areas that are currently under study. One of these areas is diffraction [33]-[34]. One can show that the same formalism which gives deep inelastic scattering also gives diffraction and that there is a simple relation between diffractive structure functions and deep inelastic scattering. I have also not developed a formal treatment of deep inelastic scattering within the Color Glass Condensate picture [35]-[36].

The last chapter is very sketchy, and should mostly provide an introduction to the literature on this problem. The derivation of the results discussed in that chapter are onerous, and all the details have been omitted in this article. In some sense this is good, since the most interesting part of this problem is to understand and solve the renormalization group equations, and at least this problem is clearly stated, and free from the technical details from which it arises.

An area which should be better understood from the perspective presented above is the nature of shadowing for nuclei at small x. This relates deep inelastic scattering and diffraction in a non-trivial way, and the Color Glass Condensate is one of the few theories available which pretends to treat both consistently.

The other area where there is much potential is the production of quarks in hadron-hadron collisions. In particular, the charm quark may provide us a real clue about non trivial dynamics since its mass is very close to the scale of the Color Glass Condensate for large nuclei at accessible energies.

Acknowledgments

These lectures were delivered at the 40th Schladming Winter School: DENSE MATTER, March 3-10, 2001. I thank the organizers for making this a very informative meeting, with an excellent atmosphere for interaction.

I thank my colleagues Alejandro Ayala-Mercado, Adrian Dumitru, Elena Ferreiro, Miklos Gyulassy, Edmond Iancu, Yuri Kovchegov, Alex Kovner, Jamal Jalilian-Marian, Andrei Leonidov, Raju Venugopalan, and Heribert Weigert with whom the ideas presented in this talk were developed. I particularly thank Kazunori Itakura for a careful and critical reading of the manuscript.

This manuscript has been authorized under Contract No. DE-AC02-98H10886 with the U. S. Department of Energy.

References

1. L.N. Lipatov: Sov. J. Nucl. Phys. **23**, 338 (1976); E.A. Kuraev, L.N. Lipatov, Y.S. Fadin: Zh. Eksp. Teor. Fiz. **72**, 3 (1977); (Sov. Phys. JETP **45**, 1 (1977)); Ya.Ya. Balitsky, L.N. Lipatov: Sov. J. Nucl. Phys. **28**, 822 (1978)
2. V. Gribov, L. Lipatov: Sov. J. Nucl. Phys. **15**, 438 (1972); G. Altarelli, G. Parisi: Nucl. Phys. **B126**, 298 (1977); Yu.L. Dokshitser: Sov. Phys. JETP **46**, 641 (1977)
3. J. Breitweg et al.: Eur. Phys. J. **67**, 609 (1999) and references therein
4. L.V. Gribov, E.M. Levin, M. G. Ryskin: Phys. Rep. **100**, 1 (1983)
5. A.H. Mueller, Jian-wei Qiu: Nucl. Phys. **B268**, 427 (1986)
6. J.D. Bjorken: Phys. Rev. **D27**, 140 (1983)
7. L. McLerran, R. Venugopalan: Phys. Rev. **D49**, 2233 (1994); ibid. **D49**, 3352 (1994)
8. A.H. Mueller: Nucl. Phys. **B558**, 285 (1999)
9. Y. Kovchegov: Phys. Rev. **D54**, 5463 (1996); ibid. **D55**, 5445 (1997)
10. R. Venugopalan: nucl-th/9808023
11. L. McLerran, R. Venugopalan: Phys. Rev. **D50**, 2225 (1994)
12. J. Jalilian-Marian, A. Kovner, L. McLerran, H. Weigert: Phys. Rev. **D55**, 5414 (1997)
13. C.S. Lam, G. Mahlon: Phys. Rev. **D62**, 114023 (2000)
14. K. Kajantie, P. Landshoff, J. Lindfors: Phys. Rev. Lett. **59**, 2527 (1987); K. Eskola, K. Kajnatie, J. Lindfors: Nucl. Phys. **B323**, 327 (1989)
15. M. Gyulassy, Xin-Nian Wang: Phys. Rev. **D43**, 104 (1991)
16. R. Baier, A. Mueller, D. Schiff, D. Son: Phys. Lett. **B502**, 51 (2001)
17. M. Gyulassy, L. McLerran: Phys. Rev. **C56**, 2219 (1997)
18. A. Kovner, L. McLerran, H. Weigert: Phys. Rev. **D52**, 3809 (1995); ibid. **D52**, 6231 (1995)
19. A. Krasnitz, R. Venugopalan: Phys. Rev. Lett. **84**, 4309 (2000); Nucl. Phys. **B557**, 237 (1999)
20. S. Bass, B. Muller, W. Poschl: J. Phys. **G25**, L109 (1999)
21. Y. Kovchegov: hep-ph/0011252
22. Y. Kovchegov, A. H. Mueller: Nucl. Phys. **B529**, 451 (1998)
23. D. Rischke, Y. Kovchegov: Phys. Rev. **C56**, 1084 (1997)
24. A. Dumitru, L. McLerran, in preparation
25. J. Jalilian-Marian, A. Kovner, A. Leonidov and H. Weigert: Nucl. Phys. **B504**, 415 (1997); Phys. Rev. **D59**, 014014 (1999); J. Jalilian-Marian, A. Kovner, H. Weigert: Phys. Rev. **D59**, 014015 (1999)
26. E. Iancu, L. McLerran: hep-ph/0103032
27. E. Iancu, A. Leonidov, L. McLerran: hep-ph/0012009; hep-ph/0011241; E. Ferreiro, E. Iancu, A. Leonidov, L. McLerran: in preparation
28. I. Balitsky: Nucl. Phys. **B463**, 99 (1996)
29. A. Kovner, J. G. Milhano, H. Weigert: Phys. **D62**, 114005 (2000)
30. H. Weigert: hep-ph/0004044
31. Y. Kovchegov: Phys. Rev. **D60**, 034008 (1999); ibid. **D61**, 074018 (2000)
32. M. Braun: Eur. Phys. J. **C16**, 337 (2000)
33. W. Buchmuller, T. Gehrmann, A. Hebecker: Nucl. Phys. **B537**, 477 (1999)
34. L. McLerran, Y. Kovchegov: Phys. Rev. **D60**, 054025 (1999)
35. A. H. Mueller: Nucl. Phys. **B307**, 34 (1988); ibid. **B317**, 573 (1989); ibid. **B335**, 115 (1990)
36. L. McLerran, R. Venugopalan: Phys. Rev. **D59**, 094002 (1999)

Lecture Notes in Physics

For information about Vols. 1–546
please contact your bookseller or Springer-Verlag

Monographs

For information about Vols. 1–29 please contact your bookseller or Springer-Verlag

Vol. m 30: A. J. Greer, W. J. Kossler, Low Magnetic Fields in Anisotropic Superconductors. VII, 161 pages. 1995.

Vol. m 31 (Corr. Second Printing): P. Busch, M. Grabowski, P.J. Lahti, Operational Quantum Physics. XII, 230 pages. 1997.

Vol. m 32: L. de Broglie, Diverses questions de mécanique et de thermodynamique classiques et relativistes. XII, 198 pages. 1995.

Vol. m 33: R. Alkofer, H. Reinhardt, Chiral Quark Dynamics. VIII, 115 pages. 1995.

Vol. m 34: R. Jost, Das Märchen vom Elfenbeinernen Turm. VIII, 286 pages. 1995.

Vol. m 35: E. Elizalde, Ten Physical Applications of Spectral Zeta Functions. XIV, 224 pages. 1995.

Vol. m 36: G. Dunne, Self-Dual Chern-Simons Theories. X, 217 pages. 1995.

Vol. m 37: S. Childress, A.D. Gilbert, Stretch, Twist, Fold: The Fast Dynamo. XI, 406 pages. 1995.

Vol. m 38: J. González, M. A. Martín-Delgado, G. Sierra, A. H. Vozmediano, Quantum Electron Liquids and High-Tc Superconductivity. X, 299 pages. 1995.

Vol. m 39: L. Pittner, Algebraic Foundations of Non-Com-mutative Differential Geometry and Quantum Groups. XII, 469 pages. 1996.

Vol. m 40: H.-J. Borchers, Translation Group and Particle Representations in Quantum Field Theory. VII, 131 pages. 1996.

Vol. m 41: B. K. Chakrabarti, A. Dutta, P. Sen, Quantum Ising Phases and Transitions in Transverse Ising Models. X, 204 pages. 1996.

Vol. m 42: P. Bouwknegt, J. McCarthy, K. Pilch, The W3 Algebra. Modules, Semi-infinite Cohomology and BV Algebras. XI, 204 pages. 1996.

Vol. m 43: M. Schottenloher, A Mathematical Introduction to Conformal Field Theory. VIII, 142 pages. 1997.

Vol. m 44: A. Bach, Indistinguishable Classical Particles. VIII, 157 pages. 1997.

Vol. m 45: M. Ferrari, V. T. Granik, A. Imam, J. C. Nadeau (Eds.), Advances in Doublet Mechanics. XVI, 214 pages. 1997.

Vol. m 46: M. Camenzind, Les noyaux actifs de galaxies. XVIII, 218 pages. 1997.

Vol. m 47: L. M. Zubov, Nonlinear Theory of Dislocations and Disclinations in Elastic Body. VI, 205 pages. 1997.

Vol. m 48: P. Kopietz, Bosonization of Interacting Fermions in Arbitrary Dimensions. XII, 259 pages. 1997.

Vol. m 49: M. Zak, J. B. Zbilut, R. E. Meyers, From Instability to Intelligence. Complexity and Predictability in Nonlinear Dynamics. XIV, 552 pages. 1997.

Vol. m 50: J. Ambjørn, M. Carfora, A. Marzuoli, The Geometry of Dynamical Triangulations. VI, 197 pages. 1997.

Vol. m 51: G. Landi, An Introduction to Noncommutative Spaces and Their Geometries. XI, 200 pages. 1997.

Vol. m 52: M. Hénon, Generating Families in the Restricted Three-Body Problem. XI, 278 pages. 1997.

Vol. m 53: M. Gad-el-Hak, A. Pollard, J.-P. Bonnet (Eds.), Flow Control. Fundamentals and Practices. XII, 527 pages. 1998.

Vol. m 54: Y. Suzuki, K. Varga, Stochastic Variational Approach to Quantum-Mechanical Few-Body Problems. XIV, 324 pages. 1998.

Vol. m 55: F. Busse, S. C. Müller, Evolution of Spontaneous Structures in Dissipative Continuous Systems. X, 559 pages. 1998.

Vol. m 56: R. Haussmann, Self-consistent Quantum Field Theory and Bosonization for Strongly Correlated Electron Systems. VIII, 173 pages. 1999.

Vol. m 57: G. Cicogna, G. Gaeta, Symmetry and Perturbation Theory in Nonlinear Dynamics. XI, 208 pages. 1999.

Vol. m 58: J. Daillant, A. Gibaud (Eds.), X-Ray and Neutron Reflectivity: Principles and Applications. XVIII, 331 pages. 1999.

Vol. m 59: M. Kriele, Spacetime. Foundations of General Relativity and Differential Geometry. XV, 432 pages. 1999.

Vol. m 60: J. T. Londergan, J. P. Carini, D. P. Murdock, Binding and Scattering in Two-Dimensional Systems. Applications to Quantum Wires, Waveguides and Photonic Crystals. X, 222 pages. 1999.

Vol. m 61: V. Perlick, Ray Optics, Fermat's Principle, and Applications to General Relativity. X, 220 pages. 2000.

Vol. m 62: J. Berger, J. Rubinstein, Connectivity and Superconductivity. XI, 246 pages. 2000.

Vol. m 63: R. J. Szabo, Ray Optics, Equivariant Cohomology and Localization of Path Integrals. XII, 315 pages. 2000.

Vol. m 64: I. G. Avramidi, Heat Kernel and Quantum Gravity. X, 143 pages. 2000.

Vol. m 65: M. Hénon, Generating Families in the Restricted Three-Body Problem. Quantitative Study of Bifurcations. XII, 301 pages. 2001.

Vol. m 66: F. Calogero, Classical Many-Body Problems Amenable to Exact Treatments. XIX, 749 pages. 2001.

Vol. m 67: A. S. Holevo, Statistical Structure of Quantum Theory. IX, 159 pages. 2001.

Vol. m 68: N. Polonsky, Supersymmetry: Structure and Phenomena. Extensions of the Standard Model. XV, 169 pages. 2001.

Vol. m 69: W. Staude, Laser-Strophometry. High-Resolution Techniques for Velocity Gradient Measurements in Fluid Flows. XV, 178 pages. 2001.

Vol. m 70: P. T. Chruściel, J. Jezierski, J. Kijowski, Hamiltonian Field Theory in the Radiating Regime. VI, 172 pages. 2002.

Vol. m 71: S. Odenbach, Magnetoviscous Effects in Ferrofluids. X, 151 pages. 2002.

Vol. m 72: J. G. Muga, R. Sala Mayato, I. L. Egusquiza, Time in Quantum Mechanics. XII, 419 pages. 2002.